# Lecture Notes in Control and Information Sciences

Edited by M. Thoma and A. Wyner



90

# Singular Perturbations and Asymptotic Analysis in Control Systems

Edited by
P. Kokotovic, A. Bensoussan, G. Blankenship

Springer-Verlag
Berlin Heidelberg New York
London Paris Tokyo

**Editors**
Prof. Petar V. Kokotovic
Coordinated Sciences Laboratory
University of Illinois
Urbana, IL 61801

Prof. Alain Bensoussan
University of Paris Dauphine
President of INRIA

Prof. Gilmer L. Blankenship
Electrical Engineering Department
University of Maryland
College Park, MD 20742

ISBN 3-540-17362-5 Springer-Verlag Berlin Heidelberg New York
ISBN 0-387-17362-5 Springer-Verlag New York Berlin Heidelberg

Library of Congress Cataloging-in-Publication Data
Singular perturbations and asymptotic analysis
in control systems.
(Lecture notes in control and information sciences ; 90)
1. Control theory. 2. Perturbation (Mathematics) 3. Approximation theory.
I. Kokotovic, Petar V. II. Bensoussan, Alain. III. Blankenship, G. (Gilmer) IV. Series.
QA402.3.S545 1986 629.8'312 86-31608
ISBN 0-387-17362-5 (U.S.)


Printed in Germany

Offsetprinting: Mercedes-Druck, Berlin
Binding: B. Helm, Berlin
2161/3020-543210

## PREFACE

This collection of papers deals with the general role of singular perturbation techniques in control systems analysis and design problems. These methods have proven useful in the construction of "reduced order models" and the evaluation of control system designs based on those models. We have collected here a representative sampling of the use of these techniques which will be informative to those readers interested in acquiring a taste for the theory and its applications. We have also addressed those doing research in the subject matter by including some new results and methods not published elsewhere.

The first paper in this collection, *Singular Perturbation Techniques in Control Theory* is a survey of the role of singular perturbation ideas in engineering control system design. The analysis and examples which it contains summarizes much of the work in the field prior to this volume. It sets the stage for the detailed treatment of more specialized topics in the subsequent papers.

In Part I we treat optimal control problems with small parameters. The paper *Singular Perturbations for Deterministic Control Problems* provides a comprehensive treatment of deterministic optimal control problems with "fast" and "slow" states. It is based on the asymptotic analysis of both necessary conditions and the associated Hamilton-Jacobi-Bellman equation - that is, direct evaluation of the optimal cost function. The treatment using a *duality* for this equation is new. As a consequence, one can extend the concept of composite feedback involving "separation" of controls for fast and slow states which had been derived earlier for quasi-linear systems the full nonlinear case.

The treatment of optimal cost functions is continued in the stochastic case in *Singular Perturbations in Stochastic Control.* In this case the simplification of the optimal cost in the "reduced order" model is not as complete as in the deterministic case. The class of feedbacks for the reduced order system - the slow states - must retain its dependence on the fast state variables. The long time behavior of the fast state variables which permits the definition of the reduced order model is based on ergodicity conditions. Most of the material in this Chapter has not been published before.

The papers in Part II are devoted to the role of singular perturbation methods in the reduction of models of large scale systems. The first two papers in this part are concerned with the analysis of singularly perturbed models for Markov chains. In *Singular Perturbations of Markov Chains* methods for aggregation and time scale analysis are developed. An application to the reliability analysis of large scale repairable systems illustrates the results. The paper *Optimal Control of Perturbed Markov Chains* applies the methods to the analysis of the associated Bellman equation. The reader may wish to compare the treatment of the Bellman equation in this paper with the corresponding analysis in Part I.

The third paper in Part II, *Time Scale Modeling of Dynamic Networks with Sparse and Weak Coupling,* uses time scale analysis and aggregation methods to deduce sparsity patterns in large scale networks. The grouping algorithms developed in the course of this analysis are applied to synthesis of simplified models for large electric energy systems. The sparsity property exploited in this paper is directly related to the weak coupling property of Markov chains used in the previous papers in this section.

The two papers in Part III deal with the role of singular perturbation analysis in the derivation of stability criteria for nonlinear systems. In *Stability Analysis of Singu-*

*larly Perturbed Systems* nonlinear, non-autonomous singularly perturbed systems are considered at the outset. The methods are then extended to treat multiparameter perturbation problems. In *New Stability Theorems for Averaging and Their Application to the Convergence Analysis of Adaptive Identification and Control Schemes* multi-time scale methods are used to treat time varying nonlinear systems with applications to estimation of the rates of convergence of adaptive identification and control algorithms.

These papers provide just a sampling of the methods available in this rich area of applied mathematics. Some of the papers indicate the broader range of methods and applications which lie outside control theory. We trust that those readers who have found the papers in this volume interesting will be motivated to explore the many important contributions which treat related applications in engineering and applied physics.

P.V. Kokotovic
A. Bensoussan
G.L. Blankenship

Contents

# SINGULAR PERTURBATION TECHNIQUES IN CONTROL THEORY

*P.V. Kokotovic*[†]

Abstract

This paper discusses typical applications of singular perturbation techniques to control problems in the last fifteen years. The first three sections are devoted to the standard model and its convergence, stability and controllability properties. The next two sections deal with linear-quadratic optimal control and one with cheap (near-singular) control. Then the composite control and trajectory optimization are considered in two sections, and stochastic control in one section. The last section returns to the problem of modeling, this time in the context of large scale systems. The bibliography contains more than 250 titles.

## Introduction

For the control engineer, singular perturbations legitimize his ad hoc simplifications of dynamic models. One of them is to neglect some "small" time constants, masses, capacitances, and similar "parasitic" parameters which increase the dynamic order of the model. However, the design based on a simplified model may result in a system far from its desired performance or even an unstable system. If this happens, the control engineer needs a tool which will help him to improve his oversimplified design. He wants to treat the simplified design as a first step, which captures the dominant phenomena. The disregarded phenomena, if important, are to be treated in the second step.

It turns out that asymptotic expansions into reduced ("outer") and boundary layer ("inner") series, which are the main characteristic of singular perturbation techniques, coincide with the outlined design stages. Because most control systems are dynamic, the decomposition into stages is dictated by a separation of time scales. Typically, the reduced model represents the slowest (average) phenomena which in most applications are dominant. Boundary layer (and sublayer) models evolve in faster

[†]Coordinated Sciences Laboratory and Electrical Engineering Department, University of Illinois, 1101 W. Springfield Avenue, Urbana, IL 61801. This paper is based on the author's survey in the *SIAM Review*, Vol. 6, No. 4, October 1984, pp. 501-550.

time scales and represent deviations from the predicted slow behavior. The goal of the second, third, etc., design stages is to make the boundary layers and sublayers asymptotically stable, so that the deviations rapidly decay. The separation of time scales also eliminates the stiffness difficulties and prepares for a more efficient hardware and software implementation of the controller.

This paper is a tutorial presentation of typical, but not all, applications of singular perturbation techniques to control problems. The focus is on systems modeled by ordinary differential equations and most topics discussed are deterministic. Only one out of ten sections is dedicated to stochastic problems because of the existence of two excellent surveys of singular perturbation methods in stochastic differential equations, Blankenship (1979) and Schuss (1980). Sections 1 and 2 introduce a standard model and discuss its properties. Sections 3, 4, 5, and 6 deal with linear control problems in open-loop and feedback form. Sections 7 and 8 are dedicated to nonlinear, and Section 9 to stochastic problems. In Section 10 we return to the issue of modeling by examining nonstandard models common in networks and other large scale systems. Although some results are quoted as theorems, they are spelled out in a less technical form than that in the referenced works, which should be consulted for more rigorous formulations. Whenever convenient, simple examples are inserted to illustrate basic concepts.

## 1. The Standard Singular Perturbation Model

The singular perurbation model of finite dimensional dynamic systems extensively studied in mathematical literature by Tichonov (1948, 1952), Levinson (1950), Vasileva (1963), Wasow (1965), Hoppensteadt (1967, 1971), O'Malley (1971, 1973), etc., was also the first model to be used in control and systems theory. This model is in the explicit state variable form in which the derivatives of some of the states are multiplied by a small positive scalar $\varepsilon$, that is,

$$\dot{x} = f(x,z,u,\varepsilon,t), \qquad x\in R^n \tag{1.1}$$

$$\varepsilon\dot{z} = g(x,z,u,\varepsilon,t), \qquad z\in R^m \tag{1.2}$$

where $u = u(t)$ is the control vector and a dot denotes a derivative with respect to time $t$. It is assumed that $f$ and $g$ are sufficiently many times continuously differentiable functions of their arguments $x,z,u,\varepsilon,t$. The scalar $\varepsilon$ represents all the small parameters to be neglected. In most applications having a single parameter is not a restriction. For example, if $T_1$ and $T_2$ are small time constants of the same order of magnitude, $O(T_1) = O(T_2)$, then one of them can be taken as $\varepsilon$ and the other expressed as its multiple, say $T_1 = \varepsilon$, $T_2 = \alpha\varepsilon$, where $\alpha = T_2/T_1$ is fixed.

In control and systems theory the model (1.1), (1.2) is a convenient tool for "reduced order modeling," a common engineering task. The order reduction is converted into a parameter perturbation, called "singular." When we set $\varepsilon = 0$ the dimension of the state space of (1), (2) reduces from $n + m$ to $n$ because the differential equation (1.2) degenerates into an algebraic or a transcendental equation

$$0 = g(\bar{x},\bar{z},u,0,t), \tag{1.3}$$

where the bar indicates that the variables belong to a system with $\varepsilon = 0$. We will say that the model (1.1), (1.2) is in the *standard form* if and only if the following crucial assumption concerning (1.3) is satisfied.

Assumption 1.1

In a domain of interest equation (1.3) has $k \geq 1$ distinct ("isolated") real roots

$$\bar{z} = \phi_i(\bar{x},\bar{u},t), \qquad i = 1,2,\ldots,k. \tag{1.4}$$

This assumption assures that a well defined n-dimensional reduced model will correspond to each root (1.4). To obtain the i-th reduced model we substitute (1.4) into (1.1),

$$\dot{\bar{x}} = f(\bar{x},\phi_i(\bar{x},\bar{u},t),\bar{u},0,t). \tag{1.5}$$

In the sequel we will drop the subscript i and rewrite (1.5) more compactly as

$$\dot{\bar{x}} = \bar{f}(\bar{x},\bar{u},t). \tag{1.6}$$

This model is sometimes called quasi-steady-state model, because z, whose velocity $\dot{z} = \frac{g}{\varepsilon}$ is large when $\varepsilon$ is small, may rapidly converge to a root of (1.3), which is the quasi-steady-state form of (1.2). We will discuss this two-time-scale property of (1.1), (1.2) in the next section.

The convenience of using a parameter to achieve order reduction has also a drawback: it is not always clear how to pick the parameters to be considered as small. Fortunately, in many applications our knowledge of physical processes and components of the system suffice to be on the right track. Let us illustrate this by examples.

Example 1.1

A well-known model of an armature controlled DC-motor is

$$\dot{x} = ax \tag{1.7}$$

$$L\dot{z} = bx - Rz + u \tag{1.8}$$

where x, z, and u are respectively, speed, current, and voltage, R and L are armature resistance and inductance, and a and b are some motor constants. In most DC-motors L is a "small parameter" which is often neglected, $\varepsilon$=L. In this case equation (1.3) is

$$0 = b\bar{x} - R\bar{z} + u \tag{1.9}$$

and has only one root

$$\bar{z} = (\bar{u} - b\bar{x})/R. \tag{1.10}$$

Thus the reduced model (1.6) is

$$\dot{\bar{x}} = \frac{a}{R}(\bar{u} - b\bar{x}). \tag{1.11}$$

It is frequently used in the design of servosystems.

Example 1.2

In a feedback system, Fig. 1a, with a high-gain amplifier K, where the nonlinear block N is tan z, the choice of $\varepsilon$ is not as obvious. However, any student of feedback systems would pick $\varepsilon = \frac{1}{K}$, where K is the amplifier gain, and obtain

$$\dot{x} = z \tag{1.12}$$

$$\varepsilon\dot{z} = -x - \varepsilon z - \tan z + u. \tag{1.13}$$

In this case (1.3) and (1.4) yield

$$0 = -\bar{x} - 0 - \tan\bar{z} + \bar{u} \tag{1.14}$$

$$\bar{z} = \tan^{-1}(\bar{u} - \bar{x}) \tag{1.15}$$

and hence the reduced model (1.6) is

$$\dot{\bar{x}} = \tan^{-1}(\bar{u} - \bar{x}). \tag{1.16}$$

This model is represented by the block diagram in Fig. 1b in which the loop with infinite gain $\varepsilon = 0$ is replaced by the inverse of the operator in the feedback path.

It is easily seen that both (1.9) and (1.14) satify Assumption 1.1, that is, both models (1.7), (1.8) and (1.12), (1.13) appear in the *standard form* and their reduced models can be obtained by the singular perturbation $\varepsilon = 0$. To avoid a misleading conclusion that this is always the case, let us consider another simple example in which the original model is not in the standard form.

Example 1.3

In the RC-network in Fig. 1.2a the capacitances are equal, $C_1 = C_2 = 1$, while the resistance r is much smaller than R. Letting $r = \varepsilon$, using the capacitor voltages as the state variables and the input voltage u as the control, the model of this network is

$$\varepsilon\dot{v}_1 = -v_1 + v_2 \tag{1.17}$$

$$\varepsilon\dot{v}_2 = v_1 - (1 + \frac{\varepsilon}{R})v_2 + \frac{\varepsilon}{R}u. \tag{1.18}$$

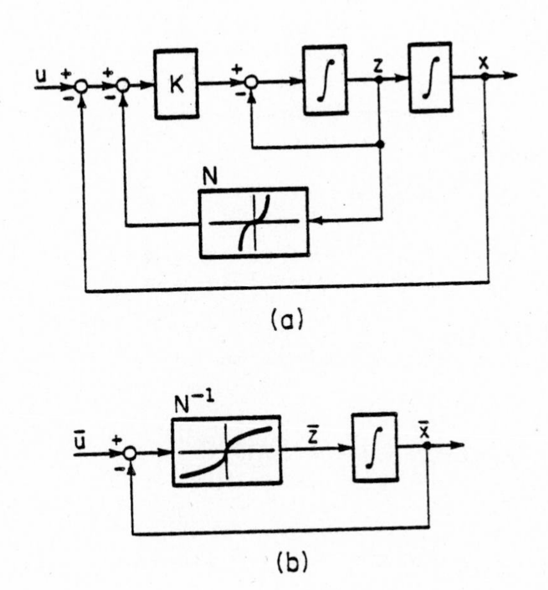


Fig. 1. System with a high gain amplifier: (a) full model, (b) reduced model.

If this model were in the form (1.1), (1.2), both $v_1$ and $v_2$ would be considered as z-variables and (1.3) would be

$$0 = -\bar{v}_1 + \bar{v}_2 \tag{1.19}$$

$$0 = \bar{v}_1 - \bar{v}_2 . \tag{1.20}$$

However, Assumption 1.1 would then be violated because the roots of (1.3), in this case $\bar{v}_1 = \bar{v}_2$, are not distinct. The question remains whether the model of this RC-network can be simplified by singular perturbation $\varepsilon = 0$, that is, by neglecting the small parasitic resistance r? Without hesitation the answer of the electrical engineer is yes, and his simplified model is given in Fig. 1.2b. To justify this simplified model a choice of state variables must be found such that Assumption 1.1 be satisfied. As will be explained in Section 10 a good choice of the x-variable is the "aggregate" voltage

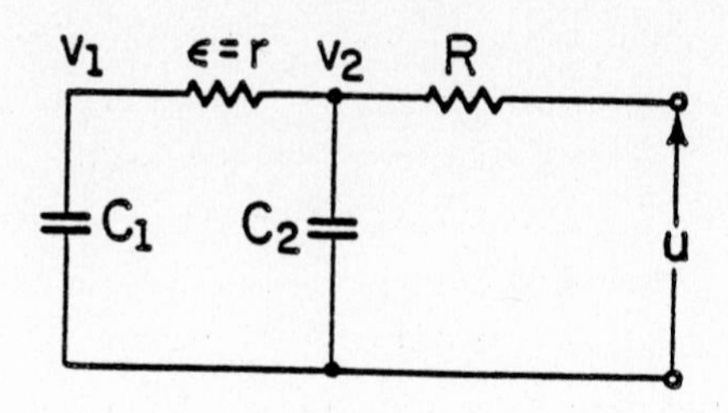


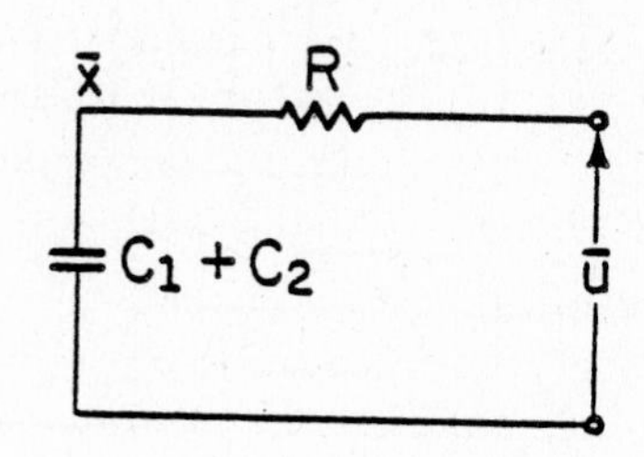


Fig. 1.2. (a) full model, (b) reduced model.

$$x = \frac{C_1 v_1 + C_2 v_2}{C_1 + C_2} \tag{1.21}$$

which, along with $v_2 = z$, transforms (1.17), (1.18) into

$$\dot{x} = -\frac{1}{2R} z + \frac{1}{2R} u \tag{1.22}$$

$$\varepsilon\dot{z} = 2x - (2 + \frac{\varepsilon}{R})z + \frac{\varepsilon}{R} u. \tag{1.23}$$

Now (1.3) becomes

$$0 = 2\bar{x} - 2\bar{z} \tag{1.24}$$

and it satisfies Assumption 1.1. The substitution of $\bar{z} = \bar{x}$ into (1.22) with $C_1 + C_2 = 1 + 1 = 2$ indeed results in the reduced model

$$\dot{\bar{x}} = -\frac{1}{2R} \bar{x} + \frac{1}{2R} \bar{u} \tag{1.25}$$

describing the circuit in Fig. 1.2b.

Most of the quoted singular perturbation literature assumes that model (1.1), (1.2) is in the standard form, that is, it satisfies Assumption 1.1. The importance of Example 1.3 is that it points out the dependence of Assumption 1.1 on the choice of state variables. In most applications a goal of modeling is to remain close to original "physical" variables. This was possible in our Examples 1.1 and 1.2, but not in Example 1.3, where a new voltage variable (1.21) had to be introduced. However, few engineers, accustomed to the simplified "equivalent" circuit in Fig. 1.2b, would question the "physicalness" of this new variable. On the contrary, physical properties of the circuit in Fig. 1.2a are more clearly displayed by the standard form (1.22), (1.23). Nevertheless the problem of presenting and analyzing singular perturbation properties in a coordinate-free form is of fundamental importance. A geometric approach to this problem has recently been developed by Fenichel (1979) Kopell (1979) and Sobolev (1984). More common are indirect approaches which deal with singular singularly perturbed problems, such as in O'Malley (1979), or transform the original "nonstandard" model into the standard form (1.1), (1.2), such as in Peponides, Kokotovic, and Chow (1982), or Campbell (1980, 1982). We will return to this modeling issue in Section 10.

## 2. Time Scale Properties of the Standard Model

Singular perturbations cause a multi-time-scale behavior of dynamic systems characterized by the presence of both slow and fast transients in the system response to external stimuli. Loosely speaking, the slow response, or the "quasi-steady-state," is approximated by the reduced model (1.6), while the discrepancy between the response of the reduced model (1.6) and that of the full model (1.1), (1.2) is the fast transient. To see this let us return to (1.1)-(1.6) and examine variable z which has been excluded from the reduced model (1.6) and substituted by its "quasi-steady-state" $\bar{z}$. In contrast to the original variable z, starting at $t_o$ from a prescribed $z^o$, the quasi-steady-state $\bar{z}$ is not free to start from $z^o$ and there may be a large discrepancy between its initial value

$$\bar{z}(t_o) = \phi(\bar{x}(t_o),\bar{u}(t_o),t_o) \tag{2.1}$$

and the prescribed initial condition $z^o$. Thus $\bar{z}$ cannot be a uniform approximation of z. The best we can expect is that the approximation

$$z = \bar{z}(t) + O(\varepsilon) \tag{2.2}$$

will hold on an interval excluding $t_o$, that is, for $t\in[t_1,T]$ where $t_1 > t_o$. However, we can constrain the quasi-steady-state $\bar{x}$ to start from the prescribed initial condition $x^o$ and, hence the approximation of x by $\bar{x}$ may be uniform. In other words,

$$x = \bar{x}(t) + O(\varepsilon) \tag{2.3}$$

may hold on an interval including $t_o$, that is, for all t in the interval $[t_o,T]$ on which $\bar{x}(t)$ exists.

The approximation (2.2) establishes that during an initial ("boundary layer") interval $[t_o,t_1]$ the original variable z approaches $\bar{z}$ and then, during $[t_1,T]$, remains close to $\bar{z}$. Let us remember that the speed of z is large, $\dot{z} = g/\varepsilon$. In fact, having set $\varepsilon$ equal to zero in (1.2) we have made the transient of z instantaneous. Will during this transient z escape to infinity or converge to its quasi-steady-state $\bar{z}$?

To answer this question let us analyze $\varepsilon\dot{z}$, which may remain finite, even when $\varepsilon$ tends to zero and $\dot{z}$ tends to infinity. We set

$$\varepsilon \frac{dz}{dt} = \frac{dz}{d\tau}, \qquad \text{hence } \frac{d\tau}{dt} = \frac{1}{\varepsilon}, \tag{2.4}$$

and use $\tau = 0$ as the initial value at $t = t_o$. The new time variable

$$\tau = \frac{t-t_o}{\varepsilon}; \qquad \tau = 0 \text{ at } t = t_o, \tag{2.5}$$

is "stretched," that is, if $\varepsilon$ tends to zero, $\tau$ tends to infinity even for fixed t only slightly larger than $t_o$. On the other hand, while z and $\tau$ almost instantaneously change, x remains at its initial value $x^o$. To describe the behavior of z as a function of $\tau$ we use the so-called "boundary layer system"

$$\frac{d\hat{z}}{d\tau} = g(x^o,\hat{z}(\tau),u,0,t_o), \tag{2.6}$$

with $z^o$ as the initial condition for $\hat{z}(\tau)$, and $x^o$, $t_o$ as fixed parameters. The solution $\hat{z}(\tau)$ of this initial value problem is used as a "boundary layer" correction of (2.2) to form a possibly uniform approximation of z,

$$z = \bar{z}(t) + \hat{z}(\tau) - \bar{z}(t_o) + O(\varepsilon). \tag{2.7}$$

Clearly $\bar{z}(t)$ is the slow, and $\hat{z}(\tau) - \bar{z}(t_o)$ is the fast transient of z.

To control these two transients the control u can also be composed of a slow control $\bar{u}(t)$, already assumed in the reduced model (1.6), and a fast control $\hat{u}(\tau)$ in the boundary layer system (2.6). The design of such a two-time-scale composite control is the main topic of several subsequent sections. In this section we concentrate on the assumptions under which the approximations (2.3) and (2.7) are valid.

Assumption 2.1

The equilibrium $\bar{z}(t_o)$ of (2.6) is asymptotically stable uniformly in $x^o$ and $t^o$, and $\hat{z}(\tau)$ starts from $z^o$ which belongs to the domain of attraction of the equilibrium $\bar{z}(t_o)$.

If this assumption is satisfied, that is, if

$$\lim_{t\to\infty} \hat{z}(\tau) = \bar{z}(t_o), \tag{2.8}$$

uniformly in $x^o$, $t_o$, then z will come close to its quasi-steady-state $\bar{z}$ at some time $t_1 > t_o$. Interval $[t_o,t_1]$ can be made arbitrarily short by making $\varepsilon$ sufficiently small. To assure that z stays close to $\bar{z}$, we think as if any instant $t \in [t_1,T]$ can be the initial instant. At such an instant z is already close to $\bar{z}$, which motivates the following assumption about the linearization of (2.6).

Assumption 2.2

The eigenvalues of $\partial g/\partial z$ evaluated along $\bar{x}(t)$, $\bar{z}(t)$, $\bar{u}(t)$ for all $t \in [t_o,T]$ have real parts smaller than a fixed negative number

$$\mathrm{Re}\lambda\{\frac{\partial g}{\partial z}\} <-c < 0. \tag{2.9}$$

Both assumptions describe a strong stability property of the boundary layer system (2.6). If $z^o$ is assumed to be sufficiently close to $\bar{z}(t_o)$, then Assumption 2.2 encompasses Assumption 2.1. We also note from (2.9) that the non-singularity of $\partial g/\partial z$ along $\bar{x}(t)$, $\bar{z}(t)$, $\bar{z}(t)$ implies that the root $\bar{z}(t)$ is distinct as required by Assymption 1.1. These assumptions are common in much of the singular perturbation literature (Tichonov (1948,1952), Levinson (1950), Vasileva (1963), Hoppensteadt (1971), et al.). These references contain the proof and refinements of the following result, frequently referred to as Tichnov's theorem.

Theorem 2.1:

If Assumptions 2.1 and 2.2 are satisfied, then (2.3) and (2.7) hold for all $t \in [t_o,T]$, while (2.2) holds for all $t \in [t_1,T]$, where the "thickness of the boundary layer" $t_1-t_o$ can be made arbitrarily small by choosing small enough $\varepsilon$.

As we shall see, many control applications of singular perturbations make use of this theorem. Let us first specialize (1.1), (1.2) to linear systems

$$\dot{x} = Ax + Bz, \quad x \in R^n \tag{2.10}$$

$$\varepsilon\dot{z} = Cx + Dz, \quad z \in R^m \tag{2.11}$$

assuming first that A, B, C, and D are constant matrices. Clearly, Theorem 2.1 holds if $\mathrm{Re}\lambda\{D\} < 0$. The root of (1.3),

$$\bar{z} = -D^{-1}C\bar{x} \tag{2.12}$$

substituted in (2.10) yields the reduced model

$$\dot{\bar{x}} = (A-BD^{-1}C)\bar{x} \ . \tag{2.13}$$

Introducing the fast variable $\eta$ as the difference between z and its quasi-steady state $\bar{z}$,

$$\eta = z + D^{-1}Cx \tag{2.14}$$

the boundary layer system of (2.11) is simply

$$\frac{d\eta}{d\tau} = D\eta, \qquad \eta(0) = z^o + D^{-1}Cx^o \tag{2.15}$$

and its initial condition $\eta(0)$ is sometimes called the "boundary layer jump." In terms of $\eta$ the approximation (2.7) becomes

$$z = \bar{z}(t) + \eta(\tau) + O(\varepsilon) \tag{2.16}$$

The two-time-scale property of the linear time invariant systems caused by the singular perturbation $\varepsilon \to 0$ is equivalent to the separation of the spectrum into its fast and slow parts. It is of interest to obtain this result algebraically. This is done by using a definition of the fast variable $\eta$ more general than (2.14), namely

$$\eta = z - Lx \tag{2.17}$$

and requiring that L be a root of the matrix equation

$$DL - \varepsilon LA + \varepsilon LBL - C = 0. \tag{2.18}$$

Then (2.17) transforms the original singular perturbation system (2.10), (2.11) into a block-triangular form

$$\dot{x} = (A-BL)x + B\eta \tag{2.19}$$

$$\varepsilon\dot{\eta} = (D + \varepsilon LB)\eta \tag{2.20}$$

where eigenvalues are the eigenvalues of the blocks. An application of the implicit function theorem to (2.18) shows that

$$L = D^{-1}C + O(\varepsilon) \ . \tag{2.21}$$

More details on this algebraic approach to time scale modeling can be found in Kokotovic (1975), Anderson (1978), O'Malley and Anderson (1978), and Avramovic (1979).

In linear-time-varying systems the time-scale properties also depend on the speed of parameter variations. A well-known example is the stability of

$$\varepsilon \frac{dz}{dt} = D(t)z. \tag{2.22}$$

For $\varepsilon = 1$, even if

$$\mathrm{Re}\lambda\{D(t)\} \leq -c_1 < 0, \quad \forall t \geq t_o \tag{2.23}$$

system (2.22) can be unstable. However, when $\varepsilon$ is small, the following result holds.

Theorem 2.2

If, in addition to (2.23), the derivative $\dot{D}(t)$ of $D(t)$ is bounded, $\|\dot{D}(t)\| \leq c_2$ for all $t \geq t_o$, then there exists $\varepsilon_1 > 0$ such that for all $0 < \varepsilon \leq \varepsilon_1$ the system (2.22) is uniformly asymptotically stable.

To prove this theorem we define $M(t)$ for all $t \geq t_o$ by

$$D'(t)M(t) + M(t)D(t) = -I. \tag{2.24}$$

In view of (2.23) $M(t)$ is positive definite and its derivative $M(t)$ is bounded, that is,

$$z'\dot{M}(t)z \leq c_3 z'z. \tag{2.25}$$

Theorem 2.2 follows from the fact that the derivative $\dot{v}$ of the Lyapunov function

$$v = z'M(t)z \tag{2.26}$$

for (2.22) is bounded by

$$\dot{v} \leq -\left(\frac{1}{\varepsilon} - c_3\right)z'z$$

This analysis reveals the meaning of the boundary layer stability assumption of Theorem 2.1. For $\varepsilon$ sufficiently small, the "frozen" spectrum of $\frac{1}{\varepsilon}\frac{\partial g}{\partial z}$, in this case $\frac{1}{\varepsilon}D(t)$, is sufficiently faster than the variations of the entries of $\frac{\partial g}{\partial z}$ and the "frozen" stability condition (2.23) applies.

We are now in the position to generalize the transformation (2.17) to the time-varying system (2.10), (2.11) that is when

$$A = A(t), \quad B = B(t), \quad C = C(t), \quad D = D(t). \tag{2.28}$$

If the transformation matrix $L = L(t)$ in (2.17) satisfies the matrix differential equation

$$\varepsilon\dot{L} = D(t)L - \varepsilon LA(t) + \varepsilon LB(t)L - C(t) \tag{2.29}$$

then the time-varying system is in the form (2.19), (2.20). Equation (2.29) has been analyzed by Chang (1969, 1972) who proved the following result.

Theorem 2.3

If the matrices (2.28) are bounded and (2.23) holds for all $t \in [t_o,T]$ then there exists $\varepsilon_2 > 0$ such that for all $t \in [t_o,T]$, $\varepsilon \in (0,\varepsilon_2]$ a bounded, continuously differentiable solution $L = L(t)$ of (2.29) exists and can be uniformly approximated by

$$L(t) = D^{-1}(t)C(t) + O(\varepsilon) \tag{2.40}$$

This theorem furnishes a simple proof of Theorem 2.1 for linear and linearizable problems. The validity of the approximation (2.3) of x by $\bar{x}$ follows from replacing $L(t)$ by $-D^{-1}(t)C(t)$ in (2.19) and neglecting $B(t)\eta$, because $\|\eta\| \leq c_4 \exp(-c_5\frac{t-t_o}{\varepsilon})$, where $c_4,c_5 > 0$. The approximation of z by (2.16) follows by the same argument.

While the approximations (2.3) and 2.7) are within an $O(\varepsilon)$ error, expressions in two-time scale asymptotic series can improve the accuracy up to any desired order. The details of construction and validation of asymptotic series are presented in Vasileva (1963), Hoppensteadt (1971), Vasileva and Butuzov (1973), and O'Malley (1974). In addition to these direct expansions, formal series can also be formed indirectly by expanding the transformation matrix L in (2.17), that is its defining equation (2.18) or (2.19). This leads to a convenient numerical procedure, because L can be computed iteratively, as in Kokotovic (1975), Anderson (1978), and Avramovic (1979). An alternative procedure for the expansion of the state equation was presented in Kokotovic Allemong, Winkelman, and Chow (1980). The validation of indirect and iterative procedures was given by Phillips (1983) who proved that they produce the terms of the asymptotic series in Vasileva and Butozov (1973).

## 3. Controllability and Stability

It is of conceptual and practical importance that many properties of singular perturbation systems can be deduced from the same properties of simpler slow and fast subsystems defined in separate time scales. In this section we concentrate on controllability and stability properties. We begin with the linear time varying control system

$$\dot{x} = A_{11}(t)x + A_{12}(t)z + B_1(t)u \tag{3.1}$$

$$\varepsilon\dot{z} = A_{12}(t)x + A_{22}(t)z + B_2(t)u \tag{3.2}$$

with a change of notation suitable for control applications. Following Chang (1972), we let $L(t)$ satisfy (2.29) in the new notation and we also define $H(t)$ as a solution of

$$-\varepsilon\dot{H} = H(A_{22} + \varepsilon LA_{12}) - \varepsilon(A_{11} - A_{12}L)H - A_{12} \tag{3.3}$$

which can be approximated by

$$H(t) = A_{12}(t)A_{22}^{-1}(t) + O(\varepsilon). \tag{3.4}$$

Denoting by $I_k$ a $k\times k$ identity we introduce the transformation

$$\begin{bmatrix} \xi \\ \eta \end{bmatrix} = \begin{bmatrix} I_n-\varepsilon HL & -\varepsilon H \\ L & I_m \end{bmatrix} \begin{bmatrix} x \\ z \end{bmatrix} \tag{3.5}$$

whose inverse is

$$\begin{bmatrix} x \\ z \end{bmatrix} = \begin{bmatrix} I_n & \varepsilon H \\ -L & I_m-\varepsilon LH \end{bmatrix} \begin{bmatrix} \xi \\ \eta \end{bmatrix}. \tag{3.6}$$

In the new coordinates $\xi$, $\eta$, the system (3.1), (3.2) separates into two subsystems

$$\dot{\xi} = (A_{11} - A_{12}L)\xi + (B_1 - HB_2 - \varepsilon LB_1)u \tag{3.7}$$

$$\varepsilon\dot{\eta} = (A_{22} + \varepsilon LA_{12})\eta + (B_2 + \varepsilon LB_1)u \tag{3.8}$$

Taking into account (2.29) and (3.4) we readily obtain the following result:

Theorem 3.1

For $\varepsilon$ small a sufficient condition for the controllability of the full system (3.1), (3.2) is the controllability of the slow (reduced) subsystem

$$\dot{\bar{x}} = (A_{11} - A_{12}A_{22}^{-1}A_{21})\bar{x} + (B_1 - A_{12}A_{22}^{-1}B_2)u \tag{3.9}$$

and the fast (boundary layer) subsystem

$$\frac{d\bar{\eta}}{d\tau} = A_{22}(t)\bar{\eta} + B_2(t)\, u, \tag{3.10}$$

where the bar indicates that $\varepsilon = 0$ and $\bar{x} = \bar{\xi}$, as in (2.21).

In (3.10) the slow time t appears as a parameter. Thus the boundary layer controllability condition is

$$\text{rank}[B_2(t), A_{22}(t)B_2(t), \ldots, A_{22}^{m-1}(t)B_2(t)] = m, \quad \forall t \geq t_o. \tag{3.11}$$

This condition appeared in Kokotovic and Yackel (1972) and has been extended by Sannuti (1977) and used for time-optimal control in Kokotovic and Haddad (1975), Javid and Kokotovic (1977), and Javid (1978). We see once more that the boundary layer system can be treated as a time-invariant system.

The sufficient condition of Theorem 3.1 is not necessary. As shown by Chow (1977) this condition excludes the controllable case when the fast subsystem is controlled through the slow subsystem. From a practical point of view, when the fast modes are neglected as "parasitics," their weak controllability and observability contribute to the robustness of the simplified design. Although the problems of observability and robust observer design for singularly perturbed systems have attracted the attention of several authors, Porter (1974,1977), Balas (1978), O'Reilly (1979,1980), Javid (1980, 1982), Khalil (1981), Saksena and Cruz (1981), more work remains to be done on this important problem.

We proceed to the stability properties. Being invariant with respect to the transformation (3.5), these properties can be inferred from the properties of the two separate systems (3.7) and (3.8). Since the reduced system (3.9) and the boundary layer system (3.10) are the regular perturbations of (3.7) and (3.8), respectively, the following result is immediate.

Theorem 3.2

If Theorem 2.2 holds for $A_{22}(t) = D(t)$ and the reduced system (3.9) is uniformly asymptotically stable, then there exists $\varepsilon^* > 0$ such that the original system (3.1), (3.2) is uniformly asymptotically stable for all $\varepsilon \in (0, \varepsilon^*]$.

This theorem also follows as a corollary from more general results by Klimshev and Krasovski (1962), Wilde and Kokotovic (1972), Hoppensteadt (1974), and Habets (1974). The time-invariant version of the Theorem 3.2 was applied to networks with parasitics by Desoer and Shensa (1970) and to control systems by B. Porter (1974). A more detailed stability analysis leads to an estimate of $\varepsilon^*$ in terms of bounds on system matrices and their derivatives. For linear time-invariant case a bound was derived by Zien (1973) and for the time-varying case by Javid (1978). A robustness bound for linear time-invariant systems, uses the Laplace transform of (3.1), (3.2) expressed in a feedback form with $u = 0$ as

$$x(s) = (sI - A_{11})^{-1}A_{12}z(s) \tag{3.12}$$

$$z(s) = [I - \varepsilon s(\varepsilon sI - A_{22})^{-1}](-A_{22}^{-1}A_{21})x(s). \tag{3.13}$$

Defining the transfer function matrices G and $\Delta G$.

$$G(s) = A_{22}^{-1}A_{21}(sI - A_{11})^{-1}A_{12} \tag{3.14}$$

$$\Delta G(s,\varepsilon) = -\varepsilon s(\varepsilon sI - A_{22})^{-1} \tag{3.15}$$

and denoting by $\bar{\sigma}$ and $\underline{\sigma}$ the largest and the smallest singular values, respectively, the robustness conditions due to Sandell (1979) is stated as follows.

Theorem 3.3

If the reduced system (3.9) is stable, the full system (3.1), (3.2) remains stable for all $\varepsilon > 0$ satisfying

$$\bar{\sigma}(\Delta G(j\omega,\varepsilon)) \leq \underline{\sigma}(I + G^{-1}(j\omega)) \tag{3.16}$$

for all $\omega > 0$.

For nonlinear singularly perturbed systems the stability is frequently analyzed using separate Lyapunov functions for the reduced system and the boundary layer system and composing them into a single Lyapunov function for the full system. Let us first illustrate this on a nonlinear system which is linear in z,

$$\dot{x} = f(x) + F(x)z \tag{3.17}$$

$$\varepsilon\dot{z} = g(x) + G(x)z \tag{3.18}$$

where $G^{-1}(x)$ exists for all x. Lyapunov function introduced by Chow (1978) consists of two functions. The first function

$$v = a'(x)Q(x)a(x) \tag{3.18}$$

establishes the asymptotic stability of the reduced system $\dot{\bar{x}} = a(\bar{x})$, where

$$a(x) = f(x) - F(x)G^{-1}(x)g(x) \tag{3.19}$$

and $Q(x) > 0$ satisfies, for some differentiable $C(x) > 0$,

$$Q(x)a_x(x) + a_x'(x)Q(x) = -C(x) \quad , \quad a_x = \frac{\partial a}{\partial x} \tag{3.20}$$

where prime indicates a transpose. The second function

$$w = (z + \Gamma g - P^{-1}\Gamma'F'v_x')'P(z + \Gamma g - P^{-1}\Gamma'F'v_x'), \tag{3.21}$$

where $\Gamma = G^{-1}(x)$ and $P(x)$ satisfies

$$P(x)G(x) + G'(x)P(x) = - I, \tag{3.22}$$

establishes the asymptotic stability (uniform in x) of the fast (boundary layer) subsystem

$$\frac{d\eta}{d\tau} = G(x)\eta + g(x). \tag{3.23}$$

The Lyapunov function $V(x,z,\varepsilon)$ for the full system (3.17), (3.18) is composed from v and w as follows

$$V(x,z,\varepsilon) = v(x) + \frac{\varepsilon}{2} w(x,z). \tag{3.24}$$

It can be used to estimate the dependence of the domain of attraction on $\varepsilon$.

Among the stability results obtained by Klimshev and Krasovski (1962), Hoppenst (1967,1974), Habets (1974), Grujic (1979,1981), and Saberi and Khalil (1984) for the more general nonlinear system

$$\dot{x} = f(x,z,t) \tag{3.25}$$

$$\varepsilon\dot{z} = g(x,z,t) \tag{3.26}$$

we briefly outline the result by Habets. The reduced system of (3.25), (3.26) is

$$\dot{x} = f(x,\varphi(x,t),t) \tag{3.27}$$

where $\varphi(x,t)$ satisfies

$$g(x,\varphi(x,t),t) = 0, \tag{3.28}$$

while the boundary layer system is

$$\frac{dz}{d\tau} = g(x,z,t). \tag{3.29}$$

For simplicity let $f(0,0,t) = 0$, $g(0,0,t)$ and hence $\varphi(0,t) = 0$.

Theorem 3.4

Suppose that there exist Lyapunov functions $v(x,t)$ for (3.27) and $w(x,z,t)$ for

(3.29) such that

$$a(\|x\|) \leq v(t,x) \leq b(\|x\|) \tag{3.30}$$

$$a(\|z - \varphi(x,t)\|) \leq w(x,z,t) \leq b(\|z - \varphi(x,t)\|) \tag{3.31}$$

where a and b are positive nondecreasing scalar functions. Furthermore, suppose that positive constants $k_1$ and $k_2$ exist such that

$$\dot{v}(x,t) \leq - k_1\|x\|^2, \tag{3.32}$$

$$\left\|\frac{\partial v}{\partial x}\right\| \leq k_2\|x\|, \tag{3.33}$$

$$\dot{w}(x,z,t) \leq k_1\|z - \varphi(x,t)\|^2, \tag{3.34}$$

$$\left|\frac{\partial w}{\partial t}\right| \leq k_2\|z - \varphi(x,t)\|(\|z - \varphi(x,t)\| + \|x\|), \tag{3.35}$$

$$\left\|\frac{\partial w}{\partial x}\right\| \leq k_2\|z - \varphi(x,t)\|, \tag{3.36}$$

$$\|f(x,z,t) - f(x,\varphi(x,t),t)\| \leq k_2\|z - \varphi(x,t)\|, \tag{3.37}$$

$$\|f(x,z,t)\| \leq k_2(\|x\| + \|z - \varphi(x,t)\|), \tag{3.38}$$

$$\|\varphi(x,t)\| \leq b(\|x\|), \tag{3.39}$$

where $\dot{v}$ in (3.32) denotes the t-derivative for the reduced system (3.27), while $\dot{w}$ in (3.34) denotes the $\tau$-derivative for the boundary layer system (3.29). If (3.33) to (3.39) are satisfied, then there exists $\varepsilon^*$ such that for all $\varepsilon \in (0,\varepsilon^*]$ the equilibrium $x = 0$, $z = 0$ of (3.25), (3.26) is uniformly asymptotically stable.

Obtaining more easily verifiable stability conditions is an active research topic, of major interest in robustness studies of adaptive systems, Ioannou and Kokotovic (1982,1983). A possible approach to these problems is to investigate the perturbations of the aboslute stability property as in Siljak (1972), Ioannou (1981), and Saksena and Kokotovic (1981).

## 4. Optimal Linear State Regulators

One of the basic results of control theory is the solution of the optimal linear state regulator problem by Kalman (1960), which reduces the problem to the solution of a matrix Riccati equation. For the linear singularly perturbed system (3.1), (3.2) this equation is also singularly perturbed. It was investigated by Sannuti and Kokotovic (1969), Yackel (1971), Haddad and Kokotovic (1971), Kokotovic and Yackel (1972), O'Malley (1972), and O'Malley and Kung (1974). Another form of the regulator solution is obtained via a Hamiltonian boundary value problem which in this case is singularly perturbed. This approach was taken by O'Malley (1972b), Wilde (1972), Wilde and Kokotovic (1983), Assatani (1974), and others. A comparison of the two approaches was given by O'Malley and Kung (1974) and O'Malley (1975). In this sectic we give an outline of the Riccati equation approach.

The problem considered is to find a control u which, for the system (3.1), (3.2) minimizes the quadratic cost

$$J = \frac{1}{2}\int_{t_o}^{t_f} (y'y + u'R(t)u)dt \tag{4.1}$$

where $y = C_1(t)x + C_2(t)z$ is the system output and $x(t_f)$ are free. For $\varepsilon > 0$ the optimal state feedback control for this problem is

$$u(t) = -R(t)^{-1}[B_1'(t)\ \frac{1}{\varepsilon}B_2'(t)]K\begin{bmatrix} x(t) \\ z(t) \end{bmatrix} \tag{4.2}$$

where K is the positive definite solution of the Riccati equation[†]

$$\frac{dK}{dt} = -KA - A'K + KBR^{-1}B'K - C'C \quad , \quad K(t_f) = 0. \tag{4.3}$$

The singularity of (4.3) is due to the fact that the system matrices

$$A = \begin{bmatrix} A_{11}(t) & A_{12}(t) \\ \frac{A_{21}(t)}{\varepsilon} & \frac{A_{22}(t)}{\varepsilon} \end{bmatrix}, \qquad B = \begin{bmatrix} B_1(t) \\ \frac{B_2(t)}{\varepsilon} \end{bmatrix}, \tag{4.4}$$

are unbounded as $\varepsilon \to 0$. It is not obvious that (4.3) is a singularly perturbed syster in the form (1.1), (1.2). However, the search for a solution in the form

$$K = \begin{bmatrix} K_{11} & \varepsilon K_{12} \\ \varepsilon K_{12}' & \varepsilon K_{22} \end{bmatrix} \tag{4.5}$$

makes the singular perturbation form explicit, Sannuti (1968). Denoting $S_{11} = B_1^{-1}RB_1'$ $S_{22} = B_2R^{-1}B_2'$, $S_{12} = B_1R^{-1}B_2'$ and substituting (4.5) into (4.3) we obtain

---

[†] For brevity, arugment t is omitted whenever convenient.

$$\frac{dK_{11}}{dt} = -K_{11}A_{11} - A_{11}'K_{11} - K_{12}A_{21} - A_{21}'K_{12}' + K_{11}S_{11}K_{11} + K_{11}S_{12}K_{12}'$$

$$+ K_{12}S_{12}'K_{11} + K_{12}S_{22}K_{12}' - C_1'C_1 \qquad (4.6)$$

$$\varepsilon\frac{dK_{12}}{dt} = -K_{11}A_{12} - K_{12}A_{22} - \varepsilon A_{11}'K_{22} - A_{21}'K_{22} + \varepsilon K_{11}S_{11}K_{12} + K_{11}S_{12}K_{21}$$

$$+ \varepsilon K_{12}S_{12}'K_{12} + K_{12}S_{22}K_{22} - C_1'C_2 \qquad (4.7)$$

$$\varepsilon\frac{dK_{22}}{dt} = -\varepsilon K_{12}'A_{12} - \varepsilon A_{12}'K_{12} - K_{22}A_{22} - A_{22}'K_{22} + \varepsilon^2 K_{12}'S_{11}K_{12} + \varepsilon K_{12}'S_{12}K_{22}$$

$$+ \varepsilon K_{22}S_{12}'K_{12} + K_{22}S_{22}K_{22} - C_2'C_2 \qquad (4.8)$$

with the end condition

$$K_{11}(t_f) = 0, \quad K_{12}(t_f) = 0, \quad K_{22}(t_f) = 0. \qquad (4.9)$$

This is clearly a singularly perturbed system of the type (1.1), (1.2) and we can apply Theorem 1.1. When we set $\varepsilon = 0$, we get

$$\frac{d\bar{K}_{11}}{dt} = -\bar{K}_{11}(A_{11} - S_{12}\bar{K}_{12}') - (A_{11} - S_{12}\bar{K}_{12}')'\bar{K}_{11} + \bar{K}_{11}S_{11}\bar{K}_{11} - \bar{K}_{12}A_{21}$$

$$- A_{21}'\bar{K}_{12}' + \bar{K}_{12}S_{22}\bar{K}_{12}' - C_1'C_1, \; \bar{K}_{11}(t_f) = 0\,, \qquad (4.10)$$

$$0 = -\bar{K}_{12}(A_{22} - S_{22}\bar{K}_{22}) - \bar{K}_{11}A_{12} - A_{21}'\bar{K}_{22} + \bar{K}_{11}S_{12}\bar{K}_{22} - C_1'C_2 \qquad (4.11)$$

$$0 = -\bar{K}_{22}A_{22} - A_{22}'\bar{K}_{22} + \bar{K}_{22}S_{22}\bar{K}_{22} - C_2'C_2. \qquad (4.12)$$

The only end condition to be imposed on this algebraic-differential system is $\bar{K}_{11}(t_f) = 0$, while (4.11) and (4.12) now play the role of (1.3). A crucial property of this system is that (4.12) is independent of (4.10) and (4.11). To satisfy Assumption 1.1 we need a unique positive definite solution $\bar{K}_{22}$ of (4.12) to exist.

Assumption 4.1

For each fixed $t \in [t_o, t_f]$ the pair $A_{22}(t)$, $B_2(t)$ is stabilizable and pair $A_{22}(t)$, $C_2(t)$ is detectable.

For this assumption to hold it is sufficient that the controllability condition (3.11) and

$$\text{rank}[C_2'(t), A_{22}'(t)C_2'(t), \ldots, A_{22}'(t)^{m-1}C_2'(t)] = m \qquad (4.13)$$

hold for all $t \in [t_o, t_f]$. Under Assumption 4.1 eigenvalues of $A_{22} - S_{22}\bar{K}_{22}$ all have negative real parts and (4.11) can be solved for $\bar{K}_{12}$ in terms of $\bar{K}_{22}$, known from (4.12), and $\bar{K}_{11}$. Thus, the root (1.4) of interest in this case is distinct (isolated). The boundary layer system at $t_f$ corresponding to (4.11), (4.12) is

$$\frac{d\hat{K}_{12}(\tau)}{d\tau} = -\hat{K}_{12}(\tau)[A_{22}(t) - S_{22}(t)\hat{K}_{22}(\tau)] - [A_{21}(t) - S_{12}(t)\bar{K}_{11}(t)]'\hat{K}_{22}(\tau)$$

$$- \bar{K}_{11}(t)A_{12}(t) - C_1'(t)C_2(t), \qquad \tau = \frac{t-t_f}{\varepsilon}, \tag{4.14}$$

$$\frac{d\hat{K}_{22}(\tau)}{d\tau} = -\hat{K}_{22}(\tau)A_{22}(t) - (A_{22}'\hat{K}_{22}(\tau) + \hat{K}_{22}(\tau)S_{22}(t)\hat{K}_{22}(\tau) - C_2'(t)C_2(t) \tag{4.15}$$

with $\hat{K}_{12} = 0$ and $\hat{K}_{22} = 0$ at $\tau = 0$. For fixed t and $\varepsilon \to 0$ the limit (2.8) of Assumption is to be taken as $\tau \to -\infty$. It follows from the regulator theory that Assumption 4.1 guarantees that, as $\tau \to -\infty$, the solution $\hat{K}_{22}(\tau)$ of (4.15) converges uniformly to the positive definite root $\bar{K}_{22}(t)$ of (4.12), that is, to the solution of a "boundary lay regulator problem for each fixed $t \in [t_o, t_f]$. The uniform asymptotic stability of equation (4.14), which is linear in $\hat{K}_{12}(\tau)$, follows from standard stability theorems Thus (4.14) and (4.15) satisfy Assumption 2.1. Furthermore, matrix $\partial g/\partial z$ of Assumpt 2.2 for (4.14), (4.15) is block upper triangular with the eignevalues identical to t eigenvalues of $-[A_{22}(t) - S_{22}(t)\bar{K}_{22}(t)]$. Thus the uniform asymptotic stability of t boundary layer regulator also guarantees that Assumption 2.2 is satisfied. Hence th following result.

Theorem 4.1

If Assumption 4.1 is satisfied then for all $t \in [t_o, t_f]$ the solution of the full Riccati equation (4.3) is approximated by

$$K_{11}(t) = \bar{K}_{11}(t) + O(\varepsilon) \tag{4.16}$$

$$K_{12}(t) = \bar{K}_{12}(t) + \hat{K}_{12}(\tau) - \bar{K}_{12}(t_f) + O(\varepsilon) \tag{4.17}$$

$$K_{22}(t) = \bar{K}_{22}(t) + \hat{K}_{22}(\tau) - \bar{K}_{22}(t_f) + O(\varepsilon) \tag{4.18}$$

that is, by the separate solution of the slow ("reduced") and the fast ("boundary lay Riccati systems. Excluding the boundary layer correction terms the approximation

$$K_{11}(t) = \bar{K}_{11}(t) + O(\varepsilon) \tag{4.19}$$

$$K_{12}(t) = \bar{K}_{12}(t) + O(\varepsilon) \tag{4.20}$$

$$K_{22}(t) = \bar{K}_{22}(t) + O(\varepsilon) \tag{4.21}$$

is valid for all $t \in [t_o, t_1]$, where $t_1 < t_f$ can be made arbitrarily close to $t_f$ by choosing $\varepsilon$ small enough.

Higher order approximations are given in Yackel and Kokotovic (1973) and O'Malle and Kung (1974). Theorem 4.1 has important practical implications. First, we note that (4.15) represents the time-invariant Riccati equation depending on the fixed parameter t, which is, in fact, an independent optimality condition for the boundary layer regulator problem (3.10) in fast time scale $\tau$. Then the resulting feedback matrix $A_{22} - S_{22}K_{22}$ satisfies Theorem 2.2, that is, it guarantees the uniform asymptotic stability of the boundary layer. This is the stabilizing role of the fast regulator feedback $K_{22}$. We reiterate that the weakly controllable (stabilizable case is excluded, that is, Theorem 4.1 requires that the fast modes be controlled directly, rather than through the slow subsystem. Although not necessary, this requirement is needed for a robust design. The slow regulator is defined by the reduced system (4.10), (4.11), (4.12). At the first glance it appears that it depends on the quasi-steady-state solution $\bar{K}_{22}$ of the fast regulator. This would allow it to differ from the regulator solution for the problem in which $\varepsilon$ is neglected already in the system (3.1), (3.2) and in the cost (4.1), rather than later in the Riccati equation. The difference between the two reduced solutions would indicate nonrobustness, because the result would depend on the moment when $\varepsilon$ is neglected.

The robustness of the optimal state regulator problem with respect to singular perturbations is established by Haddad and Kokotovic (1971). The same robustness property is not automatic in other feedback designs. Khalil (1981) gives examples of non-robust feedback designs using reduced order observers or static output feedback. Gardner and Cruz (1978) show that, even with the state feedback, Nash games are non-robust with respect to singular perturbations.

Once the robustness of the optimal state regulator is established, we can proceed with the design which consists in implementing the control law (4.2) with approximate feedback gains (4.16), (4.17), (4.18). This is a two-time scale design because the feedback gains depending on t and $\tau$ are obtained separately. However, an equivalent, but more direct approach is the so-called composite control approach developed by Suzuki and Miura (1976) and Chow and Kokotovic (1976). We will present this approach in the section on nonlinear control. The singularly perturbed optimal regulator problem for linear difference (rather than differential) equations was solved by Blankenship (1981), and Litkouhi and Khalil (1983).

## 5. Linear Optimal Control

Although convenient for the feedback solution of linear optimal control problems with free endpoints, the Riccati equation approach must be modified in order to apply to problems with fixed endpoints. Two such modifications were developed by Wilde and Kokotovic (1973) and Asatani (1976). In general, endpoint constraints require the solution of Hamiltonian boundary value problems, which are in our case singularly perturbed. Various forms of singularly perturbed boundary value problems, not directly related to control applications, were studied earlier by Levin (1957), Vishik and Liusternik (1958), Harris (1960), Vasileva (1963), Wasow (1965), O'Malley (1969), Chang (1972), and others. Most of these works develop "inner" (in $\tau$ and $\sigma$) and "outer" (in t) asymptotic expansions. This approach to the boundary value problem arising in linear optimal control was taken by O'Malley (1972b,1975), O'Malley and Kung (1974), and Sannuti (1974). The results are based on hypotheses assuring the matching at both ends of the optimal trajectory.

Another approach, more in the spirit of the regulator theory, is that of Wilde (1972) and Wilde and Kokotovic (1973). It exploits the stabilizing properties of both the positive definite and the negative definite solutions of the same Riccati equation appearing in the regulator problem. These solutions split the original boundary value problem into two initial value problems, one of which is in reverse time. We present this approach by considering the same linear optimal control problem (3.1), (3.2), and (4.1), but this time with fixed endpoints

$$x(t_o) = x_o, \quad z(t_o) = z_o; \qquad x(t_f) = x_f, \quad z(t_f) = z_f. \tag{5.1}$$

Using p and $\varepsilon q$ as the adjoint variables corresponding to x and z, respectively, the optimal control is obtained as

$$u = -R^{-1}(B_1'p + B_2'q). \tag{5.2}$$

The standard necessary optimality conditions yield the singularly perturbed boundary value problem (5.1) for the Hamiltonian system

$$\begin{bmatrix} \dot{x} \\ \varepsilon\dot{z} \\ \dot{p} \\ \varepsilon\dot{q} \end{bmatrix} = \begin{bmatrix} A_{11} & A_{12} & -S_{11} & -S_{12} \\ A_{21} & A_{22} & -S_{12}' & -S_{22} \\ -C_1'C_1 & -C_1'C_2 & -A_{11}' & -A_{21}' \\ -C_2'C_1 & -C_2'C_2 & -A_{12}' & -A_{22}' \end{bmatrix} \begin{bmatrix} x \\ z \\ p \\ q \end{bmatrix}. \tag{5.3}$$

The reduced problem is

$$\begin{bmatrix} \dot{\bar{x}} \\ \dot{\bar{p}} \end{bmatrix} = \left\{ \begin{bmatrix} A_{11} & -S_{11} \\ -C_1'C_1 & -A_{11}' \end{bmatrix} - \begin{bmatrix} A_{12} & -S_{12} \\ -C_1'C_2 & -A_{21}' \end{bmatrix} \begin{bmatrix} A_{22} & -S_{22} \\ -C_2'C_2 & -A_{22}' \end{bmatrix}^{-1} \begin{bmatrix} A_{21} & -S_{12}' \\ -C_2'C_1 & -A_{12}' \end{bmatrix} \right\} \begin{bmatrix} \bar{x} \\ \bar{p} \end{bmatrix} \tag{5.4}$$

with the end conditions

$$\bar{x}(t_o) = x_o, \quad \bar{x}(t_f) = x_f. \tag{5.5}$$

The end conditions on z had to be dropped because the slow parts $\bar{z}$ and $\bar{q}$ of z and q are obtained from the linear algebraic equations when $\varepsilon\dot{z} = 0$ and $\varepsilon\dot{q} = 0$ is set in (5.3). Hence, $\bar{z}$ and $\bar{q}$ in general do not satisfy the end conditions (5.1) and "boundary layer appear at both ends of the optimal trajectory. The layer at the left end point must be uniformly asymptotically stable in the direct, and the layer at the right end point in the reverse time.

The two-time scale design of a near optimal trajectory is summarized in the following theorem.

Theorem 5.1

Suppose that Assumption 4.1 is satisfied and $\bar{x}(t)$ and $\bar{p}(t)$ uniquely satisfy (5.4) and (5.5). Denote by $P_{22}$ the positive definite root of the Riccati equation (4.12) at $t = t_o$ and by $N_{22}$ its negative definite root at $t = t_f$. Let $L(\tau)$ and $R(\sigma)$ be the solutions of two mutually independent time-invariant initial value problems

$$\frac{dL(\tau)}{d\tau} = [A_{22}(t_o) - S_{22}(t_o)P_{22}]L(\tau) \tag{5.6}$$

$$L(0) = z_o - \bar{z}(t_o) \tag{5.7}$$

and

$$\frac{dR(\sigma)}{d\sigma} = [A_{22}(t_f) - S_{22}(t_f)N_{22}]R(\sigma) \tag{5.8}$$

$$R(0) = z_f - \bar{z}(t_f) \tag{5.9}$$

where $\tau = (t-t_o)/\varepsilon$ and $\sigma = (t-t_f)/\varepsilon$ are the "stretched" time scales. Then there exists $\varepsilon^* > 0$ such that for all $t \in [t_o,t_f]$, $\varepsilon \in (0,\varepsilon^*]$

$$x(t,\varepsilon) = \bar{x}(t) + 0(\varepsilon) \tag{5.10}$$

$$z(t,\varepsilon) = \bar{z}(t) + L(\tau) + R(\sigma) + 0(\varepsilon) \tag{5.11}$$

$$p(t,\varepsilon) = \bar{p}(t) + 0(\varepsilon) \tag{5.12}$$

$$q(t,\varepsilon) = \bar{q}(t) + P_{22}L(\tau) + N_{22}R(\sigma) + 0(\varepsilon) \tag{5.13}$$

$$u(t,\varepsilon) = \bar{u}(t) + U_L(\tau) + U_R(\sigma) + 0(\varepsilon) \tag{5.14}$$

where

$$\bar{u}(t) = - R^{-1}(B_1'\bar{p} + B_2'\bar{q}) \tag{5.15}$$

$$U_L(\tau) = - R^{-1}(t_o)B_2'(t_o)P_{22}L(\tau) \tag{5.16}$$

$$U_R(\sigma) = - R^{-1}(t_f)B_2'(t_f)N_{22}R(\sigma). \tag{5.17}$$

The time scales for these two operations can be selected to be independent. For the reduced problem, a standard two point boundary value technique is used. The advantage over the original problem is that the order is lower, and the fast phenomena due to $\varepsilon$ are eliminated.

Example 5.1

We illustrate the procedure using the system and the cost

$$\dot{x} = z \tag{5.18}$$

$$\varepsilon\dot{z} = tz + u$$

$$J = \frac{1}{2}\int_1^2 [x^2 + (9-t^2)z^2 + u^2]dt \tag{5.19}$$

with end conditions as in (5.1). Since $A_{22} = t$, $B_2 = 1$, and $C_2'C_2 = 9 - t^2$, Assumption 4.1 holds for $0 < t < 3$. The exact optimal solution must satisfy

$$\dot{x} = z$$

$$\varepsilon\dot{z} = tz - q$$

$$\dot{p} = -x$$

$$\varepsilon\dot{q} = -(9-t^2)z - p - tq \tag{5.20}$$

subject to (5.1). When $\varepsilon$ is set equal to zero, the reduced problem is

$$\dot{\bar{x}} = -\frac{1}{9}\bar{p}$$

$$\dot{\bar{p}} = -\bar{x}. \tag{5.21}$$

Its solution $\bar{x}(t)$, $\bar{p}(t)$ is easily found using the eigenvalues $\frac{1}{3}$ and $-\frac{1}{3}$ of the system matrix in (5.21), while $\bar{z}$ and $\bar{q}$ are evaluated from

$$\bar{z} = -\frac{1}{9}\bar{p}, \quad \bar{q} = -\frac{t}{9}\bar{p}. \tag{5.22}$$

Then the roots of the Riccati equation

$$2K_{22}t - K_{22}^2 + (9-t^2) = 0 \tag{5.23}$$

$$P_{22} = t_o + 3 = 4, \qquad N_{22} = t_f - 3 = -1 \tag{5.24}$$

are used in (5.6), (5.8)

$$\frac{dL}{d\tau} = -3L; \qquad \frac{dR}{d\sigma} = 3R \tag{5.25}$$

to obtain the layer correction terms

$$L = [z_o - \bar{z}(1)]e^{-3(t-1)/\varepsilon}$$

$$R = [z_f - \bar{z}(2)]e^{3(t-2)/\varepsilon}. \tag{5.26}$$

Thus the corrections $L(\tau)$ and $R(\sigma)$ are the solutions of the left and the right "boundary layer regulators," respectively. It is right regulator (5.26) that allows us to automatically satisfy the end point matching condition $z_f - \bar{z}(2)$. It is totally unstable in real time t, that is, asymptotically stable in the reverse time $t_f - t$.

We can use the same example to illustrate the more common approach by O'Malley (1972b). Starting with (5.20) an asymptotic series in t, $\tau$, and $\sigma$ would be substituted for each of the variables and the terms with like powers of $\varepsilon$ are identified. The first terms $\bar{x}(t)$, $\bar{z}(t)$, $\bar{p}(t)$, $\bar{q}(t)$ in the t-series are obtained from (5.21) and (5.22), as in this approach. However, instead of using the Riccati and the boundary layer systems, (5.23) and (5.25), the first terms $z(\tau)$, $q(\tau)$, $z(\sigma)$, $q(\sigma)$ in the $\tau$- and the $\sigma$-series would be obtained from the $\tau$- and the $\sigma$-form of (5.20), subject to appropriate matching of their initial and end conditions. This approach can handle any other type of consistent initial and end-conditions. Both approaches lead to the same asymptotic solution, but under different hypotheses. The relationship of the hypotheses was investigated by O'Malley (1975).

## 6. Singular, Cheap, and High Gain Control

In our discussions thus far the singular perturbation properties of the system to be controlled were not altered by the control law. However, even if the original system is not singularly perturbed, a strong control action can force it to have fast and slow transients, that is, to behave like a singularly perturbed system. In feedback systems, the strong control action is achieved by high feedback gain. For a high gain system to emerge as a result of an optimal control problem, the control should be "cheap," that is, instead of u'Ru, its cost in (4.1) should be only $\varepsilon^2 u'Ru$, where $\varepsilon > 0$ is very small. On the other hand, an optimal control problem (3.1), (3.2), and (4.1) with det $R = 0$ is singular in the sense that the standard optimality conditions do not provide adequate information for its solution. Singular optimal controls and resulting singular arcs have been a control theory topic of considerable research interest, see for example Bell and Jacobson (1975). By formulating and analyzing the cheap control problem as a singular perturbation problem O'Malley and Jameson (1975,1977), Jameson and O'Malley (1975), and O'Malley (1976) have developed a new tool for a study of singular controls as the limits of cheap controls. The application of these results to the design of high gain and variable structure systems was discussed in Young, Kokotovic, and Utkin (1977). Here we closely follow a presentation in O'Malley (1978).

The cheap (near-singular) control problem for a linear system

$$\dot{x} = A(t)x + B(t)u, \qquad x \in R^n, \qquad u \in R^r \tag{6.1}$$

is characterized by the presence of $\varepsilon$ in the cost functional

$$J = \frac{1}{2} \int_{t_o}^{t_f} [x'Q(t)x + \varepsilon^2 u'R(t)u]dt \tag{6.2}$$

where Q and R are as usual symmetric positive definite. For $\varepsilon > 0$ the standard optimality conditions hold,

$$u = -\frac{1}{\varepsilon^2} R^{-1} B'p \tag{6.3}$$

$$\varepsilon^2 \dot{x} = \varepsilon^2 Ax - BR^{-1}B'p, \qquad x(t_o) = x^o \tag{6.4}$$

$$\dot{p} = -Qx - A'p, \qquad p(t_f) = 0 \tag{6.5}$$

but they are not defined for $\varepsilon = 0$. The singular perturbation method of the preceding sections does not apply because (6.4) is not in the standard form. On the other hand, the singular control theory establishes that the optimal singular arcs satisfy $B'p = 0$ which is consistent with the formal reduced system $BR^{-1}B'\bar{p} = 0$ obtained from (6.4). The results of O'Malley and Jameson treat a hierarchy of cases, where Case $\ell$ is defined by requiring that for $j = 0,1,\ldots,\ell-2$ and all $t \in [t_o, t_f]$

$$B_j'QB_j = 0, \qquad B_{\ell-1}'QB_{\ell-1} > 0 \tag{6.6}$$

where

$$B_o = B, \qquad B_j = AB_{j-1} - \dot{B}_{j-1}. \tag{6.7}$$

(There are also problems beyond all cases and those where the case changes with t.) For Case $\ell$ the fast time variables are defined as

$$\tau = \frac{t - t_o}{\mu}, \qquad \sigma = \frac{t_f - t}{\mu}, \qquad \mu = \varepsilon^{\frac{1}{\ell}} \tag{6.8}$$

and the control and the corresponding trajectory are of the form

$$u = \bar{u}(t,\mu) + \frac{1}{\mu^\ell} v(\tau,\mu) + w(\sigma,\mu) \tag{6.9}$$

$$x = \bar{x}(t,\mu) + \frac{1}{\mu^{\ell-1}} \eta(\tau,\mu) + \mu\rho(\sigma,\mu) \tag{6.10}$$

where the slow limiting control is $\bar{u}(t) = \bar{u}(t,0)$ and the slow trajectory $\bar{x}(t) = \bar{x}(t,0)$ lies on a manifold of dimension $n-\ell r$. A crucial property of the control (6.9) is its term $\frac{1}{\mu^\ell} v(\tau,\mu)$, which allows a rapid transfer from the given initial state to the singular arc. In the limit as $\varepsilon \to 0$ the control behavior is impulsive and can be analyzed by distributions, Francis and Glover (1978) and Francis (1979, 1982). The trajectory will feature impulsive behavior at $t = t_o$ whenever $\ell > 1$.

Applying the Riccati approach to (6.3), (6.4), (6.5), that is, setting $u = Kp$ we get

$$u = -\frac{1}{\varepsilon^2} R^{-1} B'Kx \tag{6.11}$$

where K satisfies

$$\varepsilon^2(\dot{K} + KA + A'K + Q) = KBR^{-1}B'K, \qquad K(t_f) = 0. \tag{6.12}$$

This equation is in the standard form only if $r = n$ and det $B \neq 0$ which is a very special and unlikely situation. For $r < n$ the r.h.s. term of (6.12) is singular. Hence this equation is not in the standard form and the procedure in Section 4 does not apply. We see, however, that a reduced solution $\bar{K}_o$ satisfies

$$B'\bar{K}_o = 0 \tag{6.13}$$

but this $\bar{K}_o$ is not fully defined. Since $B'KB$ can be nonsingular, we pre- and post-multiply (6.12) by $B'$ and $B$, respectively,

$$\varepsilon^2[B'(\dot{KB}) + B'KAB + B'A'KB + B'QB] = B'KBR^{-1}B'KB. \tag{6.14}$$

Substituting $K = \varepsilon K_1$ and letting $\varepsilon = 0$ we obtain the reduced equation

$$B'QB = (B'\bar{K}_1B)R^{-1}(B'\bar{K}_1B) \tag{6.15}$$

which in Case 1, when $B'QB > 0$, has a unique solution

$$B'\bar{K}_1B = \sqrt{R^{\frac{1}{2}}(B'QB)R^{\frac{1}{2}}} > 0. \tag{6.16}$$

Such an analysis suggests that $K$ be sought in the form

$$K = \bar{K}_o(t) + \varepsilon\bar{K}_1(t) + \varepsilon\hat{K}_1(\sigma) + O(\varepsilon^2) \tag{6.17}$$

where $\sigma = (t_f - t)/\varepsilon$ and $\hat{K}_1(\sigma)$ is the boundary layer correction at $t = t_f$. Substituting (6.17) into (6.12) and equating the terms of like powers in $\varepsilon$, we obtain at $\varepsilon = 0$

$$\dot{\bar{K}}_o + \bar{K}_oA_1 + A_1'\bar{K}_o - \bar{K}_oS_1\bar{K}_o + Q_1 = 0, \quad \bar{K}_o(t_f) = 0 \tag{6.18}$$

where

$$A_1 = A - B_1(B'QB)^{-1}B'Q \tag{6.19}$$

$$Q_1 = Q - QB(B'QB)^{-1}B'Q \tag{6.20}$$

$$S_1 = B_1(B'QB)^{-1}B_1' \geq 0 \tag{6.21}$$

and

$$\frac{d\hat{K}_1}{d\sigma} = -\hat{K}_1S_f\bar{K}_1(t_f) - \bar{K}_1(t_f)S_f\hat{K}_1 - \hat{K}_1S_f\hat{K}_1 \tag{6.22}$$

where $S_f = B(t_f)R^{-1}(t_f)B'(t_f)$. It can be shown that $\bar{K}_o(t)$ is defined by (6.13) and (6.18) and that (6.22) and $B'(t_f)\hat{K}_1(0) + B'(t_f)\bar{K}_1(t_f) = 0$ uniquely define $\hat{K}_1(\sigma)$ in terms of $\bar{K}_1(t_f)$. These facts and (6.16) allows us to form the control law (6.11) with the approximation (6.17) which, in view of $B'\bar{K}_o = 0$, becomes

$$u = -\frac{1}{\varepsilon} R^{-1} B'(\bar{K}_1 + \hat{K}_1)x. \tag{6.23}$$

With this high-gain feedback control the system (6.1) is

$$\varepsilon \dot{x} = [\varepsilon A - BR^{-1}B'(\bar{K}_1 + \hat{K}_1)]x. \tag{6.24}$$

Although it is not in the standard form we can expect that the reduced solution $\bar{x}$ satisfies $B'\bar{K}_1\bar{x} = 0$, that is, the corresponding singular arc is in the null space of $B'\bar{K}_1$. Since the prescribed initial condition $x(0) = x^o$ in general does not satisfy $B'\bar{K}_1 x^o = 0$, there will be a boundary layer at $t = t_o$, the rapid transition of $x$ from $x^o$ to $\bar{x}$. Another boundary layer will exist at $t = t_f$ because of the presence of $\hat{K}_1(\sigma)$ in (6.23).

We see that the analysis of singular perturbation problems which are not in the standard form is more complex than those which are. It is often useful to transform the problem into the standard form. The time-invariant problem (6.1) can, after a change of variables, always be written as

$$\begin{bmatrix} \dot{y} \\ \dot{z} \end{bmatrix} = \begin{bmatrix} A_{11} & A_{12} \\ A_{21} & A_{22} \end{bmatrix} \begin{bmatrix} y \\ z \end{bmatrix} + \begin{bmatrix} 0 \\ B_2 \end{bmatrix} u \tag{6.25}$$

where $y \in R^{n-r}$, $z \in R^r$, and $B_2$ is a nonsingular $r \times r$ matrix. With a high-gain feedback control

$$u = \frac{1}{\varepsilon} (F_1 y + F_2 z) \tag{6.26}$$

where $F_1$ and $F_2$ are constant matrices, the system (6.25) becomes

$$\dot{y} = A_{11} y + A_{12} z \tag{6.27}$$

$$\varepsilon \dot{z} = (\varepsilon A_{21} + B_2 F_1)y + (\varepsilon A_{22} + B_2 F_2)z. \tag{6.28}$$

If $F_2$ is chosen such that

$$\mathrm{Re}\lambda\{B_2 F_2\} < 0 \tag{6.29}$$

Theorem 2.1 holds and a two-time scale design is possible by designing the reduced (slow) subsystem

$$\dot{\bar{x}} = [A_{11} - A_{12}(B_2F_2)^{-1}B_2F_1]\bar{x} \tag{6.30}$$

and the boundary layer (fast) subsystem

$$\frac{d\eta}{d\tau} = B_2F_2\eta. \tag{6.31}$$

Taking $F_1 = F_2F_s$ (6.30) becomes

$$\dot{\bar{x}} = (A_{11} - A_{12}F_s)\bar{x}. \tag{6.32}$$

It can be shown, Kokotovic (1984), that the feedback matrices can be separately chosen, $F_2$ to place the eigenvalues of $B_2F_2$ and $F_s$ to place eigenvalues of $A_{11} - A_{12}F_2$. Such a design procedure was proposed by Young, Kokotovic, and Utkin (1977).

High gain systems have good disturbance rejection properties. They have been extensively studied in control literature, most recently by Sastry and Desoer (1983) and, using a geometric approach, by Willems (1981, 1982). Insensitivity and disturbance decoupling properties are analyzed by Young (1976, 1982a,b). High-gain systems may suffer because of neglected high frequency parasitics. This aspect was addressed by Young and Kokotovic (1982).

## 7. Composite Feedback Control of Nonlinear Systems

In the preceding three sections approximations of both the optimal feedback control and the optimal trajectory consisted of slow and fast parts. They are obtained from singularly perturbed Riccati equations or two-point boundary value problems. These optimality conditions also consisted of slow and fast parts. A further step toward a final decomposition of the two time scale design has been made which decomposes the optimal control problem itself into a slow subproblem and a fast subproblem. Separate solutions of these subproblems are then composed into a *composite feedback control* which is applied to the original system. As an engineering tool the composite control approach has both conceptual and practical advantages. The fast and slow controllers appear as recognizable entities which can be implemented in separate hardware or software.

The composite control was first developed for time-invariant optimal linear state regulators by Suzuki and Miura (1976), Chow (1977), and Chow and Kokotovic (1976), and then for nonlinear systems by Chow and Kokotovic (1978a,b, 1981), Suzuki (1981) and Saberi and Khalil (1985). A frequency domain composite design was developed by Fossard and Magni (1980). Extensions to stochastic control problems are due to Bensoussan (1981) and Khalil and Gajic (1984). The composite control has also been applied to large scale systems, as will be discussed in a subsequent section. The composite control approach is now presented following Chow and Kokotovic (1981).

The optimal control problems in the preceding sections were linear and over a finite time interval. We consider now a nonlinear infinite interval problem in which the system is

$$\dot{x} = a_1(x) + A_1(x)z + B_1(x)u, \qquad x(0) = x_o \tag{7.1}$$

$$\varepsilon\dot{z} = a_2(x) + A_2(x)z + B_2(x)u, \qquad z(0) = z_o \tag{7.2}$$

where $x \in R^n$, $z \in R^m$, $u \in R^r$ and the cost to be optimized is

$$J = \int_0^\infty [p(x) + s'(x)z + z'Q(x)z + u'R(x)u]dt. \tag{7.3}$$

Assumption 7.1

There exists a domain $D \subset R^n$, containing $x = 0$ as an interior point, such that for all $x \in D$ functions $a_1$, $a_2$, $A_1$, $A_2$, $A_2^{-1}$, $B_1$, $B_2$, p, s, r, and Q are differentiable with respect to x; $a_1$, $a_2$, p, and s are zero only at $x = 0$; Q and R are positive-definite matrices for all $x \in D$; the scalar $p+s'z + z'Qz$ is a positive-definite function of its arguments x and z, that is, it is positive except for $x = 0$, $z = 0$ where it is zero.

The usual approach would be to assume that a differentiable optimal value function $V(x,z,\varepsilon)$ exists satisfying

$$0 = \min_u [p + s'z + z'Qz + u'Ru + V_x(a_1 + A_1 z + B_1 u) + \frac{1}{\varepsilon} V_z(a_2 + A_2 z + B_2 u)] \tag{7.4}$$

where $V_x$, $V_z$ denote the partial derivatives of V. Since the control minimizing (7.4) is

$$u = -\frac{1}{2} R^{-1}(B_1'V_x' + \frac{1}{\varepsilon} B_2'V_2'), \tag{7.5}$$

the problem would consist of solving the Hamilton-Jacobi equation

$$0 = p + s'z + z'Qz + V_x(a_1 + A_1 z) + \frac{1}{\varepsilon} V_z(a_2 + A_2 z)$$

$$- \frac{1}{4} (V_x B_1 + \frac{1}{\varepsilon} V_z B_2)R^{-1}(B_1'V_x' + \frac{1}{\varepsilon} B_2'V_z'), \qquad V(0,0,\varepsilon) = 0. \tag{7.6}$$

To solve (7.6) is difficutl even for well-behaved nonlinear systems. The presence of $1/\varepsilon$ terms increases the difficulties. To avoid the difficulties we do not deal with the full problem directly. In contrast, we take advantage of the fact that as $\varepsilon \to 0$ the slow and the fast phenomena separate, and define two separate lower dimensional subproblems. The solutions of the two subproblems are combined into a composite control whose stabilizing and near optimal properties can be guaranteed.

For the slow subproblem, denoted by subscript "s," the fast transient is neglected, that is

$$\dot{x}_s = a_1(x_s) + A_1(x_s)z_s + B_1(x_s)u_s, \qquad x_s(0) = x_o \tag{7.7}$$

$$0 = a_2(x_s) + A_2(x_s)z_s + B_2(x_s)u_s \tag{7.8}$$

and, since $A_2^{-1}$ is assumed to exist,

$$z_s(x_s) = -A_2^{-1}(a_2 + B_2u_s) \tag{7.9}$$

is eliminated from (7.7) and (7.3). Then the slow subproblem is to optimally control the "slow subsystem"

$$\dot{x}_s = a_o(x_s) + B_o(x_s)u_s, \qquad x_s(0) = x_o \tag{7.10}$$

with respect to "slow cost"

$$J_s = \int_0^\infty [p_o(x_s) + 2s_o'(x_s)u_s + u_s'R_o(x_s)u_s]dt \tag{7.11}$$

where

$$a_o = a_1 - A_1A_2^{-1}a_2, \qquad B_o = B_1 - A_1A_2^{-1}B_2$$

$$p_o = p - s'A_2^{-1}a_2 + a_2'A_2'^{-1}QA_2^{-1}a_2,$$

$$s_o = B_2'A_2'^{-1}(QA_2^{-1}a_2 - \tfrac{1}{2}s), \qquad R_o = R + B_2'A_2'^{-1}QA_2^{-1}B_2. \tag{7.12}$$

We note that, in view of Assumption 7.1, the equilibrium of the slow subsystem (7.10) for all $x_s \in D$ is $x_s = 0$, and

$$p_o(x_s) + 2s_o'(x_s)u_s + u_s'R_o(x_s)u_s > 0, \qquad \forall x_s \neq 0,\ \forall u_s \neq 0. \tag{7.13}$$

Our crucial Assumption 7.2 concerns the existence of the optimal value function $L(x_s)$ satisfying the optimality principle

$$0 = \min_{u_s}[p_o(x_s)+2s_o'(x_s)u_s + u_s'R_o(x_s)u_s + L_x(a_o(x_s) + B_o(x_s)u_s)] \tag{7.14}$$

where $L_x$ denotes the derivative of L with respect to its argument $x_s$. The elimination of the minimizing control

$$u_s = -R_o^{-1}(s_o + \tfrac{1}{2}B_o'L_x') \tag{7.15}$$

from (7.14) results in the Hamilton-Jacobi equation

$$0 = (p_o - s_o'R_o^{-1}s_o) + L_x(a_o - B_oR_o^{-1}s_o) - \tfrac{1}{2}L_xB_oR_o^{-1}B_o'L_x', \qquad L(0) = 0 \tag{7.16}$$

where $p_o - s_o' R_o^{-1} s_o$ is positive definite in D.

Assumption 7.2

For all $x_s \in D$, (7.16) has a unique differentiable positive-definite solution $L(x_s)$ with the property that positive constants $k_1$, $k_2$, $k_3$, $k_4$ exist such that

$$k_1 L_x L_x' \leq - L_x \bar{a}_o \leq k_2 L_x L_x' \tag{7.17}$$

$$k_3 \bar{a}_o' \bar{a}_o \leq - L_x \bar{a}_o \leq k_4 \bar{a}_o' \bar{a}_o . \tag{7.18}$$

Then $L(x_s)$ is a Lyapunov function guaranteeing the asymptotic stability of $x_s = 0$ for the slow subsystem (7.10) controlled by (7.15), that is, for the feedback system

$$\dot{x}_s = a_o - B_o R_o^{-1}(s_o + \frac{1}{2} B_o' L_x') = - \bar{a}_o(x_s). \tag{7.19}$$

It also guarantees that D belongs to the region of attraction of $x_s = 0$.

For the fast subproblem, denoted by subscript "f", we recall that only an $O(\varepsilon)$ error is made by replacing x with $x_s$, or z with $z_s$. Thus we subtract (7.8) from (7.2), introduce $z_f = z - z_s$, $u_f = u - u_s$, neglect $O(\varepsilon)$ terms, and define the fast subproblem as

$$\varepsilon \dot{z}_f = A_2(x) z_f + B_2(x) u_f, \qquad z_f(0) = z_o - z_s(0), \tag{7.20}$$

$$J_f = \int_0^\infty (z_f' Q(x) z_f + u_f' R(x) u_f) dt. \tag{7.21}$$

This problem is to be solved for every fixed $x \in D$. It has the familiar linear quadratic form and a controllability assumption is natural.

Assumption 7.3

For every fixed $x \in D$,

$$\text{rand } [B_2, A_2 B_2, \ldots, A_2^{m-1} B_2] = m. \tag{7.22}$$

Alternatively, a less demanding stabilizability assumption can be made. For each $x \in D$ the optimal solution of the fast subproblem is

$$u_f(z_f, x) = - R^{-1}(x) B_2'(x) K_f(x) z_f \tag{7.23}$$

where $K_f(x)$ is the positive-definite solution of the x-dependent Riccati equation

$$0 = K_f A_2 + A_2' K_f - K_f B_2 R^{-1} B_2' K_f + Q. \tag{7.24}$$

The control (7.23) is stabilizing in the sense that the fast feedback system

$$\varepsilon \dot{z}_f = (A_2 - B_2 R^{-1} B_2' K) z_f \triangleq \bar{A}_2(x) z_f \tag{7.25}$$

has the property that $\mathrm{Re}\lambda[\bar{A}_2(x)] < 0, \quad \forall x \in D$.

We now form a "composite" control $u_c = u_s + u_f$, in which $x_s$ is replaced by x and $z_f$ by $z + A_2^{-1}(a_2 + B_2 u_s(x))$, that is

$$u_c(x,z) = u_s(x) - R^{-1} B_2' K_f (z + A_2^{-1}(a_2 + B_2 u_s(x))$$

$$= - R_o^{-1}(s_o + \frac{1}{2} B_o' L_x') - R^{-1} B_2' K_f (z + \bar{A}_2^{-1} \bar{a}_2) \tag{7.26}$$

where

$$\bar{a}_2(x) = a_2 - \frac{1}{2} B_2 R^{-1}(b_1' L_x' + B_2' V_1), \qquad \bar{a}_2(o) = 0,$$

$$V_1' = - (s' + 2a_2' K_f + L_x \bar{A}_1) \bar{A}_2^{-1}, \qquad \bar{A}_1 = A_1 - B_1 R^{-1} B_2' K_f. \tag{7.27}$$

The properties of the system controlled by the composite control are summarized in the following theorem.

Theorem 7.1

When Assumptions 7.1, 7.2, and 7.3 are satisfied then there exists $\varepsilon^*$ such that $\forall \varepsilon \in (0,\varepsilon^*]$, the composite control $u_c$ defined by (7.26) stabilizes the full system (7.1), (7.2) in a sphere centered at $x = 0$, $z = 0$. The corresponding cost $J_c$ is bounded. Moreover, $J_c$ is near optimal in the sense that $J_c \to J_s$ as $\varepsilon \to 0$.

This theorem shows that the considered nonlinear regulator problem is well-posed with respect to $\varepsilon$. It is the basis for a two-time scale design procedure developed by Chow and Kokotovic (1981) and Saberi and Khalil (1985).

## 8. Nonlinear Trajectory Optimization

We now consider a more general class of nonlinear optimal control problems on a finite interval $[t_o, t_f]$, frequently encountered in flight dynamics and start-up or shut-down operations for industrial plants. In Section 5 we have discussed such problems for linear systems and quadratic functionals. In this section we deal with nonlinear systems in the form

$$\dot{x} = f(x,z,u), \qquad x \in R^n \tag{8.1}$$

$$\varepsilon \dot{z} = g(x,z,u), \qquad z \in R^m \tag{8.2}$$

and the functional to be minimized

$$J = \int_0^{t_f} v(x,z,u) dt \tag{8.3}$$

where for simplicity of notation we do not show the dependence of f, g, and v on $\varepsilon$ and t. The Hamiltonian function for this problem is defined as

$$H = v + p'f + q'g \tag{8.4}$$

that is, the second adjoint variable $\varepsilon q$ is scaled for $g/\varepsilon$. This was the problem that attracted control engineers to singular perturbations, Kokotovic and Sannuti (1968), Sannuti and Kokotovic (1969), Kelley and Edelbaum (1970), and Kelley (1970a,b,c, 1971a,b), and singular perturbationists to control, Bagirova, Vasileva, Imanaliev (1967), O'Malley (1972, 1974). Papers by Kelley (1970a,b,c, 1971a,b, 1973) demonstrated the relevance of singular perturbations and boundary layer approximation for aircraft maneuver optimization and similar flight dynamics problems. These applications were further advanced by Ardema (1976, 1979, 1980), Calise (1976, 1978, 1979, 1980, 1981), Mehra, et al. (1979), Sridhar and Gupta (1980), and Shinar (1981, 1983). An application to nuclear reactors was reported in Reddy and Sannuti (1975). Asymptotic expansions and their validity were investigated by Hadlock (1970, 1973), O'Malley (1974), Sannuti (1974a,b, 1975), Freedman and Granoff (1976), Freedman and Kaplan (1976), Kurina (1977), Vasileva and Dmitriev (1980), Vasileva and Faminskaya (1981). A methodology similar to that of Sections 5 and 7 was developed by Chow (1979).

A different methodology was developed for linear time-optimal controls by Collins (1973), Kokotovic and Haddad (1975a,b), Javid and Kokotovic (1977), Javid (1978), and Halanay and Mirica (1979), in which case the bang-bang control exhibits outer low frequency and inner high frequency switches.

Using the Pontryagin's principle, or $\frac{\partial H}{\partial u} = 0$ if the control is unconstrained, u is eliminated in terms of the state and adjoint variables. The result is a nonlinear singularly perturbed $(2n+2m)$-dimensional boundary value problem

$$\dot{x} = \frac{\partial H}{\partial p} \qquad \dot{p} = -\frac{\partial H}{\partial x} \tag{8.5}$$

$$\varepsilon\dot{z} = \frac{\partial H}{\partial q} \qquad \varepsilon\dot{q} = -\frac{\partial H}{\partial z}\,. \tag{8.6}$$

In general, the initial and final states are required to be on some lower dimensional manifolds $M_o$ at $t = t_o$ and $M_f$ at $t = t_f$, that is, the boundary conditions for (8.5), (8.6) are

$$x(t_o), z(t_o) \in M_o, \qquad p(t_o), q(t_o) \perp M_o \tag{8.7}$$

$$x(t_f), z(t_f) \in M_f, \qquad p(t_f), q(t_f) \perp M_f. \tag{8.8}$$

From general properties of singularly perturbed boundary value problems, Wasow (1965), Chang (1972), Vasileva and Butuzov (1973), we know that an optimal trajector consists of a slow "outer" part with "boundary layers" at the ends. In the limit as $\varepsilon \to 0$ the problem decomposes into one slow and two fast subproblems. The slow ("outer") subproblem

$$\dot{x}_s = -\frac{\partial H_s}{\partial p_s}, \qquad \dot{p}_s = -\frac{\partial H_s}{\partial x_s} \tag{8.9}$$

is 2n-dimensional. To satisfy the remaining 2m boundary conditions, the layer ("inner") corrections $z_L(\tau_L)$, $z_R(\tau_R)$ for z, and $q_L(\tau_L)$, $q_R(\tau_R)$ for q are determined from the initial (L) and final (R) boundary layer systems with appropriately defined Hamiltonians $H^L$ and $H^R$, that is,

$$\frac{dz_L}{d\tau_L} = \frac{\partial H^L}{\partial q_L}, \qquad \frac{dq_L}{d\tau_L} = -\frac{\partial H^L}{\partial z_L} \tag{8.10}$$

$$\frac{dz_R}{d\tau_R} = \frac{\partial H^R}{\partial q_R}, \qquad \frac{dq_R}{d\tau_R} = -\frac{\partial H^R}{\partial z_R} \tag{8.11}$$

where $\tau_L = \frac{t-t_o}{\varepsilon}$, while $\tau_R = \frac{t_f-t}{\varepsilon}$ is the reversed fast time scale. The results of these subproblems are used to form approximations

$$u = u_s(t) + u_L(\tau_L) + u_R(\tau_R) + O(\varepsilon) \tag{8.12}$$

$$x = x_s(t) + O(\varepsilon) \tag{8.13}$$

$$z = z_s(t) + z_L(\tau_L) + z_R(\tau_R) + O(\varepsilon). \tag{8.14}$$

As was already discussed in Section 5, the L-layer must asymptotically decay, that is, the initial condition at $\tau_L = 0$ for (8.10) must be on a stable manifold. The endlayer $z_R(\tau_R)$ must asymptotically decay as $\tau_R \to \infty$, that is, as $t \to -\infty$ and hence $z_R(0)$ must lie on a totally unstable manifold of (8.11).

In realistic nonlinear problems the matching of layers and reduced solutions is not an easy task. It is more complex if the control is constrained and if singular arcs occur. For this reason practical approaches are problem-dependent and based on prior experience. This is particularly true in flight dynamics, where reduced order approximations based on "energy state" or "point mass" and "rigid body" models are common. In flight dynamics singular perturbations facilitate numerical computations in two ways: first, they reduce the number of costate initial values that must be determined simultaneously, and, second, they improve the conditioning of the boundary value problem. For example, the wild, undamped phugoid-like oscillations characteristic of the system (8.5), (8.6) for lifting atmospheric flight is avoided for the most part, being relegated to boundary layer corrections. Kelley, Calise and Ardema contain details of several applications containing the layers not only at the ends, but also at some inner points where the reduced trajectory is permitted to be discontinuous. Another difficulty in these applications is a proper choice of fast and slow variables, and the selection of one or several small parameters. Time scales differ in low thrust (aircraft) and high thrust (missle) applications.

## 9. Stochastic Filtering and Control

Research in singular perturbation of filtering and stochastic control problems with white noise inputs has revealed difficulties not present in deterministic problems. This is due to the fact that the input white noise process "fluctuates" faster than the dynamic variables, which as $\varepsilon \to 0$, themselves tend to white noise processes. In their surveys of stochastic differential equations and diffusion models Blankenship (1979) and Shuss (1980) and Kushner (1982) in a note, stress the importance of attaching clear probabilistic meaning to time scales.

To illustrate the problems arising in the singularly perturbed formulation of systems with white noise problems, we note that setting $\varepsilon = 0$ in the linear system

$$\dot{x} = A_{11}x + A_{12}z + B_1u + G_1w \tag{9.1}$$

$$\varepsilon\dot{z} = A_{21}x + A_{22}z + B_2u + G_2w \tag{9.2}$$

where w(t) is white Gaussian noise, is inadequate, since

$$\bar{z} = -A_{22}^{-1}(A_{21}\bar{x} + B_2\bar{u} + G_2w) \tag{9.3}$$

has a white noise component and, therefore, has infinite variance. As shown by Haddad (1976), variable $\bar{z}$ from (9.3) may be substituted for z in defining a reduced (slow) subsystem, but $\bar{z}$ cannot serve as an approximation for z in the mean square sense.

For the linear filtering of (9.1), (9.2) with respect to the observations

$$y = C_1x + C_2z + v \tag{9.4}$$

where v(t) is a white Gaussian noise independent of w(t), Haddad (1976) demonstrated that the Kalman filter can be approximately decomposed into two filters in different time scales.

For the control problem of the system (9.1), (9.2) and (9.4) with respect to the cost functional

$$J = E\{x'(T)\Gamma_1x(T) + 2\varepsilon x'(T)\Gamma_{12}z(T) + \varepsilon z'(T)\Gamma_2z(T)$$
$$+ \int_0^T (x'L_1x + 2x'L_{12}z + z'L_2z + u'Ru)dt\} \tag{9.5}$$

it was demonstrated by Teneketzis and Sandell (1977) and Haddad and Kokotovic (1977) that the optimal solution may be approximated by the solutions of two reduced order stochastic control problems in the slow and fast time scale. However, to avoid divergence $J \sim \frac{1}{\varepsilon}$, it is required that $L_2 \sim \varepsilon$, and $\Gamma_2 \sim \varepsilon^{\frac{1}{2}}$. More recently Khalil and Gajic

(1984) approached this problem via singularly perturbed Lyapunov equations. Razevig (1978) and Singh and Ram-Nandan (1982) have established the weak convergence, as $\varepsilon \to 0$, of the fast stochastic variable z which satisfies the Ito equation

$$\varepsilon dz = Az\, dt + \sqrt{\varepsilon}\, G dw; \qquad \mathrm{Re}\lambda(A) < 0 \tag{9.6}$$

where $\dot{w}(t)$ is Gaussian white noise with covariance W, that is,

$$\lim_{\varepsilon \to 0} z(t;\varepsilon) = \bar{z} \text{ weakly} \tag{9.7}$$

where $\bar{z}$ is a constant Gaussian random vector with covariance P satisfying the Lyapunov equation

$$AP + PA' + GWG' = 0. \tag{9.8}$$

Khalil (1978) assumes a colored noise disturbance in the fast subsystem to account for situations when the correlation time of the input stochastic process is longer than the time constants of fast variables. Thus the optimal solution to the stochastic regulator problem can be approximated by the optimal solution of the slow subproblem and optimal cost J does not diverge.

A composite control approach to a class of nonlinear systems driven by white noise disturbances, as a stochastic version of the results reviewed in Section 8, was developed by Bensoussan (1981). He considered

$$dx = (c(x)z + d(x) + 2\beta(x)u)dt + \sqrt{2}\, dw_1 \tag{9.9}$$

$$\varepsilon dz = (a(x)z + b(x) + 2\alpha(x)u)dt + \varepsilon\sqrt{2}\, dw_2 \tag{9.10}$$

$$J^{\varepsilon}_{x,z}(u(\cdot)) = E\int_0^{\infty} e^{-\nu t}[(f(x) + h(x)z)^2 + u^2]dt \tag{9.11}$$

where $w_1(t)$, $w_2(t)$ are standard Wiener processes independent of each other.

The optimal feedback law is obtained as

$$u^{\varepsilon}(x,z) = -\beta(x)V^{\varepsilon}_x(x,z) - \frac{\alpha(x)V^{\varepsilon}_z(x,z)}{\varepsilon} \tag{9.12}$$

where $V^{\varepsilon}(x,z)$ is the Bellman function. As $\varepsilon \to 0$. the optimal solution converges to the solutions of the two subproblems. The slow subproblem is

$$dx = (-\frac{c}{a}(b + 2\alpha u_s) + d + 2\beta u_s)dt + \sqrt{2}\, dw \tag{9.13}$$

$$J^{o}_{x}(u_s(\cdot)) = E\int_0^{\infty} e^{-\nu t}[(f - \frac{h}{a}(b + 2\alpha u_s)^2 + u_s^2]dt. \tag{9.14}$$

The fast subproblem is an x-dependent deterministic optimal control problem given by

$$\varepsilon \dot{z}_f = az_f + 2\alpha u_f \tag{9.15}$$

$$J^o_{z_f}(u_f(\cdot)) = \int_0^\infty (h^2 z_f^2 + u_f^2)dt. \tag{9.16}$$

The composite control is formed as in Section 8, namely,

$$u_c(x,z) = u_s(x) + u_f(x,z) \tag{9.17}$$

where $u_s(x)$ is the optimal control for (9.13), (9.14) and $u_f(x,z)$ is the optimal control for (9.15), (9.16).

Singular perturbations of quasi-variational inequalities arising in optimal stochastic scheduling problems are investigated by Hopkins and Blankenship (1981). Results for wide-band noise disturbances are obtained by Blankenship and Meyer (1977), Blankenship and Papanicolaou (1978), and El-Ansary and Khalil (1982). Time scales in stochastic differential equations are studied by Blankenship and Sachs (1977) and Blankenship (1978). Singular perturbations of stochastic filtering and control are an active research topic which has some common features with problems of mathematical physics, surveyed by Blankenship (1979) and Schuss (1980).

## 10. Time Scale Modeling of Networks

In the last several years time scale modeling and singular perturbation techniques have been extensively used in the study of large scale systems. We first give an overview of main topics and then concentrate on modeling issues. A time scale modeling methodology was developed for Markov chains with weak interactions by Gaitsgori and Pervozvanski (1975, 1979, 1980), Delebecque and Quadrat (1981), Phillips and Kokotovic (1981), Delebecque (1983), and Coderich et al. (1983) and for networks with weak connections by Avramovic, et al. (1980), Kokotovic (1981), Kokotovic, et al. (1982), Peponides, et al. (1982), and Peponides and Kokotovic (1983), summarized in a monograph by Chow, et al. (1982). This methodology has been applied to energy and power systems for management of dams, as in Delebecque and Quadrat (1978), and for network equivalencing as in Chow, et al. (1982). The models of large scale systems obtained by this methodology consist of a slow "core" which represents the only coupling of otherwise decoupled fast models of local subsystems. This model structure motivated a "multimodeling" approach to the decentralized control by Khalil and Kokotovic (1978, 1979a,b) further developed by Khalil (1979, 1980, 1981), Saksena and Cruz (1981, 1982), and Saksena and Başar (1983). The characteristic of multimodeling approach is that each local controller has a different model of the same large scale system which agrees with the models of other controllers only in the model of the slow core. A multi-parameter singular perturbation model with one slow core and N local fast subsystems captures this situation

$$\dot{x} = A_{00}x + \sum_{i=1}^{N} A_{0i}z_i + \sum_{i=1}^{N} B_{0i}u_i \tag{10.1}$$

$$\varepsilon_i\dot{z}_i = A_{i0}x + A_{ii}z_i + \sum_{\substack{j=1\\ j\neq i}}^{N} \varepsilon_{ij}A_{ij}z_j + B_{ii}u_i \ . \tag{10.2}$$

This model allows us to assume that each controller neglects all other fast subsystems and concentrates on its own subsystem, plus the interaction with others through the slow core. For the i-th controller, this is simply effected by setting $\varepsilon$-parameters to zero, except for $\varepsilon_i$. The i-th controller's simplified model is then

$$\dot{x}^i = A_i x^i + A_{0i}z_i + B_{0i}u_i + \sum_{\substack{j=1\\ j\neq i}}^{N} B_{ij}u_j \tag{10.3}$$

$$\varepsilon_i\dot{z}_i = A_{i0}x^i + A_{ii}z_i + B_{ii}u_i \tag{10.4}$$

which is often all the i-th controller knows about the whole system. The k-th controller, on the other hand, has a different model of the same large scale system. Control $u_i$ can be divided into a slow part, which contributes to the control of the core, and a fast part controlling only its own fast subsystem. The multiparameter perturbation problem has been solved under rather restrictive D-stability assumptions, Khalil and Kokotovic (1979), Ozguner (1979), and Khalil (1981). Stochastic multimodeling problems are even more complex, because of the so-called nonclassical information patterns, Saksena and Başar (1983).

Singular perturbation problems for multiple controllers with different cost functionals (e.g., differential games) are complex even with a single perturbation parameter. We have already mentioned the ill-posedness of linear Nash games with respect to singular perturbations, Gardner and Cruz (1978). Singularly perturbed differential games were further investigated by Salman and Cruz (1979), Khalil and Kokotovic (1979), Khalil and Medanic (1980). Singularly perturbed pursuit-evasion problem was studied by Farber and Shinar (1980) and Shinar (1981).

Let us conclude this section and the whole survey with a closer look at a fundamental property of large scale systems--the fact that the time scales are caused by weak connections, Kokotovic (1981). Although this is a property of a wide class of nonlinear systems, such as power systems, Peponides, Kokotovic, and Chow (1982), and multimarket economies, Peponides and Kokotovic (1983), we restrict our discussion to linear time-invariant systems in the form

$$\varepsilon\dot{v} = [A + \varepsilon B(\varepsilon)]v, \qquad v \in R^n, \tag{10.5}$$

where A represents strong *internal* connections within a subsystem while $\varepsilon B$ are weak *external* connections among subsystems. If A is singular, this is not a standard form

(1.1), (1.2) because the crucial Assumption (1.1) is violated. Of the rich literature dealing with (10.5) and its generalizations we mention only a few representative references. Vasileva in (1975, 1976) and in her monograph with Butuzov (1978) treats (10.5) as a "critical case" of singular perturbations. For O'Malley (1978), 1979) and O'Malley and O'Flaherty (1977, 1980) these are "singular singularly perturbed" problems. In the monographs by Campbell (1980, 1982) they are special cases of "singular systems" of differential equations. These terminological differences imply different approaches, or different assumptions about the structure of A and B in (10.5). Denoting by $\mathcal{R}(A)$ and $\mathcal{N}(A)$ the range space and the null space of A, respectively, let us assume, as in Peponides, et al. (1982) that

$$\mathcal{R}(A) \oplus \mathcal{N}(A) = R^n \tag{10.7}$$

$$\dim \mathcal{R}(A) = \rho, \ \dim \mathcal{N}(A) = \nu \tag{10.8}$$

and, hence $\rho + \nu = n$. Then $\mathcal{N}(A)$ is the *equilibrium manifold* of

$$\frac{dv(\tau)}{d\tau} = Av(\tau) \tag{10.9}$$

and there exists a $\rho \times n$ matrix Q such that

$$Qv = 0 \iff v \in \mathcal{N}(A). \tag{10.10}$$

Moreover, let the rows of an $\nu \times n$ matrix P span the left null space of A, then

$$P\frac{dv}{d\tau} = PAv = 0 \tag{10.11}$$

represents a *conservation manifold* of (10.9) because

$$Pv(0) = Pv(\tau), \qquad \forall \tau > 0. \tag{10.12}$$

The time scales of (10.5) are clear from (10.9) and (10.12) which represent the "near-equilibrium" and "near-conservation" properties of (10.5).

Theorem 10.1

The slow and fast variables of (10.5) are x and z, respectively,

$$x = Pv, \qquad z = Qv, \qquad v = Sx + Tz, \tag{10.13}$$

and this change of variables transforms (10.5) into

$$\dot{x} = PB(\varepsilon)Sx + PB(\varepsilon)Tz \tag{10.14}$$

$$\varepsilon\dot{z} = QB(\varepsilon)Sx + [QAT + \varepsilon QB(\varepsilon)T]z \tag{10.15}$$

which is a standard form because QAT is nonsingular due to (10.7).

This defines the fastest time scale $\tau = \frac{t}{\varepsilon}$ and

$$\dot{\bar{x}} = PB(0)S\bar{x} \tag{10.16}$$

is the slow (reduced) subsystem of (10.14), (10.15). If PB(0)S is singular, there will be time scales slower than t and the same procedure can be continued. This is the essence of a sequential determination of time scales by Coderich, et al. (1983), Delebecque (1983), and Khalil (1984a).

Example 10.1

Let us re-examine the RC-network in Fig. 1.2 and its model (1.17), (1.18). In this case

$$A = \begin{bmatrix} -1 & 1 \\ 1 & -1 \end{bmatrix}, \qquad B = \begin{bmatrix} 0 & 0 \\ 0 & -\frac{1}{R} \end{bmatrix} \tag{10.17}$$

and Q and P can be defined as

$$Q = [1 \quad -1], \qquad P = [p \quad p]. \tag{10.18}$$

For (1.21) with $C_1 = C_2$ coefficient p is $\frac{1}{2}$. The near conservation property of the network in Fig. 1.2 refers to the fact that if $R = \infty$, the total charge on the capacitors and the "aggregate" voltage x is the voltage on the sum of the capacitors with that charge. During the fast transient this voltage remains essentially constant, while the actual voltages $v_1$ and $v_2$ converge to their quasi-steady state $v_1 = v_2$. Their difference

$$z = Qv = v_1 - v_2 \tag{10.19}$$

is the fast variable. Its substitution into (1.22), (1.23) would put the network model in the form (10.15).

In networks and Markov chains, A is often block-diagonal and each of its N blocks $A_i$ represents a local network or Markov chain with the property that

$$\det A_i = 0, \qquad i=1,\ldots,N. \tag{10.20}$$

The most interesting case is when $\dim N(A_i) = 1$ for all $i=1,\ldots,N$ and hence $\nu = N$. Then P is an $N \times n$ dimensional *aggregation* matrix and $x = Pv$ defines one aggregate variable for each subsystem. In Markov chains the aggregate variable $x_i$ is the probability for the Markov process to be in the class i of the strongly interacting states. For the multimodeling approach to decentralized control it is of crucial importance that QAT is block diagonal, that is, the fast subsystems are indeed "local." The variables in the same subsystem are "coherent" because their response to the excitation of system-wide slow modes is identical. This is why for slow phenomena all the variables of the same subsystem can be aggregated into one variable. Aggregation and coherency are generalized to nonlinear networks in Peponides (1982), and Peponides, et al. (1982) and extended to modeling of multimarket economies in Peponides and Kokotovic (1983). The relationship of aggregability and weak coupling was investigated in early aggregation works by Simon and Ando (1961) and Simon (1962). These concepts can now be further analyzed by singular perturbation techniques.

In applications, an inverse problem is of even greater importance. We have seen that weak connections imply the time scales. The inverse question is how to use the knowledge of time scales to find the weak connections and decompose a large network into weakly connected subnetworks ("areas"). An efficient computer algorithm was developed for power systems by Avramovic (1980), Avramovic, et al. (1980), and Chow et al. (1982). Other applications involve models of queueing networks. It appears from these first experiences that singular perturbations and time scales will play an important role in computer assisted modeling of large scale systems.

## Concluding Remarks

Several results discussed in this paper have already been extended to distributed parameter systems. Typical references are Lions (1973), Asatani (1976), Desoer (1977), and Balas (1982) and it is clear that more work will be done in this area. Averaging and homogenization, Bensoussan, Lions, and Papanicolaou (1978), Blankenship (1979) are a related class of time-scale methods which have not been discussed. We expect to see more control applications of these methods. Our discussion of stochastic control, with the help of Blankenship (1979), and Schuss (1980), indicates that most of the major problems are still open for an efficient time-scale asymptotic treatment.

This is not to say that all is quiet on the deterministic front. The composite control approach is still restricted to special classes of systems. Trajectory optimization problems with singular arcs and state and control constraints have so far been treated in a semi-heuristic way and are in need of theoretical support. Time scaling of nonlinear models is a crucial unsolved problem. Will geometric methods help?

The developments in modeling and control of large scale systems, Chow, et al. (1982) are extremely encouraging and expected to continue at a rapid rate. When the relationship between weak or sparse connections and time scales is fully understood, the time scale asymptotic methods will be one of the most powerful tools for analysis and design of large scale systems. Let us not forget that one of the advantages of time scale methods is that they do not depend on linearity and should apply to most nonlinear models.

## Acknowledgements

The author is thankful to Mrs. Dixie Murphy for her expert typing. Preparation of this text was supported in part by the Joint Services Electronics Program under Contract N00014-84-C-0419 and in part by the National Science Foundation under Grant ECS-83-11851.

REFERENCES

Allemong, J. J. and P. V. Kokotovic (1980), "Eigensensitivities in reduced order modeling," IEEE Trans. Automat. Control, AC-25, pp. 821-822.

Altshuler, D. and A. H. Haddad (1978), "Near optimal smoothing for singularly perturbed linear systems," Automatica, 14, pp. 81-87.

Anderson, B. D. O. and J. B. Moore (1971), Linear Optimal Control, Prentice Hall, Englewood Cliffs, New Jersey.

Anderson, P. M. and A. A. Fouad (1977), Power System Control and Stability, Iowa State University Press.

Anderson, L. (1978), "Decomposition of two time scale linear systems," Proc. JACC, pp. 153-163.

Andreev, Yu H. (1982), "Differential geometry method, in control theory -- a survey," Automatika Telemachnika, 10, pp. 5-46.

Ardema, M. D. (1976), "Solution of the minimum time-to-climb problem by matched asymptotic expansions," AIAAJ, 14, pp. 843-850.

Ardema, M. D. (1979), "Linearization of the boundary layer equations for the minimum time to climb problem," AIAAJ Guidance and Control, 2, pp. 434-436.

Ardema, M. D. (1980), "Nonlinear singularly perturbed optimal control problems with singular arcs," Automatica, 16, pp. 99-104.

Ardema, M. D. (1983), "Singular perturbations in Systems and Control," CISM Courses and Lectures, 180, Springer, New York.

Asatani, K. (1974), "Suboptimal control of fixed-end-point minimum energy problem via singular perturbation theory," J. Math. Anal. Appl., 45, pp. 684-697.

Asatani, K. (1976), "Near-optimum control of distributed parameter systems via singular perturbation theory," J. Math. Anal. Appl., 54, pp. 799-819.

Asatani, K., M. Shiotani, and Y. Huttoni (1977), "Suboptimal control of nuclear reactors with distributed parameters using singular perturbation theory," Nuclear Science and Engineering, 6, pp. 119-xxx.

Athans, M. and P. L. Felt (1966), Optimal Control: An Introduction to the Theory and its Applications, McGraw-Hill, New York.

Avramovic, B. (1979), "Subspace iteration approach to the time scale separation," Proc. IEEE Conference on Decision and Control, pp. 684-697.

Avramovic, B. (1980), "Time scales, coherency, and weak coupling," Ph.D. Thesis, Coordinated Science Laboratory, Report R-895, Univ. Illinois, Urbana.

Avramovic, B., P. V. Kokotovic, J. R. Winkelman and J. H. Chow (1980), "Area decomposition of electromechanical models of power systems," Automatica, 16, pp. 637-648.

Bagirova, N., A. B. Vasileva and M. I. Imanaliev (1967), "The problem of asymptotic solutions of optimal control problems," Differential Equations, 3, pp. 985-988.

Balas, M. J. (1978), "Observer stabilization of singularly perturbed systems," J. Guidance and Control, 1, pp. 93-95.

Balas, M. J. (1982), "Reduced order feedback control of distributed parameter systems via singular perturbation methods," J. Math. Anal. Appl., 87, pp. 281-294.

Bell, D. J. and D. H. Jacobson (1975), Singular Optimal Control Problems, Academic Press, New York.

Bensoussan, A., J. L. Lions and G. C. Papanicolaou (1978), Asymptotic Analysis for Periodic Structures, North-Holland, New York.

Bensoussan, A. (1981), "Singular perturbation results for a class of stochastic control problems," IEEE Trans. Automat. Control, AC-26, pp. 1071-1080.

Bensoussan, A. (1984), "On some singular perturbation probelms arising in optimal control," Stochastic Anal. and Applic., 2, pp. 13-53.

Blankenship, G. and D. Meyer (1977), "Linear filtering with wide band noise disturbances," 16th IEEE Conf. on Decision and Control, pp. 580-584.

Blankenship, G. and G. C. Papanicolaoi (1978), "Stability and control of stochastic systems with wide-band noise disturbance, SIAM J. Appl. Math., 34, pp. 437-476.

Blankenship, G. (1978), "On the separation of time scales in stochastic differential equations," Proc. 7th IFAC Congress, Helsinkii, pp. 937-944.

Blankenship, G. (1979), "Asymptotic analysis in mathematical physics and control theory: some problems with common features," Richerche di Automatica, 10, p. 2.

Blankenship, G. and S. Sachs (1979), "Singularly perturbed linear stochastic ordinary differential equations," SIAM J. Mat. Anal., 10, pp. 306-320.

Blankenship, G. (1981), "Singularly perturbed difference equations in optimal control problems," IEEE Trans. Automat. Control, AC-26, pp. 911-917.

Bogoliubov, N. N. and Y. A. Mitropolsky (1961), Asymptotic Methods in the Theory of Non-Linear Oscillations, Second Ed., Hinduston Publishing, Delhi.

Bratus, A. S. (1977), "Asymptotic solutions of some probabilistic optimal control problems," Appl. Math. Mech., (PMM), 41, pp. 13-xx.

Brauner, C. M. (1978), "Optimal control of a perturbed system in enzyme kinetics," Proc. 7th IFAC Congress, Helsinki, pp. 945-948.

Brockett, R. W. (1970), Finite Dimensional Linear Systems, Wiley, New York.

Bryson, A. E. and Y. C. Ho (1975), Applied Optimal Control, Hemisphere, Washington, DC.

Butuzov, V. F. and A. B. Vasileva (1970), "Differential and difference equation systems with a small parameter in the case when unperturbed (degenerated) system is on the spectrum," Differential Equations, 6, pp. 499-510.

Butuzov, V. F. and M. V. Fedoryuk (1970), "Asymptotic methods in theory of ordinary differential equations," Progress in Mathematics, 8, R. V. Gamkrelidze, ed., Plenum Press, New York, pp. 1-82.

Calise, A. J. (1976), "Singular perturbatin methods for variational problems in aircraft flight," IEEE Trans. Automat. Control, AC-21, pp. 345-353.

Calise, A. J. (1978), "A new boundary layer matching procedure for singularly perturbed systems," IEEE Trans. Automat. Control, AC-23, pp. 434-438.

Calise, A. J. (1979), "A singular perturbation analysis of optimal aerodynamic and thrust magnitude control, IEEE Trans. Automatic Control, AC-24, pp. 720-730.

Calise, A. J. (1980), "A singular perturbation analysis of optimal thrust control with proportional navigation guidance," AIAAJ Guidance and Control, 3, pp. 312-318.

Calise, A. J. (1981), "Singular perturbation theory for on-line optimal flight path control," AIAAJ Guidance and Control, 4, pp. 398-405.

Campbell, S. L. (1978), "Singular perturbation of autonomous linear systems II," Differential Equations, 29, pp. 362-373.

Campbell, S. L. and N. J. Rose (1978), "Singular perturbation of autonomous linear systems III," Houston J. Math., 4, pp. 527-539.

Campbell, S. L. and N. J. Rose (1979), "Singular perturbation of autonomous linear systems," SIAM J. Math. Anal., 10, pp. 542-551.

Campbell, S. L. (1980), Singular Systems of Differential Equations, Pitman, New York.

Campbell, S. L. (1981), "A more singular singularly perturbed linear system," IEEE Trans. Automat. Control, AC-26, pp. 507-510.

Campbell, S. L. (1981), "On an assumption guaranteeing boundary layer convergence of singularly perturbed systems," Automatica, 17, pp. 645-646.

Campbell, S. L. (1982), Singular Systems of Differential Equations II, Pitman, New York.

Carr, Jack (1981), Applications of Centre Manifold Theory, Lecture Notes in Applied Mathematical Sciences, 35, Springer-Verlag, New York.

Chang, K. W. (1969), "Remarks on a certain hypothesis in singular perturbations," Proc. Amer. Math. Soc., 23, pp. 41-45.

Chang, K. Wand and W. A. Coppel (1969), "Singular perturbations of initial value problems over a finite interval," Arch. Rational Mech. Anal., 32, pp. 268-280.

Chang, K. W. (1972), "Singular perturbations of a general boundary problem," SIAM J. Math. Anal., 3, pp. 520-526.

Chow, J. H. and P. V. Kokotovic (1976a), "A decomposition of near optimum regulators for systems with slow and fast modes," IEEE Trans. Automat. Control, AC-21, pp. 701-705.

Chow, J. H. and P. V. Kokotovic (1976b), "Eigenvalue placement in two time scale systems," Proc. IFAC Symp. on Large Scale Systems, Udine, Italy, pp. 321-326.

Chow, J. H. (1977a), "Singular perturbation of nonlinear regulators and systems with oscillatory modes," Ph.D. Thesis, Coordinated Science Laboratory, Report R-

801, Univ. Illinois, Urbana.

Chow, J. H. (1977b), "Preservation of controllability in linear time invariant perturbed systems," Int. J. Control, 25, pp. 697-704.

Chow, J. H. (1978), "Asymptotic stability of a class of nonlinear singularly perturbed systems," J. Franklin Inst., 306, pp. 275-278.

Chow, J. H. (1978), "Pole-placement design of multiple controllers via weak and strong controllability," Int. J. Syst. Sci., 9, pp. 129-135.

Chow, J. H. and P. V. Kokotovic (1978a), "Near-optimal feedback stabilization of a class of nonlinear singularly perturbed systems," SIAM J. Control Optim., 16, pp. 756-770.

Chow, J. H. (1978b), "Two time scale feedback design of a class of nonlinear systems," IEEE Trans. Automat. Control, AC-23, pp. 438-443.

Chow, J. H., J. J. Allemong and P. V. Kokotovic (1978), "Singular perturbation analysis of systems with sustained high frequency oscillations, Automatica, 14, pp. 271-279.

Chow, J. H. (1979), "A class of singularly perturbed nonlinear, fixed endpoint control problems," J. Optim. Theory Appl., 29, pp. 231-251.

Chow, J. H. and P. V. Kokotovic (1981), "A two-stage Lyapunov-Bellman feedback design of a class of nonlinear systems," IEEE Trans. Automat. Control, AC-26, pp. 656-663.

Chow, J. H., Ed. (1982), Time Scale Modeling of Dynamic Networks, Lecture Notes in Control Information Sciences, 47, Springer-Verlag, New York.

Chow, J. H. and P. V. Kokotovic (1983), "Sparsity and time scales," Proc. of 1983 American Control Conference, 2, San Francisco, pp. 656-661.

Chow, J. H., P. V. Kokotovic and Y. K. Hwang (1983), "Aggregate modelling of dynamic networks with sparse interconnections," Proc. 22nd Decision and Control Conference, San Antonio, pp. 223-229.

Chow, J. H., J. Cullum and R. A. Willoughby (1984), "A sparity-based technique for identifying slow-coherent areas in large power systems," IEEE Trans. Power Appar. Systems, PAS-103, pp. 463-471.

Cobb, D. (1984), "Slowest fast stability eigenvalues systems," Proc. 23rd Decision and Control Conference, Las Vegas, Nevada, pp. 280-282.

Coderch, M., A. S. Willsky, S. S. Sastry and D. A. Castanon (1983), "Hierarchical aggregation of linear systems with multiple time scales," IEEE Trans. Automat. Control, AC-28, pp. 1017-1030.

Collins, W. B. (1973), "Singular perturbations of linear time-optimal control," Recent Mathematical Developments in Control, D. J. Bell, ed., Academic Press, New York, pp. 123-136.

Coppel, W. A. (1965), Stability and Asymptotic Behavior of Differential Equations, D. C. Heath and Company, Boston.

Coppel, W. A. (1967), "Diochotomies and reducibility," J. Differentialll Equations, 3, pp. 500-521.

Cori, R. and C. Maffezzoni (1984), "Practical optimal control of a drum boiler power plant," Automatica, 20, pp. 163-173.

Delebecque, F. and J. P. Quadrat (1978), "Contribution of stochastic control, singular perturbation averaging and team theories to an example of large scale systems: management of hydropower production," IEEE Trans. Automat. Control, AC-23, pp. 209-222.

Delebecque, F. and J. P. Quadrat (1981), "Optimal control of Markov chains admitting strong and weak interactions," Automatica, 17, pp. 281-296.

Delebecque, F. (1983), "A reduction process for perturbed Markov chains," SIAM J. Appl. Math, 43, pp. 325-350.

Delebecque, F., J. P. Quadrat and P. V. Kokotovic (1984), " unified view of aggregation and coherency in networks and Markov chains," Int. J. Control, 40, pp. 939-952.

Demello, F. P. and C. Concordia (1969), "Concepts of synchronous machine stability as affected by excitation control," IEEE Trans. Power Applications and Systems, PAS-88, pp. 316-329.

Desoer, C. A. (1970), "Singular perturbation and bounded input bounded state stability," Electronic Letters, 6, pp. 16-17.

Desoer, C. A. and M. J. Shensa (1970), "Network with very small and very large parasitics: natural frequencies and stability," Proc. IEEE, 58, pp. 1933-1938.

Desoer, C. A. (1977), "Distributed networks with small parasitic elements: input-output stability," IEEE Trans. Circuits and Systems, CAS-24, pp. 1-8.

Dmitriev, M. G. (1978), "On a class of singularly perturbed problems of optimal control," J. Appl. Math. Mech., PMM, 42, pp. 238-242.

Dontchev, A. L. (1983), Perturbations. Approximations and Sensitivity Analysis of Optimal Control Systems. Lecture notes in Control and Information Science, 52, Springer-Verlag, New York.

Dontchev, A. L. and V. M. Veliov (1983), "Singular perturbtion in Mayer's problem for linear systems," SIAM J. Control Optim., 21, pp. 566-581.

Dragan, V. and A. Hakanay (1982), "Suboptimal stabilization of linear systems with several time scales," Int. J. Control, 36, pp. 109-126.

Eckhaus, W. (1973), Matched Asymptotic Expansions and Singular Perturbations, North-Holland/American Elsevier, New York.

Eckhaus, W. (1977), "Formal approximation and singular perturbations," SIAM Review, 19, pp. 593-633.

El-Ansary, M. and H. Khalil (1982), "Reduced-order modeling of nonlinear singularly perturbed systems driven by wide-band noise," Proc. 21st IEEE Conference on Decision and Control, Orlando, FL.

Elliott, J. R. (1977), "NASA's advanced control low program for the F-8 digital fly-by-wire aircraft," IEEE Trans. Automat. Control, 22, pp. 753-757.

Etkin, B. (1972), Dynamics of Atmospheric Flight, Wiley, New York.

Farber, N. and J. Shinar (1980), "Approximate solution of singularly perturbed nonlinear pursuit-evasion games," J. Optim. Theory Appl., 32, pp. 39-73.

Fenichel, N. (1979), "Geometric singular perturbation theory for ordinary differential equations," J. Differential Equations, 31, pp. 53-98.

Ficola, A., R. Marino and S. Nicosic (1983), "A singular perturbation approach to the dynamic control of elastic robots," Proc. 21st Allerton Conf. Comm., Control, Comput., University of Illinois, pp. 335-342.

Fossard, A. G. and J. S. Magni (1980), "Frequential analysis of singularly perturbed systems with state or output control," J. Large Scale Systems, 1, pp. 223-228.

Fossard, A., J. M. Berthelot and J. F. Magni (1983), "On coherency-based decomposition algorithms," Automatica, 19, pp. 247-253.

Francis, B. A. and K. Glover (1978), "Bounded peaking in the optimal linear regulator with cheap control," IEEE Trans. Automat. Control, AC-23, pp. 608-617.

Francis, B. A. (1979), "The optimal linear-quadratic time-invariant regulator with cheap control," IEEE Trans. Automat. Control, AC-24, pp. 616-621.

Francis, B. A. (1982), "Convergence in the boundary layer for singularly perturbed equations," Automatica, 18, pp. 57-62.

Freedman, M. I. and B. Granoff (1976), "Formal asymptotic solution of a singularly perturbed nonlinear optimal control problem," J. Optim. Theory Appl., 19, pp. 301-325.

Freedman, M. I. and J. L. Kaplan (1976), "Singular perturbations of two point boundary value problems arising in optimal control," SIAM J. Control Optim., 14, pp. 189-215.

Freedman, M. I. (1977), "Perturbation analysis of an optimal control problem involving bang-bang-controls," J. Differential Equations, 25, pp. 11-29.

Gaitsgori, V. G. and A. A. Pervozvanskii (1975), "Aggregation of states in a Markov chain with weak interactions," Kibernetika, 3, pp. 91-98. (In Russian.)

Gaitsgori, V. G. (1979), "Perturbation method in optimal control problems," J. Systems Sci., 5, pp. 91-102.

Gaitsgori, V. G. (1980), "On the optimization of weakly controlled stochastic systems," Sov. Math. Dokl., 21, pp. 408-410.

Gardner, B. F., Jr. and J. B. Cruz, Jr. (1978), "Well-posedness of singularly perturbed Nash games," J. Franklin Inst., 306, 5, pp. 355-374.

Gicev, T. R. and A. L. Dontchev (1979), "Convergence of the solutions of the singularly perturbed time optimal problem," Appl. Math. Mech., PMM, 43, pp. 466-474.

Glizer, V. J. and M. G. Dmitriev (1975), "Singular perturbations in a linear optimal control problem with quadratic functional," Sov. Math. Dokl., 16, pp. 1555-1558.

Glizer, V. J. (1976), "On a connection of singular peturbations with the penalty function method," Sov. Math. Dokl., 17, pp. 1503-1505.

Glizer, V. J. (1977), "On the continuity of the regulator problem with respect to singular perturbations," Appl. Math. Mech., PMM, 41, pp. 573-576.

Glizer, V. J. (1978), "Asymptotic solution of a singularly perturbed Cauchy problem in optimal control," Differential Equations, 14, pp. 601-612.

Glizer, V. J. (1979), "Singular perturbations and generalized functions," Sov. Math. Dokl., 20, pp. 1360-1364.

Grasman, J. (1982), "On a class of optimal control problems with an almost cost-free solution," IEEE Trans. Automat. Control, AC-27, pp. 441-445.

Grishin, S. A. and V. I. Utkin (1980), "On redefinition of discontinuous systems," Differential Equations, 16, pp. 227-235.

Grujic, L. T. (1979), "Singular perturbations, large scale systems and asymptotic stability of invariant sets," Int. J. Systems Science, 12, pp. 1323-1341.

Grujic, L. T. (1981), "Uniform asymptotic stability of nonlinear singularly perturbed large-scale systems," Int. J. Control, 33, pp. 481-504.

Habets, P. (1974), "Stabilite asymptotique pour des problemes de perturbations singulieres," in Bressanone, Edizioni Cremonese, Rome, Italy, pp. 3-18.

Haddad, A. H. and P. V. Kokotovic (1971), "Note on singular perturbation of linear state regulators," IEEE Trans. Automat. Control, AC-16, 3, pp. 279-281.

Haddad, A. H. (1976), "Linear filtering of singularly perturbed systems," IEEE Trans. Automat. Control, AC-31, pp. 515-519.

Haddad, A. H. and P. V. Kokotovic (1977), "Stochastic control of linear singularly perturbed systems," IEEE Trans. Automat. Control, AC-22, pp. 815-821.

Hadlock, C. R. (1970), "Singular perturbations of a class of two point boundary value problems arising in optimal control," Ph.D. Thesis, Coordinated Science Laboratory, Report R-481, Univ. Illinois, Urbana.

Hadlock, C. A. (1973), "Existence and dependence on a parameter of solutions of a nonlinear two-point boundary value problem," J. Differential Equations, 14, pp. 498-517.

Halanay, A. and St. Mirica (1979), "The time optimal feedback control for singularly perturbed linear systems," Rev. Roum. Mat. Pures et Appl., 24, pp. 585-596.

Hale, J. K. (1980), Ordinary Differential Equations, Krieger Publishing Company.

Harris, W. A., Jr. (1960), "Singular perturbations of two-point boundary problems for systems of ordinary differential equations," Arch. Rat. Mech. Anal., 5, pp. 212-225.

Hopkins, W. E., Jr. and G. L. Blankenship (1981), "Perturbation analysis of a system of quasi-variational inequalities for optimal stochastic scheduling," IEEE Trans. Automat. Control, AC-26, pp. 1054-1070.

Hoppensteadt, F. (1967), "Stability in systems with parameters," J. Math. Anal. Appl., 18, pp. 129-134.

Hoppensteadt, F. (1971), "Properties of solutions of ordinary differential equations with small parameters," Comm. Pure Appl. Math., 34, pp. 807-840.

Hoppensteadt, F. (1974), "Asymptotic stability in singular perturbation problems, II," J. Differential Equations, 15, pp. 510-521.

Howes, F. A. (1976), "Effective characterization of the asymptotic behaviour of solutions of singularly perturbed boundary value problems," SIAM J. Appl. Math., 30, pp. 296-306.

Ioannou, P. (1981), "Robustness of absolute stability," Int. J. Control, 34, pp. 1027-1033.

Ioannou, P. A. (1982), "Robustness of model reference adaptive schemes with respect to modeling errors," Ph.D. Thesis, Coordinated Science Laboratory, Report R-955, Univ. Illinois, Urbana.

Ioannou, P. and P. V. Kokotovic (1982), "An asymptotic error analysis of identifiers and adaptive observers in the presence of parasitics," IEEE Trans. Automat. Control, AC-27, pp. 921-927.

Ioannou, P. A. and P. V. Kokotovic (1983), Adaptive Systems with Reduced Models, Lecture Notes in Control and Information Sciences 47, Springer-Verlag, New York.

Ioannou, P. A. (1984), "Robust direct adaptive controller," Proc. 23rd IEEE Conf. on Decision and Control, Las Vegas, Nevada, pp. 1015-1019.

Ioannou, P. A. and P. V. Kokotovic (1984), "Robust redesign of adaptive control," IEEE Trans. Automat. Control, AC-29, pp. 202-211.

Ioannou, P. A. and P. V. Kokotovic (1985), "Decentralized adaptive control of interconnected systems with reduced-order models," Automatica, 21, pp. xxx-xxx.

Jameson, A. and R. E. O'Malley, Jr. (1975), "Cheap control of the time-invariant regulator," Appl. Math. Optim., 1, pp. 337-354.

Jamshidi, M. (1974), "Three-stage near-optimum design of nonlinear control processes," Proc. IEEE, 121, pp. 886-892.

Javid, S. H. (1977), "The time-optimal control of singularly perturbed systems," Ph.D. Thesis, Coordinated Science Laboratory, Report R-794, Univ. Illinois, Urbana.

Javid, S. H. and P. V. Kokotovic (1977), "A decomposition of time scales for iterative computation of time optimal controls," J. Optim. Theory Appl., 21, pp. 459-468.

Javid, S. H. (1978a), "The time optimal control of a class of nonlinear singularly perturbed systems," Int. J. Control, 27, pp. 831-836.

Javid, S. H. (1978b), "Uniform asymptotic stability of linear time varying singularly perturbed systems," J. Franklin Inst., 305, pp. 27-37.

Javid, S. H. (1980), "Observing the slow states of a singularly perturbed system," IEEE Trans. Automat. Control, AC-25, pp. 277-280.

Javid, S. H. (1982), "Stabilization of time varying singularly perturbed systems by observer based slow state feedback," IEEE Trans. Automat. Control, AC-27, pp. 702-704.

Kailath, T. (1980), Linear Systems, Prentice-Hall, Englewood Cliffs, NJ.

Kalman, R. E. (1960), "Contributions to the theory of optimal control," Bol. Soc. Mat. Mexicana, 5, pp. 102-119.

Kelley, H. J. and T. N. Edelbaum (1970), "Energy climbs, energy turns and asymptotic expansions," J. Aircraft, 7, pp. 93-95.

Kelley, H. J. (1970a), "Boundary layer approximations to powered-flight attitude transients," J. Spacecraft and Rockets, 7, p. 879.

Kelley, H. J. (1970b), "Singular perturbations for a Mayer variational problem," AIAAJ, 8, pp. 1177-1178.

Kelley, H. J. (1971a), "Flight path optimization with multiple time scales," J. Aircraft, 8, p. 238.

Kelley, H. J. (1971b), "Reduced-order modeling in aircraft mission analysis," AIAAJ, 9, p. 349.

Kelley, H. J. (1973), "Aircraft maneuver optimization by reduced-order approximations," Control and Dynamic Systems, C. T. Leonides, ed., Academic Press, New York, pp. 131-178.

Khalil, H. K. and P. V. Kokotovic (1978), "Control strategies for decision makers using different models of the same system," IEEE Trans. Automat. Control, AC-23, pp. 289-298.

Khalil, H. K., A. Haddad and G. Blankenship (1978), "Parameter scaling and well-posedness of stochastic singularly perturbed control systems," Proc. 12th Asilomar Conferences, Pacific Grove, CA, pp. 407-411.

Khalil, H. K. (1978a), "Multimodeling and multiparameter singular perturbation in control and game theory," Ph.D. Thesis, Coordinated Science Laboratory, Report T-65, Univ. Illinoi, Urbana.

Khalil, H. K. (1978b), "Control of linear singularly perturbed systems with colored noise disturbances," Automatica, 14, pp. 153-156.

Khalil, H. K. (1979), "Stabilization of multiparameter singularly perturbed systems," IEEE Trans. Automat. Control, AC-24, pp. 790-791.

Khalil, H. K. and P. V. Kokotovic (1979a), "D-stability and multiparameter singular perturbations," SIAM J. Control Optim., 17, pp. 56-65.

Khalil, H. K. (1979b), "Control of linear systems with multiparameter singular perturbations," Automatica, 15, pp. 197-207.

Khalil, H. K. (1979c), "Feedback and well-posedness of singularly perturbed Nash games," IEEE Trans. Automat. Control,

Khalil, H. K. (1980), "Multimodel design of a Nash strategy," J. Optim. Theory Appl., 31, pp. 553-564.

Khalil, H. K. and J. V. Medanic (1980), "Closed-loop Stackelberg strategies for singularly perturbed linear quadratic problems," IEEE Trans. Automat. Control, AC-25, pp. 66-71.

Khalil, H. K. (1981a), "Asymptotic stability of a class of nonlinear multiparameter

singularly perturbed systems," Automatica, 17, pp. 797-804.
Khalil, H. K. (1981b), "On the robustness of output feedback control methods to modeling errors," IEEE Trans. Automat. Control, AC-28, pp. 524-528.
Khalil, H. K. and Z. Gajic (1984), "Near optimal regulators for stochastic linear singularly perturbed systems," IEEE Trans. Automat. Control, AC-29, pp. 531-541.
Khalil, H. (1984a), "A further note on the robustness of output feedback control to modeling errors," IEEE Trans. Automat. Control, AC-29, pp. 861-862.
Khalil, H. K. (1984b), "Time scale decompositin of linear implict singularly perturbed systems," IEEE Trans. Automat. Control, AC-29, pp. 1054-1056.
Khalil, H. K. (1984c), "Feedback control of implicit singularly perturbed systems," Proc. 23rd IEEE Conference on Decision and Control, Las Vegas, Nevada, pp. 1219-1223.
Khalil, H. K. (1985), "Output feedback control of linear two time scale systems," 1985 American Control Conference, Boston.
Klimushev, A. I. and N. N. Krasovskii (1962), "Uniform asymptotic stability of systems of differential equations with a small parameter in the derivative terms," J. Appl. Math. Mech., 25, pp. 1011-1025.
Koda, M. (1982), "Sensitivity analysis of singularly perturbed systems," Int. J. Systems Science, 13, pp. 909-919.
Kokotovic, P. V. and P. Sannuti (1968), "Singular perturbation method for reducing the model order in optimal control design," IEEE Trans. Automat. Control, AC-13, pp. 377-384.
Kokotovic, P. V. and R. A. Yackel (1972), "Singular perturbation of linear regulators: basic theorems," IEEE Trans. Automat. Control, AC-17, pp. 29-37.
Kokotovic, P. V. (1975), "A Riccati equation for block-diagonalization of ill-conditioned systems," IEEE Trans. Automat. Control, AC-20, pp. 812-814.
Kokotovic, P. V. and A. H. Haddad (1975a), "Controllability and time-optimal control of systems with slow and fast modes," IEEE Trans. Automat. Control, AC-20, pp. 111-113.
Kokotovic, P. V. and A. H. Haddad (1975b), "Singular perturbtaions of a class of time-optimal controls," IEEE Trans. Automat. Control, AC-20, pp. 163-164.
Kokotovic, P. V., R. E. O'Malley, Jr. and P. Sannuti (1976), "Singular perturbations and order reduction in control theory--an overview," Automatica, 12, pp. 123-132.
Kokotovic, P. V., J. J. Allemong, J. R. Winkelman and J. H. Chow (1980), "Singular perturbation and iterative separation of time scales," Automatica, 16, pp. 23-33.
Kokotovic, P. V. (1981), "Subsystems, time-scales and multimodeling," Automatica, 17, pp. 789-795.
Kokotovic, P. V., B. Avramovic, J. H. Chow and J. R. Winkelman (1982), "Coherency-based decomposition and aggregation," Automatica, 17, pp. 47-56.
Kokotovic, P. V. (1984), "Applications of singular perturbation techniques to control problems," SIAM Review, 26, pp. 501-550.
Kokotovic, P. V. (1985), "Control theory in the 80's: trends in feedback design," Automatica, 21, pp. xx-xx.
Kokotovic, P. V., K. Khorasani and M. Spong (1985), "A slow manifold approach to nonlinear two time scale systems," 1985 American Control Conference, Boston.
Kopel, N. (1979), "A geometric approach to boundary layer problems exhibiting resonance," SIAM J. Appl. Math., 37, pp. 436-458.
Kouvaritakis, B. (1978), "The optimal root loci of linear multivariable systems," Int. J. Control, 28, pp. 33-62.
Kouvaritakis, B. and J. M. Edmunds (1979), "A multivariable root loci: a unified approach to finite and infinite zeros, Int. J. Control, 29, pp. 393-428.
Krtolica, R. (1984), "A singular perturbation model of reliability in systems control," Automatica, 2, pp. 51-57.
Kung, C. F. (1976), "Singular perturbation of an infinite interval linear state regulator problem in optimal control," J. Math. Anal. Appl., 55, pp. 365-374.
Kurina, G. A. (1977), "Asymptotic solution of a classical singularly perturbed optimal control problem," Sov. Math. Dokl., 18, pp. 722-726.
Kuruoghu, N., D. E. Clough and W. F. Ramirez (1981), "Distributed parameter

estimation for systems with fast and slow dynamics," Chemical Engineering Science, 3, pp. 1357-xxxx.

Kushner, H. J. (1982), "A cautionary note on the use of singular perturbation methods for 'small noise' models," Stochastics, 6, pp. 117-120.

Kushner, H. J. (1984), Approximations and Weak Convergence Methods for Random Processes with Application to Stochastic System Theory, M.I.T. Press.

Kwakernaak, H. and S. Swan (1972), Linear Optimal Control Systems, Wiley, New York.

Ladde, G. S. and D. D. Siljak (1983), "Multiparameter singular perturbations of linear systems with multiple time scales," Automatica, 19, pp. 385-394.

Lagerstrom, P. A. and R. G. Casten (1972), "Basic concepts underlying singular perturbation techniques," SIAM Review, 14, pp. 63-120.

Lakin, W. D. and P. Van der Driessche (1977), "Time-scales in population biology," SIAM J. Appl. Math., 32, pp. 694-705.

Lehtomaki, N. A., D. A. Castanon, B. C. Levy, G. Stein, N. R. Sandell, Jr. and M. Athans (1984), "Robustness and modelin error chacterization," IEEE Trans. Automat. Control, AC-29, pp. 212-220.

Levin, J. J. and N. Levinson (1954), "Singular perturbations on non-linear systems of differential equations and an associated boundary layer equation," J. Rat. Mech. Anal., 3, pp. 274-280.

Levin, J. (1957), "The asymptotic behavior of the stable initial manifold of a system of nonlinear differential equations," Trans. Am. Math. Soc., 85, pp. 357-368.

Levinson, N. (1950), "Perturbations of discontinuous solutions of non-linear systems of differential equations," Acta Math., 82, pp. 71-106.

Lions, J. L. (1983), "Perturbations singulieres dans les problemes aux limites et en controle optimal," Lecture Notes in Mathematics, 323, Spring-Verlag, New York.

Litkouhi, B. and H. K. Khalil (1983), "Infinite-time regulators for singularly perturbed difference equations," Proc. 20th Allerton Conference on Communication, Control, and Computing, Univ. Illinoi, October 6-8, 1982, pp. 843-854.

Litkouhi, B. and H. Khalil (1984), "Infinite-time regulators for singularly perturbed difference equations," Int. J. Control, 39, pp. 587-598.

Lomov, S. A. (1981), Introduction to the General Theory of Singular Perturbations, Nauka, Moscow. (In Russian.)

Lukyanov, A. G. and V. I. Utkin (1981), "Methods for reduction of dynamic system equations to a regular form," Aut. Remote Control, 4, pp. 5-13.

Luse, D. W. and H. K. Khalil (1983), "A frequency domain approach for systems with slow and fast modes," Proc. American Control Conference, San Francisco, California, pp. 443-444.

Luse, D. W. (1985), "A continuation method for hole-placement for singularly perturbed systems," American Control Conference, Boston.

Mehra, R. K., R. B. Washburn, S. Sajon and J. V. Corell (1979), "A study of the application of singular perturbation theory," NASA, CR3167.

Moiseev, N. N. and F. L. Chernousko (1981), "Asymptotic methods in the theory of optimal control," IEEE Trans. Automat. Control, AC-26, pp. 993-1000.

O'Malley, R. E., Jr. (1971), "Boundary layer methods for nonlinear initial value problems," SIAM Review, 13, pp. 425-434.

O'Malley, R. E., Jr. (1972a), "The singularly perturbed linear state regulator problem," SIAM J. Control, 10, pp. 399-413.

O'Malley, R. E., Jr. (1972b), "Singular perturbation of the time invariant linear state regulator problem," J. Differential Equations, 12, pp. 117-128.

O'Malley, R. E., Jr. (1974), "Boundary layer methods for certain nonlinear singularly perturbed optimal control problems," J. Math. Anal. Appl., 45, pp. 468-484.

O'Malley, R. E., Jr. and C. F. Kung (1974), "The matrix Riccati approach to a singularly perturbed regulator problem," J. Differential Equations, 17, pp. 413-427.

O'Malley, R. E., Jr. (1974), "The singularly perturbed linear state regulator problem, II," SIAM J. Control, 13, pp. 327-337.

O'Malley, R. E., Jr. (1974), Introduction to Singular Perturbations, Academic Press, New York.

O'Malley, R. E., Jr. (1975), "On two methods of solution for a singularly perturbed

linear state regulator problem," SIAM Review, 17, pp. 16-37.
O'Malley, R. E., Jr. and A. Jameson (1975), "Singular perturbations and singular arcs--part I," IEEE Trans. Automat. Control, 20, pp. 218-226.
O'Malley, R. E., Jr. (1976), "A more direct solution of the nearly singular linear regulator problem," SIAM J. Control Option, 14, pp. 1063-1077.
O'Malley, R. E., Jr. and A. Jameson (1977), "Singular perturbations and singular arcs--part II," IEEE Trans. Automat. Control, 22, pp. 328-337.
O'Malley, R. E., Jr. and J. E. Flaherty (1977), "Singular singular perturbation problems," Singular Perturbations and Boundary Layer Theory, Lecture Notes in Mathematics, 594, Springer-Verlag, New York, pp. 422-436.
O'Malley, R. E., Jr. (1978), "Singular perturbations and optimal control," in Mathematical Control Theory, Lecture Notes in Mathematics, 680, Springer-Verlag, New York.
O'Malley, R. E., Jr. (1978), "On singular singularly-perturbed initial value problems," Applicable Anal., 8, pp. 71-81.
O'Malley, R. E., Jr. and R. L. Anderson (1978), "Singular perturbations and slow mode approximation for large scale linear systems," Proc. of IFAC/IRIA Workshop on Singular Perturbations in Control, France, pp. 113-121.
O'Malley, R. E., Jr. (1979), "A singular singularly-perturbed linear boundary value problem," SIAM J. Math. Anal., 10, pp. 695-708.
O'Malley, R. E., Jr. and J. E. Flaherty (1980), "Analytical and numerical methods for nonlinear singular singularly perturbed initial value problems," SIAM J. Appl. Math., 38, pp. 225-248.
O'Malley, R. E., Jr. (1982), Book Reviews, Bull. (New Series), Amer. Math. Soc., 7, 2, pp. 414-420.
O'Malley, R. E., Jr. (1983), "Slow/fast decoupling - analytical and numerical aspects," CISM Courses and Lectures, 280, M. Ardema, ed., Springer-Verlag, pp. 143-159.
O'Reilly, J. (1979a), "Two time scale feedback stabilization of linear time varying singularly perturbed systems," J. Franklin Inst., 308, pp. 465-474.
O'Reilly, J. (1979b), "Full order observers for a class of singularly perturbed linear time varying systems," Int. J. Control, 30, pp. 745-756.
O'Reilly, J. (1980), "Dynamical feedback control for a class of singularly perturbed linear systems using a full order observer," Int. J. Control, 31, pp. 1-10.
O'Reilly, J. (1983a), "Partial cheap control of the time-invariant regulator," Int. J. Control, 37, pp. 909-927.
O'Reilly, J. (1983b), Observer for Linear Systems, Academic Press, London.
O'Reilly, J. (1985), "The robustness of linear feedback control systems to unmodeled high frequency dynamics," Proc. IEE 'Control 85' Conference, Cambridge.
Ozguner, U. (1979), "Near-optimal control of composite systems: the multi-time scale approach," IEEE Trans. Automat. Control, AC-24, pp. 652-655.
Peponides, G., P. V. Kokotovic and J. H. Chow (1982), "Singular perturbations and time scales in nonlinear models of power systems," IEEE Trans. Circuits and Systems, CAS-29, pp. 758-767.
Peponides, G. M. (1982), "Nonexplicit singular perturbations and interconnected systems," Ph.D. Thesis, Coordinated Science Laboratory, Report R-960, Univ. Illinois, Urbana.
Peponides, G. and P. Kokotovic (1983), "Weak connections, time scales and aggregation of nonlinear systems," IEEE Trans. Automat. Control, AC-28, pp. 729-735.
Pervozvanskii, A. A. and V. G. Gaitsgori (1978), "Suboptimization, decomposition and aggregation," 7th IFAC World Congress, Helsinki.
Pervozvanskii, A. A. (1979a), Decomposition, Aggregation and Suboptimization, Nauka, Moscow. (In Russian.)
Pervozvanskii, A. A. (1979b), "Perturbation method for LQ problems: duality in degerate cases," Proc. 2nd Warsaw Workshop on Multilevel Control, Warsaw.
Pervozvanskii, A. A. (1980), "On aggregation of linear control systems," Autom. and Remote Control, 8, pp. 88-95.
Pervozvanskii, A. A. (1981), "Degeneracy in LQ and LQG problems of optimal control; possibilities to simplify the synthesis," Proc. 8th IFAC Congress, Kyoto.
Phillips, R. G. (1980), "Reduced order modeling and control of two time scale discrete systems," Int. J. Control, 31, pp. 765-780.
Phillips, R. G. (1980), "Decomposition of time scales in linear systems and

Markovian decision processes," Ph.D. Thesis, Coordinated Science Laboratory, Report R-902, Univ. Illinois, Urbana.
Phillips, R. G. and P. V. Kokotovic (1981), "A singular perturbation approach to modelling and control of Markov chains," IEEE Trans. Automat. Control, AC-26, pp. 1087-1094.
Phillips, R. G. (1983), "The equivalence of time-scale decomposition techniques used in the analysis and design of linear systems," Int. J. Control, 37, pp. 1239-1257.
Porter, B. (1974), "Singular perturbation methods in the design of stabilizing feedback controllers for multivariable linear systems," Int. J. Control, 20, pp. 689-692.
Razevig, V. D. (1978), "Reduction of stochastic differential equations with small parameters and stochastic integrals," Int. J. Control, 28, pp. 707-720.
Reddy, P. B. and Sanuti, P. (1975), "Optimal control of a coupled-core nuclear reactor by a singular perturbation method," IEEE Trans. Automat. Control, 20, pp. 766-769.
Saberi, A. and H. Khalil (1984), "Quadratic-type Lyapunov functions for singularly perturbed systems," IEEE Trans. Automat. Control, AC-29, pp. 542-550.
Saberi, A. and H. Khalil (1985), "Stabilization and regulation of nonlinear singularly perturbed systems-composit control," IEEE Trans. Automat. Control, AC-30, to appear.
Saksena, V. R. and P. V. Kokotovic (1981), "Singular perturbation of the Popov-Kalman-Yakubovich lemma," Systems and Control Letters, 1, pp. 65-68.
Saksena, V. R. and J. B. Cruz, Jr. (1981), "Nash strategies in decentralized control of multiparameter singularly perturbed large-scale systems," J. Large Scale Systems, 2, pp. 219-234.
Saksena, V. R. and J. B. Cruz, Jr.(1982), "A multimodel approach to stochastic Nash games," Automatica, 18, pp. 295-305.
Saksena, V. R. and T. Basar (1982), "A multimodel approach to stochastic team problems," Automatica, 18, pp. 713-720.
Saksena, V. R. and J. B. Cruz, Jr. (1984), "Robust Nash strategies for a class of non-linear singularly perturbed problems," Int. J. Control, 39, pp. 293-310.
Salman, M. A. and J. B. Cruz, Jr. (1979), "Well posedness of linear closed Stackelberg strategies for singularly perturbed systems," J. Franklin Inst., 308, 1, pp. 25-37.
Sandell, N. R., Jr. (1979), "Robust stability of systems with applications to singular perturbation," Automatica, 15, pp. 467-470.
Sannuti, P. (1968), "Singular perturbation method in the theory of optimal control," Ph.D. Thesis, Coordinated Science Laboratory, Report R-379, Univ. Illinois, Urbana.
Sannuti, P. and P. V. Kokotovic (1969), "Near optimum design of linear systems by a singular perturbation method," IEEE Trans. Automat. Control, AC-14, pp. 15-22.
Sannuti, P. (1969), "Singular perturbation method for near-optimum design of high order nonlinear systems," Automatica, 5, pp. 773-779.
Sannuti, P. (1974a), "A note on obtaining reduced order optimal control problems by singular perturbations," IEEE Trans. Automat. Control, AC-19, p. 256.
Sannuti, P. (1974b), "Asymptotic solution of singularly perturbed optimal control problems," Automatica, 10, pp. 183-194.
Sannuti, P. (1975), "Asymptotic expansions of singularly perturbed quasi-linear optimal systems," SIAM J. Control, 13, 3, pp. 572-592.
Sannuti, P. (1977), "On the controllability of singularly perturbed systems," IEEE Trans. Automat. Control, AC-22, pp. 622-624.
Sannuti, P. (1978), "On the controllabilty of some singularly perturbed nonlinear systems," J. Math. Anal. Appl., 64, pp. 579-591.
Sannuti, P. (1981), "Singular perturbations in the state space approach of linear electrical networks," Circuit Theory and Appl., 9, pp. 47-57.
Sannuti, P. (1983), "Direct singular perturbation analysis of high-gain and cheap control problems," Automatica, 19, pp. 41-51.
Sannuti, P. and H. Wason (1983), "Singular perturbation analysis of cheap control problems," Proc. 22nd Decision and Control Conference, San Antonio, TX, pp. 231-236.
Sannuti, P. (1984), "Determination of multivariable root-loci," Proc. 18th

Conference Inform. Science, Princeton.
Sannuti, P. (1985), "Multi-time scale decomposition in cheap control problems," IEEE Trans. Automat. Control, AC-30, pp. xxx-xxx.
Sastry, S. S. and C. A. Desoer (1981), "Jump behaviour of circuits and systems," IEEE Trans. Circuits and Systems, CAS-28, pp. 1109-1124.
Sastry, S. S. (1983), "Asymptotic unbounded root loci-formulas and computation," IEEE Trans. Automat. Control, AC-28, 5, pp. 557-568.
Schuss, Z. (1980), "Singular perturbation methods in stochastic differential equations of mathematical physics," SIAM Review, 22, pp. 119-155.
Sebald, A. V. and A. H. Haddad (1978), "State estimation for singularly perturbed systems with uncertain perturbation parameter," IEEE Trans. Automat. Control, AC-23, pp. 464-469.
Shaked, U. (1976), "Design techniques for high-feedback gain stability," Int. J. Control, 24, pp. 137-144.
Shaked, U. (1978), "The asymptotic behaviour of the root loci of multivariable optimal regulators," IEEE Trans. Automat. Control, AC-23, pp. 425-430.
Shinar, J. (1981), "Solution techniques for realistic pursuit-evasion games," Advances in Control and Dynamic Systems, C. T. Leondes, ed., 17, Academic Press, New York, pp. 63-124.
Shinar, Jr. (1983), "On applications of singular perturbation techniques in nonlinear optimal control," Automatica, 19, pp. 203-211.
Shinar, J. and N. Farber (1984), "Horizontal variable-speed interception game solved by forced singular perturbation technique, J. Optim. Theory and Control, 42, pp. 603-636.
Siljak, D. D. (1972), "Singular perturbation of absolute stability," IEEE Trans. Automat. Control, AC-17, p. 720.
Simon, H. A. and A. Ando (1961), "Aggregation of variables in dynamic systems," Econometrica, 29, pp. 111-138.
Simon, H. A. (1962), "The architecture of complexity," Proc. American Philosophical Society, 104, pp. 467-482.
Singh, R.-N. P. (1982), "The linear-quadratic-Gaussian problem for singularly perturbed systems," Int. J. Systems Science, 13, pp. 93-100.
Slater, G. L. (1984), "Perturbation analysis of optimal integral controls," ASME Trans., J. Dyn. Systems, Meas. and Control, 106, pp. 114-116.
Sobolev, V. (1984), "Integral manifolds and decomposition of singularly perturbed systems," Systems and Control Letters, 4, pp.169-179.
Sridhar, B. and N. K. Gupta (1980), "Missile guidance laws based on singular perturbation methodology," J. Guidance and Control, 3, pp. 158-165.
Suzuki, M. and M. Miura (1976), "Stabilizing feedback controllers for singularly perturbed linear constant systems," IEEE Trans. Automat. Control, AC-21, pp. 123-124.
Suzuki, M. (1981), "Composite controls for singularly perturbed systems," IEEE Trans. Automat. Control, AC-26, pp. 505-507.
Syrcos, G. P. and P. Sannuti (1984), "Near-optimum regulator design of singularly perturbed systems via Chandrasekhar equations," Int. J. Control, 39, pp. 1083-1102.
Teneketzis, D. and N. R. Sandell, Jr. (1977), "Linear regulator design for stochastic systems by a multiple time scale method," IEEE Trans. Automat. Control, AC-22, pp. 615-621.
Tikhonov, A. (1948), "On the dependence of the solutions of differential equations on a small parameter," Mat. Sb., 22, pp. 193-204. (In Russian.)
Tikhonov, A. N. (1952), "Systems of differential equations containing a small parameter multiplying the derivative," Mat. Sb., 31, 73, pp. 575-586. (In Russian.)
Tsai, E. P. (1978), "Perturbed stochastic linear regulator problems," SIAM J. Control, 16, pp. 396-410.
Utkin, V. I. (1977a), Sliding Modes and Their Application to Variable Structure Systems, Mir Moscow. (In English.)
Utkin, V. I. (1977b), "Variable structure systems with sliding modes: a survey," IEEE Trans. Automat. Control, AC-22, pp. 212-222.
Utkin, V. I. (1983), "Variable structure systems: state of the art and perspectives," Automation and Remote Control, 9, pp. 5-25.
Van Harten, A. (1984), "Singularly perturbed systems of diffusion type and feedback

control," Automatica, 20, pp. 79-91.

Vasileva, A. B. (1963), "Asymptotic behavior of solutions to certain problems involving nonlinear differential equations containing a small parameter multiplying the highest derivatives," Russian Math. Surveys, 18, 3, pp. 13-81.

Vasileva, A. B. and V. F. Butuzov (1973), Asymptotic Expansions of Solutions of Singularly Perturbed Differential Equations, Nauka, Moscow. (In Russian.)

Vasileva, A. B. (1975), "Singularly perturbed systems containing indeterminancy in the case of degeneracy," Sov. Math. Dokl., 16, pp. 1121-1125.

Vasileva, A. B. (1976), "Singularly perturbed systems with an indeterminacy in their degenerate equations," J. Differential Equations, 12, pp. 1227-1235.

Vasileva, A. B. and V. A. Anikeeva (1976), "Asymptotic expansions of solutions of non-linear problems with singular boundary conditions," J. Differential Equations, 12, pp. 1235-1244.

Vasileva, A. B. and M. V. Faminskaya (1977), "A boundary-value problem for singularly perturbed differential and difference systems when the unperturbed system is on a spectrum," J. Differential Equations, 13, pp. 738-742.

Vasileva, A. B. and M. Dmitriev (1978), "Singular perturbations and some optimal control problems," Proc. 7th IFAC World Congress, Paper 23.6.

Vasileva, A. B. and V. F. Butuzov (1978), Singularly Perturbed Systems in Critical Cases, Moscow University Press. (In Russian.)

Vasileva, A. B. and M. G. Dmitriev (1980), "Determination of the structure of generalized solutions of nonlinear optimal control problems," Sov. Math. Dokl., 21, pp. 104-109.

Vasileva, A. B. and M. V. Faminskaya (1981), "An investigation of a nonlinear optimal control problem by the methods of singular perturbation theory," Sov. Math. Dokl., 21, pp. 104-108.

Vishik, M. I. and L. A. Liusternik (1958), "On the asymptotic behavior of the solutions of boundary problems for quasi-linear differential equations," Dokl. Akad. Nauk SSSR, 121, pp. 778-781. (In Russian.)

Wasow, W. (1965), Asymptotic Expansions for Ordinary Diffential Equations, Wiley-Interscience, New York.

Wilde, R. R. and P. V. Kokotovic (1972a), "Stability of singularly perturbed systems and networks with parasitics," IEEE Trans. Automat. Control, AC-17, pp. 245-246.

Wilde, R. R. (1972), "A boundary layer method for optimal control of singularly perturbed systems," Ph.D. Thesis, Coordinated Science Laboratory, Report R-547, Univ. Illinois, Urbana.

Wilde, R. R. and P. V. Kokotovic (1972b), "A dichotomy in linear control theory," IEEE Trans. Automat. Control, AC-17, pp. 382-383.

Wilde, R. R. (1973), "Optimal open and closed-loop control of singularly perturbed linear systems," IEEE Trans. Automat. Control, AC-18, pp. 616-625.

Willems, J. C. (1981), "Almost invariant subspaces; an approach to high gain feedback design, part I, almost controlled invariant subspaces," IEEE Trans. Automat. Control, AC-26, pp. 235-252.

Willems, J. C. (1982), "Almost invariant subspaces: an aproach to high gain feedback design, part II, almost conditionally invariant subspaces," IEEE Trans. Automat. Control, AC-27, pp. 1071-1085.

Winkelman, J. R., J. H. Chow, J. H. Allemong and P. V. Kokotovic (1980), "Multi-time-scale analysis of power systems," Automatica, 16, pp. 35-43.

Womble, M. E., J. E. Potter and J. L. Speyer (1976), "Approximations to Riccati equations having slow and fast modes," IEEE Trans. Automat. Control, AC-21, pp. 846-855.

Yackel, R. A. (1971), "Singular perturbation of the linear state regulator," Ph.D. Thesis, Coordinated Science Laboratory, Report R-532, Univ. Illinois, Urbana.

Yackel, R. A. and P. V. Kokotovic (1973), "A boundary layer method for the matirx Riccati equation," IEEE Trans. Automat. Control, AC-18, 1, pp. 17-24.

Yeung, K. S. (1985), "A note on a Riccati equation for block-diagonalization of ill-conditioned systems," IEEE Trans. Automat. Control, AC-30, pp. xxx-xxx.

Young, K. D., P. V. Kokotovic and V. I. Utkin (1977), "A singular perturbation analysis of high gain feedback systems," IEEE Trans. Automat. Control, AC-

22, pp. 931-938.

Young, K. D. (1977), "Analysis and synthesis of high gain and variable structure feedback systems," Ph.D. Thesis, Coordinated Science Laboratory, Report R-800, Univ. Illinois, Urbana.

Young, K. D. (1978), "Multiple time-scales in single-input single-output high-gain feedback systems," J. Franklin Inst., 306, pp. 293-301.

Young, K. D. (1982), "Near insensitivity of linear feedback systems," J. Franklin Inst., 314, pp. 129-142.

Young, K. D. (1982), "Disturbance decoupling by high gain feedback," IEEE Trans. Automat. Control, AC-27, pp. 970-971.

Young, K. D. and P. V. Kokotovic (1982), "Analysis of feedback loop interactions with actuator and sensor parasitics," Automatica, 18, pp. 577-582.

Zien, L. (1973), "An upper bound for the singular parameter in a stable, singularly perturbed system," J. Franklin Inst., 295, pp. 373-381.

# Part I: OPTIMAL CONTROL

# SINGULAR PERTURBATIONS FOR DETERMINISTIC CONTROL PROBLEMS

*A. Bensoussan*[†]

INTRODUCTION.

The problems considered in this article are of the following type. Consider a dynamic system whose evolution is governed by

$$(1)\qquad \begin{aligned} \frac{dx^\varepsilon}{dt} &= f(x^\varepsilon,y^\varepsilon,v) \qquad & x^\varepsilon(o) &= x_0 \\ \varepsilon\frac{dy^\varepsilon}{dt} &= g(x^\varepsilon,y^\varepsilon,v) \qquad & y^\varepsilon(o) &= y_0 \end{aligned}$$

in which $v(t)$ represents a control. The parameter $\varepsilon$ tends to 0. The state of the system $(x^\varepsilon(t), y^\varepsilon(t))$ contains one part $x^\varepsilon(t)$ which varies slowly, and one part $y^\varepsilon(t)$ which varies fastly.

Such a situation is common in the applications. It appears for example in economic models to take into account long term and short term variations, but also in many problems of engineering, biology, mechanics...

The terminology "singular" explains as follows : the problem corresponding to $\varepsilon = 0$, namely

$$(2)\qquad \begin{aligned} \frac{dx}{dt} &= f(x,y,v) \\ g(x,y,v) &= 0 \end{aligned}$$

is of a type different from the case $\varepsilon > 0$ (an algebraic equation replaces a differential equation).

[†]INRIA, Domaine de Voluceau, Rocquencourt, B.P. 105, 78150 LE CHESNAY CEDEX, France and *Universite de Paris - Dauphine.*

An other way of expressing the same idea is to say that, in the limit, the state's size shrinks to x, the slow system.

The control problem consists in minimizing the cost

$$J^{\varepsilon}(v(.)) = \int_0^T \ell(x^{\varepsilon}(t), y^{\varepsilon}(t), v(t))dt + h(x^{\varepsilon}(T)). \tag{3}$$

The problems of interest are two fold. We want to study the behaviour of the quantity $\inf_{v(.)} J^{\varepsilon}(v(.))$, as $\varepsilon \to 0$. Moreover, we want to construct "good" if not optimal controls for the $\varepsilon$ problem.

The general philosophy of the approach developed in this article, is that the limit problem is simpler than the $\varepsilon$ problem. Note that this may not be the case, notwithstanding the reduction of the size, since the $\varepsilon$ problem is more regular than the limit problem.

This underlying philosophy legitimates the assumption that the algebraic equation entering in (2) can be solved in y, in a unique way. We also assume that the limit problem is well posed, meaning that it admits a unique optimal control $u_0$. This situation occurs when the limit control problem is not too far from convexity. Note that everything can be localized, hence just local optimality is sufficient.

One the other hand, a minimum of assumptions is made on the $\varepsilon$ problem itself, in particular we do not assume the existence of an optimal control for the $\varepsilon$ problem.

Therefore, a natural "good" control to use is $u_0$ itself. We prove the convergence of $\inf_{v(.)} J^{\varepsilon}(v(.))$ to $J(u_0)$, in a general case, including constraints. At this stage, no estimate of the error is given. When more regularity on the limit problem is available, an estimate or order $\varepsilon$ is given. This is the object of section 1.

It is possible to improve this estimate. To build a control which is better than $u_0$ (i.e which yields an approximation of $\inf_{v(.)} J^{\varepsilon}(v(.))$ which is of higher order) requires the introduction of boundary layer terms.

We develop in sections 2, 3, 4 in full details an expansion $u^{\varepsilon}$ which approximates the infimum as accurately as desired.

Two expansions are needed (inner and outer expansions), namely one of regular perturbation type and one of boundary layer type (at 0 and T). The improvement of accuracy is as follows : to obtain $\varepsilon^2$ one needs to add to $u_0$, boundary layer terms (cf. P. FAURRE, M. CLERGET, F. GERMAIN [1]) and the structure of the set of solutions is interesting. We have presented it beyond what is strictly necessary to solve the boundary layer problems.

The non linear case (often referred as the trajectory optimization in the litterature) has been considered in particular by P. SANNUTI [1], [2], R.E. O'MALLEY [3], [5], P. SANNUTI - P.V. KOKOTOVIC [1], C.R. HADLOCK [1], M.I. FREEDMAN, B. GRANOFF [1], M.I. FREEDMAN - J. KAPLAN [1], A.B. VASILEVA, V.A. ANIKEEVA [1], P. HABETS [1], M. ARDEMA [1], [2] ...).

In general the point of view is to write the necessary conditions of optimality and to find expansions. A problem which is considered is to solve the necessary conditions of optimality for the $\varepsilon$ problem by perturbation techniques. We do not treat this problem here. On the other hand the evaluation of the cost function for "good" controls does not seem very much considered in the litterature, nor the expansion of the optimal cost. The fact that the control $u_0$ itself yields an approximation of order $\varepsilon$ was known at least in the L.Q case, although the proof given relies on the boundary layer analysis. We show this fact in general without using the boundary layer.

The presentation of the convergence in the "constraints" case (lack of regularity) has not either appeared in the litterature.

The study of Bellman equations in duality seems also original. It should be interesting to study the complete structure of the set of solutions.

In the Dynamic Programming approach, the main concept is that of composite feedback, due to J. CHOW - P.V. KOKOTOVIC [1]. We extend this work and prove in particular that the decomposition of the composite feedback as the sum of the limit feedback and a complementary term involving the fast state is general, and not restricted to a quasi linear structure of the dynamics.

# 1. OPEN LOOP CONTROL PROBLEMS.

## 1.1. Setting of the problem

Let us consider functions f,g,l,h such that

(1.1) $f(x,y,v) : R^n \times R^m \times R^k \to R^n$

$g(x,y,v) : R^n \times R^m \times R^k \to R^m$

$l(x,y,v) : R^n \times R^m \times R^k \to R$

$h(x) : R^n \to R.$

All these functions are twice continuously differentiable in x,y,v. All derivatives of f,g are bounded. The second derivatives of l,h are bounded.

(1.2) $g_y(x,y,v) \leq - \mu I \qquad , \qquad \mu > 0.$

Let $v(.) \in L^2(0,T;R^k)$. For $\varepsilon$ given, one solves the differential equations

(1.3) $\frac{dx^\varepsilon}{dt} = f(x^\varepsilon,y^\varepsilon,v) \qquad x^\varepsilon(0) = x_0$

$\varepsilon\frac{dy^\varepsilon}{dt} = g(x^\varepsilon,y^\varepsilon,v) \qquad y^\varepsilon(0) = y_0.$

There is one and only one solution of (1.3), such that $x^\varepsilon \in H^1(0,T;R^n)$, $y^\varepsilon \in H^1(0,T;R^m)$.

One then considers the functional

(1.4) $J^\varepsilon(v(.)) = \int_0^T \ell(x^\varepsilon(t),y^\varepsilon(t),v(t))dt + h(x^\varepsilon(T))$

which is well defined since $\ell$ has quadratic growth. An admissible control satisfies the constraints

(1.5) $v(t) \in U_{ad}$ , a.e ; $U_{ad}$ convex closed non empty subset of $R^k$.

Our objective is to study the behaviour of inf $J^\varepsilon(v(.))$ as $\varepsilon$ tends to 0.

## 1.2. The limit problem

Consider first the algebraic equation

$$g(x,y,v) = 0 \tag{1.6}$$

in which x,v are parameters and we solve (1.6) in y. By virtue of (1.2), the equation (1.6) has a unique solution $\bar{y}(x,v)$. Moreover differentiating formally (1.6) with respect to x,v we obtain

$$\begin{aligned} &g_x + g_y\, \bar{y}_x = 0 \\ &g_v + g_y\, \bar{y}_v = 0. \end{aligned} \tag{1.7}$$

These formulas show that $\bar{y}_x$, $\bar{y}_v$ are continuous functions of x,v, and bounded.

Consider then the system, for $v(.) \in L^2(0,T;R^k)$

$$\left| \begin{aligned} &\frac{dx}{dt} = f(x,\bar{y}(x,v),v) \\ &x(0) = x_0. \end{aligned} \right. \tag{1.8}$$

By the properties of $\bar{y}$, (1.8) has one and only one solution x(.) in $H^1(0,T;R^n)$. The limit problem consists in minimizing

$$J(v(.)) = \int_0^T \ell(x(t),\bar{y}(t),v(t))dt + h(x(T)) \tag{1.9}$$

in which we have set

$$\bar{y}(t) = \bar{y}(x(t),v(t)). \tag{1.10}$$

We shall make assumptions on the limit problem. We shall assume basically that the necessary conditions of optimality (Pontryagin principle) are satisfied, as well as 2nd order conditions. This will imply, among other things, that the limit problem has a unique optimal solution.

We shall define the Hamiltonian

$$H(x,y,v,p,q) = \ell(x,y,v) + p.f(x,y,v) + q.g(x,y,v). \tag{1.11}$$

Let us consider the vector $w_o(t) = (x_o(t), y_o(t), u_o(t), p_o(t), q_o(t))$. The necessary conditions of optimality are

(1.12) $$\frac{dx_o}{dt} = f(w_o) \qquad x_o(0) = x_o$$

$$g(w_o) = 0$$

$$\frac{-dp_o}{dt} = H_x(w_o) \qquad p_o(T) = h_x(x_o(T))$$

$$H_y(w_o) = 0,$$

$$H_v(w_o(t)) \,.\, (v - u_o(t)) \geq 0 \qquad \forall\, v \in U_{ad}.$$

The 2nd order conditions are given by

(1.13) $$\begin{pmatrix} H_{yy} & H_{yv} \\ H_{vy} & H_{vv} \end{pmatrix} (x,y,v,p_o(t),q_o(t)) \geq \beta I, \quad \forall t,\ \forall x,y,v,\ \beta > 0$$

$$H_{xx} - (H_{xy}\ H_{xv}) \begin{pmatrix} H_{yy} & H_{yv} \\ H_{vy} & H_{vv} \end{pmatrix}^{-1} \begin{pmatrix} H_{yx} \\ H_{vx} \end{pmatrix} \geq 0$$

for the same arguments,

(1.14) $$h_{xx} \geq 0$$

Conditions (1.12), (1.13), (1.14) imply that $u_o(.)$ is an optimal control and in fact the unique one, for the problem (1.8), (1.9). The above conditions are verified in the convex case (linear dynamics and $\ell$ convex).

Other possibilities can be considered. Suppose for instance that there exists $x^*, y^*, u^*$ (define $w^* = (x^*, y^*, u^*, 0, 0)$) such that

(1.15) $$f(w^*) = 0,\ g(w^*) = 0$$

$$\ell_x(w^*) = \ell_y(w^*) = \ell_v(w^*) = 0\ ,\quad h_x(x^*) = 0$$

and

$$\begin{pmatrix} \ell_{yy} & \ell_{yv} \\ \ell_{vy} & \ell_{vv} \end{pmatrix}(x,y,v) \geq \beta I \tag{1.16}$$

,

$$\ell_{xx} - (\ell_{xy}\ \ell_{xv}) \begin{pmatrix} \ell_{yy} & \ell_{yv} \\ \ell_{vy} & \ell_{vv} \end{pmatrix}^{-1} \begin{pmatrix} \ell_{yx} \\ \ell_{vx} \end{pmatrix} \geq 0.$$

Conditions (1.15), (1.16) coïncide with (1.12), (1.13), when $x_0 = x^*$. Therefore the constant control $u^*$ is optimal for (1.8), (1.9) when $x_0 = x^*$. It is possible to show, at least when $U_{ad} = R^k$ and for data sufficiently smooth (cf A. BENSOUSSAN [1]) that taking $x_0 - x^*$ sufficiently small, there exists a function $w_0(t)$ satisfying conditions (1.12), (1.13).

### 1.3. Convergence

We can state the following convergence result

*Theorem 1.1. Assume* (1.1), (1.2) *and the existence of* $w_0(t)$ *such that* (1.12), (1.13), (1.14) *hold. Then one has*

$$\text{Inf } J^\varepsilon(v(.)) \to \inf J(v(.)). \tag{1.17}$$

*If* $u^\varepsilon$ *satisfies*

$$J^\varepsilon(u^\varepsilon) \leq J^\varepsilon(u_0) \tag{1.18}$$

*then*

$$\begin{aligned} u^\varepsilon - u_0 &\to 0 \quad \text{in} \quad L^2(0,T;R^k) \\ y^\varepsilon - y_0 &\to 0 \quad \text{in} \quad L^2(0,T;R^m) \\ x^\varepsilon - x_0 &\to 0 \quad \text{in} \quad H^1(0,T;R^n) \end{aligned} \tag{1.19}$$

□

The proof of Theorem 1.1 is done in several Lemmas

Lemma 1.1. $J^\varepsilon(u_o) \to J(u_o)$ , as $\varepsilon \to 0$

Proof.

Consider $\bar{x}^\varepsilon$, $\bar{y}^\varepsilon$ which are the solution of

$$(1.20) \qquad \frac{dx^\varepsilon}{dt} = f(\bar{x}^\varepsilon,\bar{y}^\varepsilon,u_o) \qquad \bar{x}^\varepsilon(0) = x_o$$

$$\varepsilon\frac{dy^\varepsilon}{dt} = g(\bar{x}^\varepsilon,\bar{y}^\varepsilon,u_o) \qquad \bar{y}^\varepsilon(0) = y_o.$$

We start with proving a priori estimates. We have

$$\frac{1}{2}\frac{d}{dt}|\bar{x}^\varepsilon(t)|^2 \le k|\bar{x}^\varepsilon|(1 + |\bar{x}^\varepsilon| + |\bar{y}^\varepsilon| + |u_o|)$$

$$\frac{1}{2}\varepsilon\frac{d}{dt}|\bar{y}^\varepsilon(t)|^2 = g(\bar{x}^\varepsilon,\bar{y}^\varepsilon,u_o)\bar{y}^\varepsilon$$

$$= (g(\bar{x}^\varepsilon,\bar{y}^\varepsilon,u_o) - g(\bar{x}^\varepsilon,0,u_o)).\bar{y}^\varepsilon + g(\bar{x}^\varepsilon,0,u_o).\bar{y}^\varepsilon$$

$$\le -\mu|\bar{y}^\varepsilon|^2 + k|\bar{y}^\varepsilon|(1 + |\bar{x}^\varepsilon| + |u_o|).$$

Therefore integrating

$$\frac{1}{2}|\bar{x}^\varepsilon(t)|^2 \le \frac{1}{2}|x_o|^2 + k\int_0^t |\bar{x}^\varepsilon|(1 + |\bar{x}^\varepsilon| + |\bar{y}^\varepsilon| + |u_o|)ds$$

$$\frac{1}{2}\varepsilon|\bar{y}^\varepsilon(t)|^2 + \mu\int_0^t |\bar{y}^\varepsilon(s)|^2 ds \le \frac{1}{2}\varepsilon|y_o|^2 + k\int_0^t |\bar{y}^\varepsilon|(1 + |\bar{x}^\varepsilon| + |u_o|)ds$$

hence for a convenient choice of $k_o$

$$|\bar{x}^\varepsilon(t)|^2 + \varepsilon|\bar{y}^\varepsilon(t)|^2 + \mu\int_0^t |\bar{y}^\varepsilon(s)|^2 ds \le k_o\Big(1 + \int_0^t |\bar{x}^\varepsilon(s)|^2 ds + \int_0^T |u_o(s)|^2 ds\Big).$$

From Gronwall's inequality, we deduce

$$(1.21) \qquad |\bar{x}^\varepsilon(t)|^2 \le K_o$$

$$\int_0^T |\bar{y}^\varepsilon(t)|^2 dt \le K_o$$

and from the 1st differential equation (1.21), we get also

$$\int_0^T \left|\frac{d\bar{x}^\varepsilon}{dt}\right|^2 dt \le K_0. \tag{1.22}$$

From the estimates (1.21), (1.22), we can assert that, at least for a subsequence

$$\bar{x}^\varepsilon \to x^* \quad \text{in} \quad H^1(0,T;R^n) \text{ weakly, and } L^2(0,T;R^n) \text{ strongly} \tag{1.23}$$

$$\bar{y}^\varepsilon \to y^* \quad \text{in} \quad L^2(0,T;R^m) \text{ weakly.}$$

Considering now

$$\xi^\varepsilon = g(\bar{x}^\varepsilon,\bar{y}^\varepsilon,u_0)$$

from (1.23) it remains in a bounded set of $L^2(0,T;R^m)$. But from the second differential equation (1.20), taking $\phi \in C_0^\infty(0,T;R^m)$

$$-\,\varepsilon \int_0^T \bar{y}^\varepsilon\, \phi dt = \int_0^T \xi^\varepsilon\, \phi dt \to 0$$

hence

$$\xi^\varepsilon \to 0 \quad \text{in} \quad L^2(0,T;R^m) \text{ weakly.} \tag{1.24}$$

To proceed we use the classical technique of MINTY [1] (cf also J.L. LIONS [1]). Let $z \in L^2(0,T;R^m)$. We have from (1.2)

$$\int_0^T (g(\bar{x}^\varepsilon,\bar{y}^\varepsilon,u_0) - g(\bar{x}^\varepsilon,z,u_0)).(\bar{y}^\varepsilon - z)dt \le 0$$

hence

$$0 \ge \int_0^T \xi^\varepsilon(t)(\bar{y}^\varepsilon(t) - z(t))dt - \int_0^T g(\bar{x}^\varepsilon,z,u_0).(\bar{y}^\varepsilon - z)dt. \tag{1.25}$$

From the 2nd differential equation (1.20) we deduce

$$\frac{1}{2}\,\varepsilon|\bar{y}^\varepsilon(T)|^2 - \frac{1}{2}\,\varepsilon|y_0|^2 = \int_0^T \xi^\varepsilon(t)\bar{y}^\varepsilon(t)dt \ge -\frac{1}{2}\,\varepsilon|y_0|^2$$

which together with (1.25) implies

$$0 \geq -\frac{\varepsilon}{2}|y_o|^2 - \int_0^T g(\bar{x}^\varepsilon,\bar{z},u_o)(\bar{y}^\varepsilon - z)dt - \int_0^T \xi^\varepsilon(t)z(t)dt.$$

Noting that

$$g(\bar{x}^\varepsilon,z,u_o) \to g(x^*,z,u_o) \quad \text{in} \quad L^2(0,T;R^m) \text{ strongly}$$

and taking account of (1.24), obtains

$$\int_0^T g(x^*(t),z(t),u_o(t)).(y^*(t) - z(t))dt \geq 0.$$

Picking

$$z(t) = y^*(t) - \lambda\eta(t)$$

and letting $\lambda$ tend to 0, it follows

$$\int_0^T g(x^*(t),y^*(t),u_o(t)).\eta(t)dt \geq 0$$

and since $\eta$ is arbitrary, we get

(1.26) $$g(x^*(t),y^*(t),u_o(t)) = 0.$$

In fact, we can reinforce (1.25), by making full use of the assumption (1.2). Choosing $z = y^*$, yields

$$-\mu\int_0^T |\bar{y}^\varepsilon(t) - y^*(t)|^2dt \geq -\frac{\varepsilon}{2}|y_o|^2 - \int_0^T g(\bar{x}^\varepsilon,y^*,u_o)(\bar{y}^\varepsilon - y^*)dt$$
$$- \int_0^T \xi^\varepsilon(t)z(t)dt$$

and by virtue of (1.26) the right hand side tends to 0.

Therefore

$$\bar{y}^\varepsilon \to y^* \quad \text{in} \quad L^2(o,T;R^m) \text{ strongly.}$$

It is then possible to pass to the limit in the 1st differential equation (1.20) and to deduce

$$\frac{dx^*}{dt} = f(x^*,y^*,u_o) \qquad\qquad x^*(0) = x_o$$

which together with (1.26) implies $x^* = x_o$, $y^* = y_o$.

From the uniqueness of the limit we can assert that

$$\bar{x}^\varepsilon \to x_o \quad \text{in} \quad H^1(0,T;R^n)$$

$$\bar{y}^\varepsilon \to y_o \quad \text{in} \quad L^2(0,T;R^m)$$

and thus the desired result obtains □

Lemma 1.2. The functions $u^\varepsilon, y^\varepsilon$ remain bounded in $L^2(0,T;R^k)$ and $L^2(0,T;R^m)$ respectively. The function $x^\varepsilon$ remains bounded in $H^1(0,T;R^n)$.

Proof.

Let us set

$$\tilde{u}^\varepsilon = u^\varepsilon - u_o \quad , \quad \tilde{x}^\varepsilon = x^\varepsilon - x_o \quad , \quad \tilde{y}^\varepsilon = y^\varepsilon - y_o .$$

It will be convenient to use the notation $\sigma = (x,y,v)$(recalling that $w = (x,y,v,p,q)$). We thus write

$$\sigma_o = (x_o,y_o,u_o) \quad , \quad \tilde{\sigma}^\varepsilon = (\tilde{x}^\varepsilon,\tilde{y}^\varepsilon,\tilde{u}^\varepsilon).$$

Let us establish the formula

$$\text{(1.27)} \qquad J^\varepsilon(u^\varepsilon) = J(u_o) - \int_0^T q_o\cdot(g(\sigma^\varepsilon) - g(\sigma_o))dt + \int_0^T H_v(\sigma_o)\tilde{u}^\varepsilon dt$$

$$+ \int_0^T \int_0^1 \int_0^1 \lambda\, H_{\sigma\sigma}(w^\varepsilon_{\lambda\mu})\tilde{\sigma}^\varepsilon.\tilde{\sigma}^\varepsilon dt d\lambda d\mu$$

$$+ \int_0^1 \int_0^1 \lambda\, h_{xx}(x_o(T) + \lambda\mu\tilde{x}^\varepsilon(T))\tilde{x}^\varepsilon(T).\tilde{x}^\varepsilon(T) d\lambda d\mu$$

where we have set

$$w^{\varepsilon}_{\lambda\mu} = (\sigma_o + \lambda\mu\, \tilde{\sigma}^{\varepsilon}, p_o, q_o).$$

Indeed, one has

$$(1.28)\qquad J^{\varepsilon}(u^{\varepsilon}) = J(u_o) + \int_0^T \ell_{\sigma}(\sigma_o)\tilde{\sigma}^{\varepsilon}dt + h_x(x_o(T))\tilde{x}^{\varepsilon}(T) +$$

$$+ \int_0^T \int_0^1 \int_0^1 \lambda\, \ell_{\sigma\sigma}(w^{\varepsilon}_{\lambda\mu})\tilde{\sigma}^{\varepsilon}.\tilde{\sigma}^{\varepsilon}dt d\lambda d\mu +$$

$$+ \int_0^1 \int_0^1 \lambda\, h_{xx}(x_o(T) + \lambda\mu\tilde{x}^{\varepsilon}(T))\tilde{x}^{\varepsilon}(T)^2 d\lambda d\mu$$

But

$$\int_0^T \ell_{\sigma}(\sigma_o)\tilde{\sigma}^{\varepsilon}dt + h_x(x_o(T))\tilde{x}^{\varepsilon}(T) = \int_0^T (- \frac{dp_o}{dt} - f^*_x(\sigma_o)p_o - g^*_x(\sigma_o)q_o).\tilde{x}^{\varepsilon}dt$$

$$- \int_0^T (f^*_y(\sigma_o)p_o + g^*_y(\sigma_o)q_o).\tilde{y}^{\varepsilon}dt +$$

$$+ \int_0^T \ell_v(\sigma_o)\tilde{u}^{\varepsilon}dt + p_o(T)\tilde{x}^{\varepsilon}(T)$$

$$= \int_0^T p_o.(f(\sigma^{\varepsilon}) - f(\sigma_o) - f_{\sigma}(\sigma_o)\tilde{\sigma}^{\varepsilon})dt -$$

$$- \int_0^T q_o.g_{\sigma}(\sigma_o)\tilde{\sigma}^{\varepsilon}dt + \int_0^T H_v(\sigma_o)\tilde{u}^{\varepsilon}dt.$$

Adding and substracting the 2nd term to the right hand side of (1.27) and using an expansion similar to (1.28), we obtain (1.27).

We next estimate the quadratic form at the right hand side of (1.27). Let us introduce

$$z^{\varepsilon} = \begin{pmatrix} \tilde{y}^{\varepsilon} \\ \tilde{u}^{\varepsilon} \end{pmatrix} + \begin{pmatrix} H_{yy} & H_{yv} \\ H_{vy} & H_{vv} \end{pmatrix}^{-1} \begin{pmatrix} H_{yx} \\ H_{vx} \end{pmatrix} \tilde{x}^{\varepsilon}$$

where the arguments entering into the Hamiltonians are $w^{\varepsilon}_{\lambda\mu}$.

Then

$$H_{\sigma\sigma}(w^{\varepsilon}_{\lambda\mu})(\tilde{\sigma}^{\varepsilon})^2 = \begin{pmatrix} H_{yy} & H_{yv} \\ H_{vy} & H_{vv} \end{pmatrix}(Z^{\varepsilon})^2 +$$

$$+ (H_{xx} - (H_{xy}\ H_{xv}) \begin{pmatrix} H_{yy} & H_{yv} \\ H_{vy} & H_{vv} \end{pmatrix}^{-1} \begin{pmatrix} H_{yx} \\ H_{vx} \end{pmatrix} (\tilde{x}^{\varepsilon})^2 \geq \gamma|Z^{\varepsilon}|^2$$

where $\gamma$ is a positive number independant of $\lambda,\mu$.

Therefore we deduce from (1.27) and from the last condition (1.12)

$$J^{\varepsilon}(u^{\varepsilon}) \geq J(u_o) - \int_0^T q_o\cdot(g(\sigma^{\varepsilon}) - g(\sigma_o))dt + \gamma\int_0^T\int_0^1\int_0^1 \lambda|Z^{\varepsilon}|^2 dt d\lambda d\mu \tag{1.29}$$

(note that $Z^{\varepsilon}$ depends on $\lambda,\mu$).

Noting that

$$\left|\int_0^T q_o\cdot(g(\sigma^{\varepsilon}) - g(\sigma_o))dt\right| \leq \gamma_1\Big[(\int_0^T |\tilde{x}^{\varepsilon}|^2 dt)^{1/2} + (\int_0^T |\tilde{y}^{\varepsilon}|^2 dt)^{1/2} +$$

$$+ \int_0^T |\tilde{u}^{\varepsilon}|^2 dt)^{1/2}\Big] \leq \gamma_1'\Big[(\int_0^T |\tilde{x}^{\varepsilon}|^2 dt)^{1/2} + (\int_0^T\int_0^1\int_0^1 \lambda|Z^{\varepsilon}|^2 dt d\lambda d\mu)^{1/2}\Big]$$

and the assumtion (1.18) on $u^{\varepsilon}$,

we obtain

$$J^{\varepsilon}(u^{\varepsilon}) - J(u_o) \geq -\gamma_1'\Big[(\int_0^T |\tilde{x}^{\varepsilon}|^2 dt)^{1/2} + (\int_0^T\int_0^1\int_0^1 \lambda|Z^{\varepsilon}|^2 dt d\lambda d\mu)^{1/2}\Big] \tag{1.30}$$

$$+ \gamma\int_0^T\int_0^1\int_0^1 \lambda|Z^{\varepsilon}|^2 dt d\lambda d\mu.$$

On the other hand

$$\frac{d\tilde{x}^{\varepsilon}}{dt} = f(\sigma^{\varepsilon}) - f(\sigma_o) \qquad \tilde{x}^{\varepsilon}(0) = 0$$

$$\left|\frac{d\tilde{x}^\varepsilon}{dt}\right| \le k(|\tilde{x}^\varepsilon| + |\tilde{y}^\varepsilon| + |\tilde{u}^\varepsilon|)$$

$$\le k_1|\tilde{x}^\varepsilon| + k_2|Z^\varepsilon|$$

hence from Gronwall's inequality

$$\int_0^T |\tilde{x}^\varepsilon|^2 dt \le C \int_0^T |Z^\varepsilon(\lambda,\mu,t)|^2 dt \quad \forall\ \lambda,\mu \tag{1.31}$$

$$\le C_1 \int_0^T \int_0^1 \int_0^1 \lambda |Z^\varepsilon|^2 dt d\lambda d\mu.$$

This estimate used in (1.30) implies, taking account of Lemma 1.1, that

$$\int_0^T \int_0^1 \int_0^1 \lambda |Z^\varepsilon|^2 d\lambda d\lambda d\mu \le C \quad \text{independant of } \varepsilon.$$

From (1.31) it follows that $\int_0^T |\tilde{x}^\varepsilon|^2 dt \le C$, which with the definition of $Z^\varepsilon$ implies that $\tilde{y}^\varepsilon, \tilde{u}^\varepsilon$ remain bounded in $L^2$. Considering the expression of $\frac{d\tilde{x}^\varepsilon}{dt}$, we easily conclude □

Proof of Theorem 1.1.

Consider again (1.29) which can also be written as

$$J^\varepsilon(u^\varepsilon) \ge J(u_o) - \int_0^T q_o \cdot \varepsilon\frac{dy^\varepsilon}{dt}\, dt + \gamma \int_0^T \int_0^1 \int_0^1 \lambda |Z^\varepsilon|^2 dt d\lambda d\mu$$

hence

$$J^\varepsilon(u_o) - J(u_o) + \int_0^T q_o \cdot \varepsilon\frac{dy^\varepsilon}{dt}\, dt \ge \gamma \int_0^T \int_0^1 \int_0^1 \lambda |Z^\varepsilon|^2 dt d\lambda d\mu. \tag{1.32}$$

Now from Lemma 1.2, we know that $\varepsilon\frac{dy^\varepsilon}{dt}$ remains in a bounded set of $L^2(0,T;R^m)$, hence we can extract a subsequence such that

$$\varepsilon\, \frac{dy^\varepsilon}{dt} \to \eta \quad \text{in } L^2(0,T;R^m) \text{ weakly.}$$

Taking $\phi \in C_0^\infty(0,T;R^m)$ we have

$$\int_0^T \phi \cdot \varepsilon\, \frac{dy^\varepsilon}{dt}\, dt \to \int_0^T \phi \cdot \eta\, dt$$

and also

$$= - \varepsilon \int_0^T \frac{d\phi}{dt} y^\varepsilon dt \to 0$$

since $y^\varepsilon$ is bounded. Therefore $\eta = 0$. Using this fact and Lemma 1.1, we deduce from (1.32) that

$$\int_0^T \int_0^1 \int_0^1 \lambda |Z^\varepsilon|^2 dt d\lambda d\mu \to 0, \text{ as } \varepsilon \to 0.$$

Using (1.31) and the definition of $Z^\varepsilon$, we easily prove (1.19).

Since there exists always $u^\varepsilon$ such that

$$J^\varepsilon(u^\varepsilon) \le J^\varepsilon(u_0)$$

and

$$J^\varepsilon(u^\varepsilon) \le \inf J^\varepsilon(v(.)) + \varepsilon.$$

Taking account of (1.19), we have $J^\varepsilon(u^\varepsilon) \to J(u_0)$, hence

$$J(u_0) \le \underline{\lim} \inf J^\varepsilon(v(.)).$$

Since also

$$\inf J^\varepsilon(v(.)) \le J^\varepsilon(u_0) \to J(u_0)$$

we also have

$$\overline{\lim} \inf J^\varepsilon(v(.)) \le J(u_0)$$

which completes the proof of (1.7) □

1.4. Stronger convergence results in the case of regularity.

One can improve the convergence result of Theorem 1.1, when the following additional regularity is satisfied

$$\frac{dy_0}{dt}, \frac{dq_0}{dt} \in L^2(0,T;R^m) \tag{1.33}$$

In fact this holds in the unconstrained case, namely

Lemma 1.3. When $U_{ad} = R^k$, the additional regularity (1.33) holds, as well as

(1.34) $$\frac{du_o}{dt} \in L^2(0,T;R^k)$$

Proof.

We have

$$g(\sigma_o) = 0 \;,\quad H_y(w_o) = 0 \;,\quad H_v(w_o) = 0$$

Differentiating formally in t, yields

$$g_\sigma(\sigma_o)\dot{\sigma}_o = 0 \;,\quad H_{yw}(w_o)\dot{w}_o = 0 \;,\quad H_{vw}(w_o)\dot{w}_o = 0.$$

Developing these expressions, we express first

$$\dot{y}_o = - g_y^{-1}(g_x\, \dot{x}_o + g_v\, \dot{u}_o)$$

$$\dot{q}_o = - g_y^{*-1}(H_{yx}\, \dot{x}_o + H_{yy}\, \dot{y}_o + f_y^*\, \dot{p}_o + h_{yv}\dot{u}_o)$$

and using the last one we obtain an equation in $\dot{u}_o$,

(1.35) $$(H_{vv} - H_{vy}g_y^{-1}g_v - g_v^*g_y^{*-1}H_{yv} + g_v^*g_y^{*-1}H_{yy}g_y^{-1}g_v)\dot{u}_o +$$

$$+ (H_{vx} - H_{vy}g_y^{-1}g_x - g_v^*g_y^{*-1}H_{yx} + g_v^*g_y^{*-1}H_{yy}g_y^{-1}g_x)\dot{x}_o$$

$$+ (f_v^* - g_v^*g_y^{*-1}f_y^*)\dot{p}_o = 0.$$

Since $\dot{x}_o$, $\dot{p}_o \in L^2(0,T)$ and the matrices entering in (1.35) are bounded, the result will follow if the matrix entering in (1.35) as the coefficient of $\dot{u}_o$ is invertible, with bounded inverse. In fact this matrix, denoted by M can be expressed as

$$M = H_{vv} - H_{vy}\, H_{yy}^{-1}\, H_{yv} + (g_v^*g_y^{*-1} - H_{vy}\, H_{yy}^{-1})\, H_{yy}\, (g_y^{-1}g_v - H_{yy}^{-1}\, H_{yv})$$

$$\geq H_{vv} - H_{vy}\, H_{yy}^{-1}\, H_{yv}.$$

On the other hand

$$(H_{vv} - H_{vy} H_{yy}^{-1} H_{yv})v^2 = H_{vv} v^2 + H_{yy} y^2 + 2H_{yv} vy$$

with

$$y = - H_{yy}^{-1} H_{yv} v$$

hence from (1.13)

$$\geq \beta (|v|^2 + |y|^2) \geq \beta |v|^2.$$

The proof is now complete □

Our objective is to prove the following

_Theorem 1.2. We make the assumptions of Theorem_ 1.1, _and the regularity assumption_ (1.33). _Then we have_

$$(1.36) \qquad |\text{Inf } J^\varepsilon(v(.)) - \inf J(v(.))| \leq C\varepsilon.$$

_If_ u _satisfies_ (1.18) _then_

$$(1.37) \qquad |u^\varepsilon - u_0|_{L^2} \;,\; |y^\varepsilon - y_0|_{L^2} \leq C\sqrt{\varepsilon}$$

$$|x^\varepsilon - x_0|_{H^1} \leq C\sqrt{\varepsilon} \;,\; |y^\varepsilon(t)| \leq C.$$

□

The proof relies on the following improvement of Lemma 1.1,

_Lemma 1.4. We have_

$$(1.38) \qquad |J^\varepsilon(u_0) - J(u_0)| \leq C\varepsilon.$$

_Proof._

Let us improve the convergence of $\bar{x}^\varepsilon, \bar{y}^\varepsilon$ to $x_0, y_0$ (cf Lemma 1.1) and in fact simplify the proof, thanks to the regularity (1.33). We have in fact

$$(1.39) \qquad |\bar{x}^\varepsilon - x_o|_{H^1} \le C\sqrt{\varepsilon} \;,\quad |\bar{y}^\varepsilon - y_o|_{L^2} \le C\sqrt{\varepsilon}\;,\quad |\bar{y}^\varepsilon(t)| \le C.$$

Indeed set

$$x_1^\varepsilon = \bar{x}^\varepsilon - x_o \;,\quad y_1^\varepsilon = \bar{y}^\varepsilon - y_o$$

then one has

$$\frac{dx_1^\varepsilon}{dt} = f(\bar{x}^\varepsilon,\bar{y}^\varepsilon,u_o) - f(x_o,y_o,u_o) \qquad x_1^\varepsilon(0) = 0$$

$$\frac{\varepsilon dy_1^\varepsilon}{dt} = -\frac{\varepsilon dy_o}{dt} + g(\bar{x}^\varepsilon,\bar{y}^\varepsilon,u_o) - g(x_o,y_o,u_o), \qquad y_1^\varepsilon(0) = y_o - y_o(0).$$

Multiplying the first equation by $x_1^\varepsilon$ , and the $2^{nd}$ by $y_1^\varepsilon$, integrating and making use of the assumption (1.2), yields

$$\varepsilon|y_1^\varepsilon(t)|^2 + |x_1^\varepsilon(t)|^2 + \mu\int_0^t |y_1^\varepsilon(s)|^2 ds \le \varepsilon\,|y_o - y_o(0)|^2 +$$

$$+ C[\varepsilon^2\int_0^T |\frac{dy_o}{dt}|^2 dt + \int_0^t |x_1^\varepsilon(s)|^2 ds]$$

from which the desired estimates (1.39) follow easily.

Now applying (1.27) with $u^\varepsilon = u_o$, which is possible, we obtain

$$(1.40) \qquad J^\varepsilon(u_o) = J(u_o) - \int_0^T q_o\cdot\frac{\varepsilon d\bar{y}_\varepsilon}{dt}\,dt + \int_0^T\int_0^1\int_0^1 \lambda[H_{xx}(x_1^\varepsilon)^2 + H_{yy}(y_1^\varepsilon)^2 +$$

$$+ 2H_{xy}\,y_1^\varepsilon\,x_1^\varepsilon]dt d\lambda d\mu + \int_0^1\int_0^1 \lambda\, h_{xx}\, x_1^\varepsilon(T)^2 d\lambda d\mu.$$

But

$$\int_0^T q_o\cdot\frac{d\bar{y}_\varepsilon}{dt}\,dt = q_o(T)\bar{y}_\varepsilon(T) - q_o(0)y_o - \int_0^T \frac{dq_o}{dt}\,\bar{y}_\varepsilon dt$$

and the last estimate (1.39) implies

$$|\int_0^T q_o\cdot\frac{d\bar{y}_\varepsilon}{dt}\,dt| \le C.$$

This and the two first estimates (1.39) imply at once from the expression (1.40) that (1.38) holds true □

Proof of Theorem 1.2.

Consider (1.27) which is now written as

(1.41)
$$J^{\varepsilon}(u^{\varepsilon}) = J(u_o) - \varepsilon\, q_o(T).y_o(T) + \varepsilon\, q_o(0)y_o + \varepsilon \int_0^T \frac{dq_o}{dt}\, y_o dt$$
$$- \varepsilon\, q_o(T)\tilde{y}_{\varepsilon}(T) + \varepsilon \int_0^T \frac{dq_o}{dt}\, \tilde{y}_{\varepsilon} dt + \int_0^T H_v(\sigma_o)\tilde{u}^{\varepsilon} dt$$
$$+ \int_0^T \int_0^1 \int_0^1 \lambda\, H_{\sigma\sigma}(w^{\varepsilon}_{\lambda\mu})(\tilde{\sigma}^{\varepsilon})^2 dt d\lambda d\mu + \int_0^1 \int_0^1 \lambda\, h_{xx}(\tilde{x}^{\varepsilon}(T))^2 d\lambda d\mu.$$

We use next the relations

$$\frac{d\tilde{x}^{\varepsilon}}{dt} = f(\sigma^{\varepsilon}) - f(\sigma_o) \qquad \tilde{x}^{\varepsilon}(0) = 0$$

$$\frac{\varepsilon d\tilde{y}^{\varepsilon}}{dt} = g(\sigma^{\varepsilon}) - g(\sigma_o) \qquad \tilde{y}^{\varepsilon}(0) = y_o - y_o(0)$$

to deduce as in Lemma 1.4,

$$\varepsilon\, |\tilde{y}^{\varepsilon}(t)|^2 + |\tilde{x}^{\varepsilon}(t)|^2 + \mu \int_0^t |\tilde{y}^{\varepsilon}(s)|^2 ds \le \varepsilon |y_o - y_o(0)|^2 +$$
$$+ C[\varepsilon^2 \int_0^T |\frac{dy_o}{dt}|^2 dt + \int_0^t |\tilde{x}^{\varepsilon}(s)|^2 ds + \int_0^t |\tilde{u}^{\varepsilon}(s)|^2 ds].$$

Taking account of the definition of $Z^{\varepsilon}$, we derive the inequality

$$\varepsilon |\tilde{y}^{\varepsilon}(t)|^2 + |\tilde{x}^{\varepsilon}(t)|^2 + \mu \int_0^t |\tilde{y}^{\varepsilon}(s)|^2 ds \le \varepsilon |y_o - y_o(0)|^2$$
$$+ C[\varepsilon^2 \int_0^T |\frac{dy_o}{dt}|^2 dt + \int_0^t |\tilde{x}^{\varepsilon}(s)|^2 ds + \int_0^t |Z^{\varepsilon}(s,\lambda,\mu)|^2 ds]$$

hence also

(1.42)
$$\varepsilon |\tilde{y}^{\varepsilon}(t)|^2 + |\tilde{x}^{\varepsilon}(t)|^2 + \int_0^T |\tilde{y}^{\varepsilon}(s)|^2 ds \le C[\,\varepsilon + \int_0^T |Z^{\varepsilon}(s,\lambda,\mu)|^2 ds]$$

$\forall\ \lambda,\mu.$

We make use of this estimate in the expression (1.41) and minorize the quadratic form as in (1.29). We obtain

$$J^{\varepsilon}(u^{\varepsilon}) \geq J(u_0) + \gamma\int_0^T \int_0^1 \int_0^1 \lambda|Z^{\varepsilon}|^2 dtd\lambda d\mu - C\varepsilon.$$

From Lemma 1.4 and the property (1.18) qualifying $u^{\varepsilon}$, we get

$$\int_0^T \int_0^1 \int_0^1 \lambda|Z^{\varepsilon}|^2 dtd\lambda d\mu \leq C\varepsilon$$

hence also

$$\int_0^T |\tilde{y}^{\varepsilon}(t)|^2 dt, \int_0^T |\tilde{u}^{\varepsilon}(t)|^2 dt, \quad |\tilde{x}^{\varepsilon}(t)|^2 \leq C\varepsilon,$$

and (1.41) implies immediately the desired result. □

Remark 1.1. The results of Theorem 1.1 and 1.2 do not require the existence of an optimal control for the problem □

## 2. ASYMPTOTIC EXPANSIONS IN THE LINEAR QUADRATIC CASE

We shall derive in this section and the next one asymptotic expansions for $\inf J^{\varepsilon}(v(.))$. This will permit to construct controls whose corresponding cost is closer to $\inf J_{\varepsilon}(v(.))$ than $u_0$. They will yield in fact an order of approximation higher that $\varepsilon$. We first study the simpler case corresponding to a linear quadratic control problem.

### 2.1. Linear quadratic control problem.

We consider here the functions

(2.1) $$f(x,y,v) = Ax + By + Gv + a$$

(2.2) $$g(x,y,v) = Cx + Dy + Hv + b$$

(2.3) $$\ell(x,y,v) = \frac{1}{2}(Qx.x + Ry.y + Nv.v) + q.x + r.y + n.v$$

(2.4) $$h(x) = \frac{1}{2} Mx.x .$$

The assumption (1.2) means

(2.5) $$D \leq -\mu I$$

and we have

(2.6) $$H_{xx} = Q \ , \quad H_{yy} = R \ , \quad H_{vv} = N$$

$$H_{xy} = H_{xv} = H_{yv} = 0.$$

We assume

(2.7) M,Q,R,N symmetric, R,N positive definite, Q,M non negative.

The assumption (1.12), (1.13), (1.14) are clearly satisfied.

In this case we can also assert that there exists an optimal control for the payoff $J^\varepsilon(v(.))$ itself. Therefore these exists $(x^\varepsilon, y^\varepsilon, u^\varepsilon, p^\varepsilon, q^\varepsilon) = w^\varepsilon$ solution of the system

(2.8) $$\frac{dx^\varepsilon}{dt} = Ax^\varepsilon + By^\varepsilon + Gu^\varepsilon + a \ , \quad x^\varepsilon(0) = x_0$$

$$\varepsilon\frac{dy^\varepsilon}{dt} = Cx^\varepsilon + Dy^\varepsilon + Hu^\varepsilon + b \ , \quad y^\varepsilon(0) = y_0$$

(2.9) $$-\frac{dp^\varepsilon}{dt} = A^*p^\varepsilon + C^*q^\varepsilon + Qx^\varepsilon + q$$

$$p^\varepsilon(T) = Mx^\varepsilon(T)$$

$$-\varepsilon\frac{dq^\varepsilon}{dt} = B^*p^\varepsilon + D^*q^\varepsilon + Ry^\varepsilon + r$$

$$q^\varepsilon(T) = 0$$

(2.10) $$Nu^\varepsilon + n + G^*p^\varepsilon + H^*q^\varepsilon = 0$$

and eliminating $u^\varepsilon$ yields

(2.11) $$\frac{dx^\varepsilon}{dt} = Ax^\varepsilon + By^\varepsilon - GN^{-1}G^*p^\varepsilon - G\,N^{-1}H^*q^\varepsilon - G\,N^{-1}n + a$$

$$\varepsilon\frac{dy^\varepsilon}{dt} = Cx^\varepsilon + Dy^\varepsilon - H\,N^{-1}G^*p^\varepsilon - H\,N^{-1}H^*q^\varepsilon - H\,N^{-1}n + b$$

$$x^\varepsilon(0) = x_o \quad , \quad y^\varepsilon(0) = y_o$$

(2.12) $$-\frac{dp^\varepsilon}{dt} = A^* p^\varepsilon + C^* q^\varepsilon + Qx^\varepsilon + q$$

$$p^\varepsilon(T) = Mx^\varepsilon(T)$$

$$-\frac{\varepsilon dq^\varepsilon}{dt} = B^* p^\varepsilon + D^* q^\varepsilon + Ry^\varepsilon + r$$

$$q^\varepsilon(T) = 0$$

2.2. Expansion.

We consider an expansion of the form

(2.13) $$x^\varepsilon(t) = x_o(t) + \varepsilon x_1(t) + \varepsilon\ (X_o(\frac{t}{\varepsilon}) + U_o(\frac{T-t}{\varepsilon})) + \tilde{x}^\varepsilon$$

$$y^\varepsilon(t) = y_o(t) + \varepsilon y_1(t) + Y_o(\frac{t}{\varepsilon}) + Z_o(\frac{T-t}{\varepsilon}) + \varepsilon(Y_1(\frac{t}{\varepsilon}) + Z_1(\frac{T-t}{\varepsilon})) + \tilde{y}^\varepsilon$$

$$p^\varepsilon(t) = p_o(t) + \varepsilon p_1(t) + \varepsilon\ (M_o(\frac{t}{\varepsilon}) + K_o(\frac{T-t}{\varepsilon})) + \tilde{p}^\varepsilon(t)$$

$$q^\varepsilon(t) = q_o(t) + \varepsilon q_1(t) + L_o(\frac{t}{\varepsilon}) + Q_o(\frac{T-t}{\varepsilon}) + \varepsilon(L_1(\frac{t}{\varepsilon}) + Q_1(\frac{T-t}{\varepsilon})) + \tilde{q}^\varepsilon.$$

Matching expansions in (2.11), (2.12) we deduce

(2.14) $$\frac{dx_o}{dt} = Ax_o + By_o - G\,N^{-1}G^* p_o - G\,N^{-1}H^* q_o - G\,N^{-1}\ n + a$$

$$Cx_o + Dy_o - H\,N^{-1}G^* p_o - H\,N^{-1}H^* q_o - H\,N^{-1}n + b = 0$$

$$x_o(0) = x_o$$

(2.15) $$-\frac{dp_o}{dt} = A^* p_o + C^* q_o + Qx_o + q$$

$$p_o(T) = Mx_o(T)$$

$$B^* p_o + D^* q_o + Ry_o + r = 0$$

2.16) $$\frac{dY_o}{d\tau} = DY_o - H\,N^{-1}H^*L_o$$

$$Y_o(0) = y_o - y_o(0)$$

$$-\frac{dL_o}{d\tau} = D^*L_o + R\,Y_o$$

2.17) $$-\frac{dZ_o}{d\tau} = D\,Z_o - H^*N^{-1}H\,Q_o$$

$$\frac{dQ_o}{d\tau} = D^*Q_o + R\,Z_o$$

$$Q_o(0) = -\,q_o(T)$$

2.18) $$\frac{dX_o}{d\tau} = B\,Y_o - G\,N^{-1}H^*L_o$$

2.19) $$-\frac{dU_o}{d\tau} = B\,Z_o - G\,N^{-1}H^*Q_o$$

2.20) $$-\frac{dM_o}{d\tau} = C^*L_o$$

2.21) $$\frac{dK_o}{d\tau} = C^*Q_o$$

2.22) $$\frac{dx_1}{dt} = Ax_1 + By_1 - G\,N^{-1}G^*p_1 - G\,N^{-1}H^*q_1$$

$$x_1(0) = -\,X_o(0)$$

$$\frac{dy_o}{dt} = Cx_1 + Dy_1 - H\,N^{-1}G^*p_1 - H\,N^{-1}H^*q_1$$

2.23) $$-\frac{dp_1}{dt} = A^*p_1 + C^*q_1 + Qx_1$$

$$p_1(T) = -\,K_o(0) + M(x_1(T) + U_o(0))$$

$$-\frac{dq_o}{dt} = B^*p_1 + D^*q_1 + Ry_1$$

2.24) $$\frac{dY_1}{d\tau} = DY_1 - H\,N^{-1}H^*L_1 + CX_o - H\,N^{-1}G^*M_o$$

$$Y_1(0) = - y_1(0)$$

$$-\frac{dL_1}{d\tau} = D^*L_1 + R\,Y_1 + B^*M_o$$

$$(2.25) \qquad -\frac{dZ_1}{d\tau} = D\,Z_1 - H\,N^{-1}H^*Q_1 + C\,U_o - H^*N^{-1}G^*K_o$$

$$\frac{dQ_1}{d\tau} = D^*Q_1 + R\,Z_1 + B^*K_o$$

$$Q_1(0) = - q_1(T).$$

Remark 2.1. The expansion (2.13) is sufficient to explain the structure of a full expansion, which can be realized by an induction argument (cf P. BERTRAN [1]).

2.3. Solution of the systems.

We first study the systems (2.14) to (2.25) and derive properties which will play an important role for the convergence. Clearly (2.14), (2.15) correspond to the limit problem and define smooth functions of time.

The next step is to study the systems (2.16), (2.17) respectively. They are terms arising from initial layer and final layer. Note that no initial conditions are specified for $L_o$, $Z_o$. More generally, for all functions for which no initial condition is specified, one has in fact a condition at infinity, namely 0. We shall moreover look for functions of $\tau$ which tend 0 exponentially fast.

We start by giving some results on Riccati equations, along the lines of the work of P. FAURRE (see P. FAURRE, M. CLERGET, F. GERMAIN [1]).

2.3.1. Riccati equations.

Let us consider the algebraic Riccati equation

$$(2.26) \qquad \pi D + D^*\pi - \pi\,H\,N^{-1}H^*\pi + R = 0 \quad , \pi \text{ symmetric.}$$

Proposition 2.1. The set of solutions of (2.26) is not empty. It contains at least a positive definite solution, which is the maximum solution, and which is the unique non negative definite solution. Provided a controllobality condition is satisfied (cf (2.33) below), it contains also a negative definite solution, which is the minimum solution and which is the unique non positive solution.

Proof.

a) Positive solution

Let us show the existence of a non negative solution.

Consider indeed the control problem

$$\frac{dY}{d\tau} = D\,Y + H\xi \qquad\qquad Y(0) = h \tag{2.27}$$

$$\xi \in L^2(0,\infty;R^k)$$

$$\mathcal{J}_h(\xi) = \frac{1}{2}\int_0^\infty (R\,Y^2 + N\,\xi^2)d\tau.$$

Note that by virtue of (1.2), $Y \in H^1(0,\infty;R^m)$ and the control problem is well defined. The optimal control is unique. We can write the Pontryagin conditions of optimality, namely

$$\frac{dY_o}{d\tau} = D\,Y_o - HN^{-1}H^*L_o \qquad\qquad Y_o(0) = h \tag{2.28}$$

$$-\frac{dL_o}{d\tau} = D^*L_o + RY_o \quad , \quad Y_o, L_o \in L^2(0,\infty;R^m).$$

Note that the set of conditions (2.28) defines a unique pair $Y_o$, $L_o$. Writing

$$L_o(0) = \pi h$$

one easily checks that in fact

$$L_o(\tau) = \pi Y_o(\tau) \qquad \forall\ \tau$$

and using this in (2.28), necessarily $\pi$ is a solution of (2.26). Note that

$$\xi_o(\tau) = -\,N^{-1}H^*L_o(\tau)$$

is an optimal control for (2.27) and

$$\mathcal{J}_h(\xi_o) = \frac{1}{2}\,\pi h.h\ . \tag{2.29}$$

Therefore $\pi$ is non negative definite. In fact it is positive definite, otherwise for some $h \neq 0$, one has $\mathcal{J}_h(\xi_0) = 0$, hence

$$Y_0 = 0, \quad \xi_0 = 0$$

which implies $h = 0$, thus a contradiction

Let us show that this matrix, denoted by $\bar{\pi}$ is the maximum solution of (2.26) and the unique non negative solution. Indeed take any non negative $\pi$, and consider the equation

$$\frac{d\tilde{Y}_0}{d\tau} = (D - HN^{-1}H^*\pi)\tilde{Y}_0 , \qquad \tilde{Y}_0 = h$$

We have, by an easy calculation,

$$\frac{1}{2}\,\pi\tilde{Y}_0(T)^2 - \frac{1}{2}\,\pi h^2 + \frac{1}{2}\int_0^T (R + \pi HN^{-1}H^*\pi)\tilde{Y}_0^{\,2} d\tau = 0.$$

Hence necessarily $\tilde{Y}_0 \in L^2(0,\infty;R^m)$. Therefore $\tilde{\xi}_0 = - N^{-1}H^*\pi\tilde{Y}_0$ is an admissible control and setting $\tilde{L}_0 = -\pi\tilde{Y}_0$, we see that $\tilde{Y}_0$, $\tilde{L}_0$ is a solution of (2.28). Therefore necessarily

$$\tilde{Y}_0 = Y_0, \quad \tilde{L}_0 = L_0, \quad \tilde{\xi}_0 = \xi_0$$

and

$$\mathcal{J}_h(\tilde{\xi}_0) = \frac{1}{2}\,\pi h^2 = \mathcal{J}_h(\xi_0) = \frac{1}{2}\,\bar{\pi}h^2,$$

i.e, $\pi = \bar{\pi}$. Let us next show that $\bar{\pi}$ is the maximum solution, in the sense that

$$\text{(2.30)} \qquad \pi \text{ solution of (2.26)} \Rightarrow \pi h^2 \leq \bar{\pi}h^2.$$

Indeed, consider any control $\xi$ and the corresponding $Y$ solution of (2.27). Pick any symmetric $\pi$. One has, by an easy calculation

$$\text{(2.31)} \qquad \mathcal{J}_h(\xi) - \frac{1}{2}\,\pi h^2 = \frac{1}{2}\int_0^\infty N(\xi + N^{-1}H^*\pi Y)^2 d\tau$$

$$+ \frac{1}{2}\int_0^\infty (\pi D + D^*\pi + R - \pi HN^{-1}H^*\pi)Y^2 d\tau$$

and thus if $\pi$ satisfies (2.26)

$$\mathcal{J}_h(\xi) \geq \frac{1}{2}\,\pi h^2$$

hence

$$\inf \mathcal{J}_h(\xi_o) = \frac{1}{2}\,\bar{\pi} h^2 \geq \frac{1}{2}\,\pi h^2.$$

b) negative solution

This is slightly more intricate. Let us consider the dynamic system

$$(2.32) \qquad -\frac{dY}{d\tau} = DY + H\xi\ , \qquad \tau \in (0,\infty)$$

$$\xi \in L^2(0,\infty;R^k)\ , \qquad Y \in L^2(0,\infty\,;R^m).$$

Note that for $\xi \in L^2(0,\infty,R^k)$, there exists one and only one Y solution of (2.32), which is square integrable. Hence Y(0) has a unique value.

Define

$$\mathcal{J}(\xi) = \frac{1}{2}\int_0^\infty (RY^2 + N\xi^2)\,d\tau,$$

Define next

$$E(h) = \{\xi \mid Y(0) = h\}.$$

Assume the controllability condition for the pair (D,H)

$$(2.33) \qquad E(h) \text{ is not empty, } \forall\ h,$$

and consider the problem

$$(2.34) \qquad \inf_{\xi \in E(h)} \mathcal{J}(\xi) = \mathcal{L}(h).$$

It is easy to check that $\mathcal{L}(h)$ is a quadratic function and thus can be written

$$(2.35) \qquad \mathcal{L}(h) = -\frac{1}{2}\,\underline{\pi} h^2$$

where $\underline{\pi}$ is a symmetric non positive matrix. The optimality principle gives the relation

$$(2.36) \qquad -\frac{1}{2}\,\underline{\pi}h^2 = \inf_{\xi(.)} \{\frac{1}{2}\int_0^\delta (RY^2 + N\xi^2)d\tau - \frac{1}{2}\,\underline{\pi}Y(+\delta)^2\}$$

where $\delta > 0$ arbitrary, $\xi(.) \in L^2(0,\delta;R^k)$ arbitrary, and $Y(\delta)$ is the value at time of the solution of (2.32) starting in h at time 0.

From this optimality principle, it is easily checked that $\underline{\pi}$ satisfies the equation (2.26).

Now pick any $\pi$ symmetric solution of (2.26).

By a calculation similar to that of (2.31), one checks that

$$(2.37) \qquad \frac{1}{2}\,\pi h^2 + \mathcal{J}(\xi) = \frac{1}{2}\int_0^\infty N(\xi + N^{-1}H^*\pi Y)^2 d\tau$$

for any $\xi \in E(h)$. Therefore

$$\mathcal{L}(h) \geq -\frac{1}{2}\,\pi h^2$$

which implies

$$\pi \geq \underline{\pi}$$

and $\underline{\pi}$ is the minimum solution. Suppose $\pi$ non positive and consider the equation

$$-\frac{d\tilde{Y}_o}{d\tau} = (D - HN^{-1}H^*\pi)\tilde{Y}_o \;, \qquad \tilde{Y}_o(0) = h.$$

We have

$$\frac{1}{2}\,\pi h^2 - \frac{1}{2}\,\pi\tilde{Y}_o(+T)^2 + \frac{1}{2}\int_0^T (N\tilde{\xi}_o^2 + R\tilde{Y}_o^2)d\tau = 0$$

where

$$\tilde{\xi}_o = -N^{-1}H^*\pi\tilde{Y}_o$$

therefore $\pi\tilde{Y}_0(+T)^2$ increases as $T \uparrow +\infty$. Since it is negative, there is a limit, which implies that

$$\tilde{\xi}_0 \in L^2(0,\infty;R^k),\ \tilde{Y}_0 \in L^2(0,\infty;R^m),\ \text{hence } \tilde{\xi}_0 \in E(h),$$

and

$$(2.38) \qquad J(\tilde{\xi}_0) = -\frac{1}{2}\pi h^2.$$

Let us check that $\tilde{\xi}_0$ is optimal. Indeed take $\xi \in E(h)$, and set

$$\tilde{Y} = Y - \tilde{Y}_0$$

$$\tilde{\xi} = \xi + N^{-1}H^*\pi Y = \xi + N^{-1}H^*\pi\tilde{Y} - \tilde{\xi}_0.$$

We have

$$(2.39) \qquad -\frac{d\tilde{Y}}{d\tau} = (D - HN^{-1}H^*\pi)\tilde{Y} + H\tilde{\xi} \qquad \tilde{Y}(0) = 0.$$

and

$$\mathcal{J}(\xi) = \frac{1}{2}\int_0^\infty [R(\tilde{Y}_0 + \tilde{Y})^2 + N(\tilde{\xi}_0 - N^{-1}H^*\pi\tilde{Y} + \tilde{\xi})^2]d\tau$$

$$\geq \mathcal{J}(\tilde{\xi}_0) + \int_0^\infty [R\tilde{Y}_0\tilde{Y} + N\tilde{\xi}_0(\tilde{\xi} - N^{-1}H^*\pi\tilde{Y})]d\tau$$

$$= \mathcal{J}(\tilde{\xi}_0) + \int_0 [R\tilde{Y}_0\tilde{Y} - H^*\pi\tilde{Y}_0\tilde{\xi} + \pi HN^{-1}H^*\pi\tilde{Y}_0\tilde{Y}]d\tau$$

and from (2.39)

$$\int_0^\infty \pi\tilde{Y}_0\, H\tilde{\xi}\, d\tau = \int_0^\infty \pi\tilde{Y}_0(-\frac{d\tilde{Y}}{d\tau} - (D - HN^{-1}H^*\pi)\tilde{Y})d\tau$$

$$= -2\int_0^\infty \pi\tilde{Y}(D - HN^{-1}H^*\pi)\tilde{Y}_0 d\tau.$$

Collecting results we obtain

$$\mathcal{J}(\xi) \geq \mathcal{J}(\tilde{\xi}_0)$$

hence $\tilde{\xi}_0$ is optimal, which implies $\pi = \underline{\pi}$.

The proof is now complete □

Remark 2.1. In the proof of part b) we have shown that the existence of a non positive solution of (2.26) implies the controllability (2.33) □

2.3.2. Duality

There are several ways to introduce duality considerations. To be consistant with the general case treated later, we shall not make use of all the possibilities which arise from explicit calculations in the present case.

Duality will be defined by the transformation

$$\pi \to - \pi^{-1} = \Sigma$$

which transforms the equation (2.26) into

$$D\Sigma + \Sigma D^* - \Sigma R \Sigma + HN^{-1}H^* = 0. \tag{2.40}$$

This equation is of the same type as (2.26) except that the constaint term may be not invertible without assumptions on H. By analogy with (2.26) we can associate with (2.40) the control problems

$$\frac{dL}{d\tau} = D^*L + R\eta \tag{2.41}$$

$$L(0) = h$$

$$\mathcal{J}_h^*(\eta) = \frac{1}{2} \int_0 (R\eta^2 + HN^{-1}H^*L^2)d\tau$$

and

$$- \frac{dL}{d\tau} = D^*L + R\eta \tag{2.42}$$

$$\mathcal{J}^*(\eta) = \frac{1}{2} \int_0^{\infty} (R\eta^2 + HN^{-1}H^*L^2)d\tau$$

and

$$\eta \in E^*(h) = \{ \eta \mid L(0) = h\}.$$

Now defining

$$\frac{1}{2} \Sigma h^2 = \inf \mathcal{J}_h^*(\eta)$$

we obtain the unique non negative solution of (2.40). To get $\Sigma$ positive definite it is sufficient to assume that.

(2.43) $\quad H H^*$ has a left inverse.

This suffices to verify the controllability condition (2.33), by virtue of Remark 2.1.

In fact the special representations of $\Sigma$ related to the control problems (2.41), (2.42) cannot be generalized and are limited to the L.Q. case. However it is possible to give two interpretations which will be generalizable.

Consider the control problem

$$(2.44) \qquad -\frac{dZ}{d\tau} = DZ + H\eta$$

$$K_h(\eta) = \frac{1}{2} \int_0^\infty (RZ^2 + N\eta^2)d\tau - Z(0)h$$

in which $\eta \in L^2(0,\infty;R^k)$ (note that Z is uniquely defined in $H^1(0,\infty;R^m)$, and in particular Z(0) is completely determined).

Let us set

$$(2.45) \qquad \inf_\eta K_h(\eta) = -\frac{1}{2} \bar{\Sigma} h^2$$

then $\bar{\Sigma} = -\underline{\pi}^{-1}$ is the positive solution of (2.40).

Define next the problem

$$(2.46) \qquad \frac{dZ}{d\tau} = DZ + H\eta \quad , \; Z(0) = \zeta$$

$$\mathcal{L}_h(\eta;\zeta) = \frac{1}{2}\int_0^\infty (RZ^2 + N\eta^2)d\tau + h\zeta \;.$$

Clearly

$$(2.47) \qquad \inf_{\eta,\zeta} \mathcal{L}_h(\eta;\zeta) = \inf_\zeta \inf_\eta (\mathcal{D}_\xi(\eta) + h\zeta)$$

$$= \inf_\zeta \left(\frac{1}{2}\,\bar{\pi}\zeta^2 + h\zeta\right)$$

$$= -\frac{1}{2}\,\bar{\pi}^{-1}h^2 = \frac{1}{2}\,\underline{\Sigma}h^2.$$

Therefore

$$(2.48) \qquad \inf_{\eta,\zeta} \mathcal{L}_h(\eta;\zeta) = \frac{1}{2}\,\underline{\Sigma}h^2.$$

It is also possible to derive (2.44) and (2.46) as duals of (2.41), (2.42), in the sense of dualizing the constraints represented by the state equation and the initial condition ; but this approach is limited to the L.Q. case.

2.3.3. Application

From Proposition 2.1 follows that (2.16) and (2.17) have one and only one solution $(Y_0, L_0)$, $(Z_0, Q_0)$. We moreover have the property

Lemma 2.1. One has

$$(2.49) \qquad |Y_0(\tau)|,\ |L_0(\tau)| \le C\, e^{-\gamma\tau}\, |y_0 - y_0(0)|$$

$$|Z_0(\tau)|,\ |Q_0(\tau)| \le C\, e^{-\gamma\tau}\, |q_0(T)|$$

Proof :

One has

$$L_0(\tau) = \pi\, Y_0(\tau)$$

$$Z_0(\tau) = -\,\Sigma Q_0(\tau)$$

and

$$- \frac{d}{d\tau} M Y_o Y_o = Y_o(R + \pi HN^{-1}H^*\pi)Y_o$$

and from the invertibility of $\pi$, the 1st and the $2^{nd}$ estimates (2.49) follow. The two last are proved in a similar way □

Lemma 2.2 : One has

$$(2.50) \qquad |X_o(\tau)|, |M_o(\tau)| \le C e^{-\gamma\tau} |y_o - y_o(0)|$$

$$|U_o(\tau)|, |K_o(\tau)| \le C e^{-\gamma\tau} |q_o(\tau)|$$

Proof.

$$X_o(\tau) = \int_\tau^\infty (G N^{-1}H^*L_o(\sigma) - B Y_o(\sigma))d\sigma$$

and from (2.49) one deduces the 1st estimate (2.50). Similar considerations hold for the other quantities □

The system (2.22), (2.23) is similar to (2.14), (2.15) and has a unique solution related to a linear quadratic control problem.

Lemma 2.3. $Y_1$, $L_1$ satisfy the same estimates as $Y_o$,$L_o$ and $Z_1$,$Q_1$ the same as $Z_o$, $Q_o$ (cf (2.49).

Proof.

One has

$$L_1(\tau) = \pi Y_1(\tau) + \rho_1(\tau)$$

and $\rho_1$ is the solution of

$$- \frac{d\rho_1}{d\tau} = (D^* - \pi HN^{-1}H^*)\rho_1 + (B^* - \pi HN^{-1}G^*)M_o + \pi CX_o$$

$$\rho_1(\infty) = 0.$$

Therefore

$$- \frac{d}{d\tau} \pi^{-1}\rho_1\rho_1 = - (\pi^{-1}R\pi^{-1} + HN^{-1}H^*)\rho_1^2 + 2\pi^{-1}\rho_1\Theta$$

where

$$\Theta = (B^* - \pi HN^{-1}G^*)M_o + \pi \, CX_o$$

and it easily follows taking account of the exponential decay of $\Theta$ that

$$|\rho_1(\tau)| \leq C\, e^{-\gamma\tau}.$$

Since next

$$\frac{dY_1}{d\tau} = (D - HN^{-1}H^*\pi)Y_1 + CX_o - HN^{-1}G^*M_o - HN^{-1}H^*\rho_1$$

$$Y_1(0) = - y_1(0).$$

Computing $\frac{d}{d\tau} \pi Y^2$, and using the Riccati equation for $\pi$ and taking account of the exponential decay of the right hand side, one proves the exponential decay of $Y_1$. The other estimates follow □

We now write several control problems related to the relations (2.14) to (2.25). Firstly (2.14), (2.15) relate to the limit problem.

(2.51) $$\frac{dx_o}{dt} = Ax_o + By_o + Gu_o + a \qquad x_o(0) = x_o$$

$$Cx_o + Dy_o + Hu_o + b = 0$$

$$J(u_o) = \frac{1}{2} \int_0^T [Qx_o^2 + Ry_o^2 + Nu_o^2 + 2q.x_o + 2r.y_o + 2n.u_o]dt + \frac{1}{2} Mx_o(T)^2$$

Next (2.16) relate to the control problem

(2.52) $$\frac{dY_o}{d\tau} = DY_o + H\xi_o \qquad Y_o(0) = y_o - y_o(0)$$

$$\mathcal{J}_o(\xi_o) = \frac{1}{2} \int_0^\infty (RY_o^2 + N\,{}_o^2)d\tau$$

and (2.17) to the control problem

(2.53) $$-\frac{dZ_o}{d\tau} = DZ_o + H\eta_o , \qquad Z_o(\infty) = 0$$

$$\mathcal{K}_o(\eta_o) = \frac{1}{2}\int_0^\infty (RZ_o^2 + N\eta_o^2)d\tau - Z_o(0)\, q_o(T).$$

The relations (2.22), (2.23) relate to the control problem

(2.54) $$\frac{dx_1}{dt} = Ax_1 + By_1 + Gu_1$$

$$x_1(0) = - X_o(0)$$

$$\frac{dy_o}{dt} = Cx_1 + Dy_1 + Hu_1$$

$$J_1(u_1) = \frac{1}{2}\int_0^T (Qx_1^2 + Ry_1^2 + Nu_1^2 + 2\,\frac{dq_o}{dt}\,.\,y_1)dt$$

$$+ \frac{1}{2}\,(Mx_1(T)^2 + 2x_1(T)(MU_o(0) - K_o(0))$$

Finally for (2.24), (2.25) we get the problems

(2.55) $$\frac{dY_1}{d\tau} = DY_1 + H\xi_1 + CX_o$$

$$Y_1(0) = - y_1(0)$$

$$\mathcal{G}_1(\xi_1) = \frac{1}{2}\int_0^\infty [RY_1^2 + N\xi_1^2 + 2B^*M_o\, Y_1 + 2G^*M_o\xi_1]d\tau$$

(2.56) $$-\frac{dZ_1}{d\tau} = DZ_1 + H\eta_1 + CU_o \qquad Z_1(\infty) = 0$$

$$\mathcal{K}_1(\eta_1) = \frac{1}{2}\int_0^\infty [RZ_1^2 + N\eta_1^2 + 2B^*K_oZ_1 + 2G^*K_o\eta_1]d\tau - q_1(T)Z_1(0).$$

The various optimal controls are obtained in function of the adjoint variables by the formulas

(2.57) $$\xi_o(\tau) = - N^{-1}H^*L_o(\tau)$$

$$\eta_o(\tau) = - N^{-1}H^*Q_o(\tau)$$

(2.58) $u_1(t) = - N^{-1}(G^* p_1(t) + H^* q_1(t))$

(2.59) $\xi_1(\tau) = - N^{-1}(H^* L_1(\tau) + G^* M_o(\tau))$

$\eta_1(\tau) = - N^{-1}(H^* Q_1(\tau) + G^* K_o(\tau))$

## 2.4. Convergence

From now on we shall assume that

(2.60) $HH^*$ is invertible.

We then have

Theorem 2.1. *We assume* (2.5), (2.7) *and* (2.60). *Then we have*

(2.61) $|\tilde{x}^\varepsilon(t)|, \quad |\tilde{p}^\varepsilon(t)| \leq C\varepsilon^2$

(2.62) $|\tilde{y}^\varepsilon|_{L^2}, \quad |\tilde{q}^\varepsilon|_{L^2} \leq C^{\ 2}$

*these quantities being defined by* (2.13).

*Proof.*

Introduce $\phi^\varepsilon, \psi^\varepsilon$ defined as follows

(2.63) $$\frac{d\phi^\varepsilon}{dt} = A\phi^\varepsilon + \varepsilon[A(X_o(\frac{t}{\varepsilon}) + U_o(\frac{T-t}{\varepsilon})) + B(Y_1(\frac{t}{\varepsilon}) + Z_1(\frac{T-t}{\varepsilon})) -$$

$$- GN^{-1}G^*(M_o(\frac{t}{\varepsilon}) + K_o(\frac{T-t}{\varepsilon})) - GN^{-1}H^*(L_1(\frac{t}{\varepsilon}) + Q_1(\frac{T-t}{\varepsilon}))]$$

$$\phi^\varepsilon(0) = - \varepsilon\, U_o(\frac{T}{\varepsilon})$$

(2.64) $$- \frac{d\psi^\varepsilon}{dt} = A^*\psi^\varepsilon + \varepsilon[A^*(M_o(\frac{t}{\varepsilon}) + K_o(\frac{T-t}{\varepsilon})) + C^*(L_1(\frac{t}{\varepsilon}) + Q_1(\frac{T-t}{\varepsilon})) +$$

$$+ Q(X_o(\frac{t}{\varepsilon}) + U_o(\frac{T-t}{\varepsilon}))]$$

$$\psi^\varepsilon(T) = M\phi^\varepsilon(T) + \varepsilon[MX_o(\frac{T}{\varepsilon}) - M_o(\frac{T}{\varepsilon})]$$

and clearly one has

$$(2.65) \qquad |\phi^{\varepsilon}(t)|, \quad |\psi^{\varepsilon}(t)| \leq C\varepsilon^{2}.$$

Let us then define

$$\tilde{\tilde{x}}^{\varepsilon} = \tilde{x}^{\varepsilon} - \phi^{\varepsilon}, \quad \tilde{\tilde{p}}^{\varepsilon} = \tilde{p}^{\varepsilon} - \psi^{\varepsilon}$$

we derive the following equations from (2.11), (2.12)

$$(2.66) \qquad \frac{d\tilde{\tilde{x}}^{\varepsilon}}{dt} = A\tilde{\tilde{x}}^{\varepsilon} + B\tilde{y}^{\varepsilon} - GN^{-1}G^{*}\tilde{\tilde{p}}^{\varepsilon} - GN^{-1}H^{*}\tilde{q}^{\varepsilon} - GN^{-1}G^{*}\psi^{\varepsilon}$$

$$\tilde{\tilde{x}}^{\varepsilon}(0) = 0$$

$$(2.67) \qquad \varepsilon\frac{d\tilde{y}^{\varepsilon}}{dt} = C\tilde{\tilde{x}}^{\varepsilon} + D\tilde{y}^{\varepsilon} - HN^{-1}G^{*}\tilde{\tilde{p}}^{\varepsilon} - HN^{-1}H^{*}\tilde{q}^{\varepsilon} - \varepsilon^{2}\frac{dy_{1}}{dt}$$

$$+ C\phi^{\varepsilon} - HN^{-1}G^{*}\psi^{\varepsilon}$$

$$\tilde{y}^{\varepsilon}(0) = - Z_{0}(\frac{T}{\varepsilon}) - \varepsilon\, Z^{1}(\frac{T}{\varepsilon})$$

$$(2.68) \qquad - \frac{d\tilde{\tilde{p}}^{\varepsilon}}{dt} = A^{*}\tilde{\tilde{p}}^{\varepsilon} + C^{*}\tilde{q}^{\varepsilon} + Q\tilde{\tilde{x}}^{\varepsilon} + Q\phi^{\varepsilon}$$

$$\tilde{\tilde{p}}^{\varepsilon}(T) = M\tilde{\tilde{x}}^{\varepsilon}(T)$$

$$(2.69) \qquad - \varepsilon\frac{d\tilde{q}^{\varepsilon}}{dt} = B^{*}\tilde{\tilde{p}}^{\varepsilon} + D^{*}\tilde{q}^{\varepsilon} + R\tilde{y}^{\varepsilon} + B^{*}\psi^{\varepsilon} + \varepsilon^{2}\frac{dq_{1}}{dt}$$

$$\tilde{q}^{\varepsilon}(T) = - L_{0}(\frac{T}{\varepsilon}) - \varepsilon\, L_{1}(\frac{T}{\varepsilon}).$$

From (2.66), ... , (2.69) one deduces the relation

$$(2.70) \qquad M\tilde{\tilde{x}}^{\varepsilon}(T)^{2} + \int_{0}^{T} [Q(\tilde{\tilde{x}}^{\varepsilon})^{2} + R(\tilde{y}^{\varepsilon})^{2} + N^{-1}(G^{*}\tilde{\tilde{p}}^{\varepsilon} + M^{*}\tilde{q}^{\varepsilon})^{2}]dt$$

$$= \varepsilon\tilde{y}^{\varepsilon}(T)(L_{0}(\frac{T}{\varepsilon}) + \varepsilon L_{1}(\frac{T}{\varepsilon})) - \varepsilon\tilde{q}^{\varepsilon}(0)(Z_{0}(\frac{T}{\varepsilon}) + \varepsilon\, Z_{1}(\frac{T}{\varepsilon})) +$$

$$+ \int_{0}^{T} [- N^{-1}G^{*}\psi^{\varepsilon}(G^{*}\tilde{\tilde{p}}^{\varepsilon} + H^{*}\tilde{q}^{\varepsilon}) - Q\tilde{\tilde{x}}^{\varepsilon}.\phi^{\varepsilon} + C\phi^{\varepsilon}\tilde{q}^{\varepsilon} - B^{*}\psi^{\varepsilon}.\tilde{y}^{\varepsilon}]dt$$

$$- \varepsilon^{2}\int_{0}^{T} (\tilde{q}^{\varepsilon}.\frac{dy_{1}}{dt} + \tilde{y}^{\varepsilon}\frac{dq_{1}}{dt})dt.$$

Using the assumption (2.60) we may write

$$(2.71) \qquad \tilde{q}^\varepsilon = (HH^*)^{-1}H(G^*\tilde{\tilde{p}}^\varepsilon + H^*\tilde{q}^\varepsilon) - (HH^*)^{-1}HG^*\tilde{\tilde{p}}^\varepsilon$$

which used in (2.68) yields

$$-\frac{d\tilde{\tilde{p}}^\varepsilon}{dt} = (A^* - C^*(HH^*)^{-1}HG^*)\tilde{\tilde{p}}^\varepsilon + C^*(HH^*)^{-1}H(G^*\tilde{\tilde{p}}^\varepsilon + H^*\tilde{q}^\varepsilon) +$$

$$+ Q\tilde{\tilde{x}}^\varepsilon + Q\phi^\varepsilon$$

$$\tilde{\tilde{p}}^\varepsilon(T) = M\tilde{\tilde{x}}^\varepsilon(T)$$

from which the following estimate is derived

$$(2.72) \qquad |\tilde{\tilde{p}}^\varepsilon(t)| \le k_o[|M^{1/2}\tilde{\tilde{x}}^\varepsilon(T)| + |G^*\tilde{\tilde{p}}^\varepsilon + H^*\tilde{q}^\varepsilon|_{L^2} + |Q^{1/2}\tilde{\tilde{x}}^\varepsilon|_{L^2} + \varepsilon^2].$$

We use this estimate in (2.71) and next in (2.70) to estimate $\tilde{q}^\varepsilon$.

The first two terms to the right hand side of (2.70) are as small as we wish because of the exponential decay of $L_o$, $L_1$, $Z_o$, $Z_1$ and since $\varepsilon\, \tilde{y}^\varepsilon(T), \varepsilon\, \tilde{q}^\varepsilon(0)$ are bounded.

Collecting results we obtain

$$(2.73) \qquad |M^{1/2}\tilde{\tilde{x}}^\varepsilon(T)|^2 + \int_0^T [|Q^{1/2}\tilde{\tilde{x}}^\varepsilon|^2 + |\tilde{y}^\varepsilon|^2 + |G^*\tilde{\tilde{p}}^\varepsilon + H^*\tilde{q}^\varepsilon|^2]dt$$

$$\le k_1\varepsilon^2[|G^*\tilde{\tilde{p}}^\varepsilon + H^*\tilde{q}^\varepsilon|_{L^2} + |Q^{1/2}\tilde{\tilde{x}}^\varepsilon|_{L^2} + |\tilde{y}^\varepsilon|_{L^2} + |M^{1/2}\tilde{\tilde{x}}^\varepsilon(T)|] + k_1\,\varepsilon^4,$$

which implies

$$(2.74) \qquad |M^{1/2}\tilde{\tilde{x}}^\varepsilon(T)|,\ |Q^{1/2}\tilde{\tilde{x}}^\varepsilon|_{L^2},\ |\tilde{y}^\varepsilon|_{L^2},\ |G^*\tilde{\tilde{p}}^\varepsilon + H^*\tilde{q}^\varepsilon|_{L^2} \le C\varepsilon^2$$

and from (2.72), (2.71)

$$|\tilde{\tilde{p}}^\varepsilon(t)| \le C\varepsilon^2 \quad , \quad |\tilde{q}^\varepsilon|_{L^2} \le C\varepsilon^2$$

which used in (2.66) yields finally

$$|\tilde{\tilde{x}}^{\varepsilon}(t)| \leq C\varepsilon^2$$

and this completes the proof of the desired results □

We turn now to the expansion of the optimal cost $J^{\varepsilon}(u^{\varepsilon})$. We shall need the following terms

(2.75) $$X_1 = q_o(0)(y_o - y_o(0)) - \int_0^T \frac{dy_o}{dt} q_o dt + \mathcal{Y}_o(\xi_o) + \mathcal{K}_o(\eta_o)$$

$$X_2 = J_1(u_1) + \frac{1}{2} M\, U_o(0)^2 + \int_0^{\infty} C(U_o Q_o + X_o L_o) d\tau +$$

$$+ \frac{dq_o}{dt}(0) \int_0^{\infty} Y_o d\tau + \frac{dq_o}{dt}(T) \int_0^{\infty} Z_o d\tau - \frac{dy_o}{dt}(T) \int_0^{\infty} Q_o d\tau -$$

$$- \left(\frac{dy_o}{dt}(0) + CX_o(0)\right) \int_0^{\infty} L_o d\tau$$

$$X_3 = \mathcal{Y}_1(\xi_1) + \mathcal{K}_1(\eta_1) + \frac{d^2 q_o}{dt^2}(0) \int_0^{\infty} \tau Y_o d\tau - \frac{d^2 q_o}{dt^2}(T) \int_0^{\infty} \tau Z_o d\tau -$$

$$- \frac{d^2 y_o}{dt^2}(0) \int_0^{\infty} \tau L_o d\tau + \frac{d^2 y_o}{dt^2}(T) \int_0^{\infty} \tau Q_o d\tau + \frac{dp_1}{dt}(0) \int_0^{\infty} \tau (BY_o + G\xi_o) d\tau +$$

$$+ \int_0^{\infty} (M_o A X_o + K_o A U_o) d\tau - \frac{dp_1}{dt}(T) \int_0^{\infty} \tau (BZ_o + G\eta_o) d\tau +$$

$$+ \int_0^{\infty} \frac{1}{2} Q(X_o^2 + U_o^2) d\tau - \int_0^T q_1 \frac{dy_1}{dt} dt - y_1(0) q_1(0) +$$

$$+ \frac{dx_1}{dt}(0) \int_0^{\infty} M_o d\tau + \frac{dx_1}{dt}(T) \int_0^{\infty} K_o d\tau.$$

One can then state the

Theorem 2.2. *Under the assumptions of Theorem 2.1, one has the estimate*

(2.76) $$|J^{\varepsilon}(u^{\varepsilon}) - J(u_o) - \varepsilon X_1 - \varepsilon^2 X_2 - \varepsilon^3 X_3| \leq C\varepsilon^4.$$

Remark 2.3. We have the related estimates

$$|J^\varepsilon(u^\varepsilon) - J(u_o)| \le C\varepsilon$$

$$|J^\varepsilon(u^\varepsilon) - J(u_o + \xi_o + \eta_o)| \le C\varepsilon^2$$

$$|J^\varepsilon(u^\varepsilon) - J(u_o + \xi_o + \eta_o + \varepsilon\, u_1)| \le C\varepsilon^3$$

$$|J^\varepsilon(u^\varepsilon) - J(u_o + \xi_o + \eta_o + \varepsilon\, u_1 + \varepsilon\, \xi_1 + \varepsilon\, \eta_1)| \le C\varepsilon^4,$$

which indicate the improvement of the accuracy, when correcting $u_o$ by addition of further terms. □

It will be convenient to define

(2.77) $$u^\varepsilon(t) = u_o(t) + \xi_o(\tfrac{t}{\varepsilon}) + \eta_o(\tfrac{T-t}{\varepsilon}) + \varepsilon(u_1(t) + \xi_1(\tfrac{t}{\varepsilon}) + \eta_1(\tfrac{T-t}{\varepsilon})) + \tilde{u}^\varepsilon(t).$$

and it follows from Theorem 2.1, that

(2.78) $$|\tilde{u}^\varepsilon|_{L^2} \le C\varepsilon^2$$

<u>Proof of Theorem 2.2.</u>

We express

$$J^\varepsilon(u^\varepsilon) = J(u_o + \xi_o + \eta_o + \varepsilon\, u_1 + \varepsilon\, \xi_1 + \varepsilon\, \eta_1 + \tilde{u}^\varepsilon).$$

We expand and leave terms of order $\varepsilon^4$. We obtain

(2.79) $$J^\varepsilon(u^\varepsilon) = \frac{1}{2}\int_0^T [Q(x_o + \varepsilon x_1 + \varepsilon(X_o + U_o))^2 +$$

$$+ R(y_o + \varepsilon y_1 + Y_o + Z_o + \varepsilon(Y_1 + Z_1))^2 +$$

$$+ N(u_o + \varepsilon u_1 + \xi_o + \eta_o + \varepsilon(\xi_1 + \eta_1))^2 +$$

$$+ 2\, Q\tilde{x}^\varepsilon(x_o + \varepsilon x_1) + 2\, R\tilde{y}^\varepsilon(y_o + \varepsilon y_1 + Y_o + Z_o + \varepsilon(Y_1 + Z_1))$$

$$+ 2\, N\tilde{u}^\varepsilon(u_o + \varepsilon u_1 + \xi_o + \eta_o + \varepsilon(\xi_1 + \eta_1)) + 2qx^\varepsilon$$

$$+ 2r\, y^{\varepsilon} + 2n\, u^{\varepsilon}]dt + \frac{1}{2} M(x_o(T) + \varepsilon x_1(T) + \varepsilon U_o(0))^2$$

$$+ M\tilde{x}^{\varepsilon}(T)(x_o(T) + \varepsilon x_1(T) + \varepsilon U_o(0)) + O(\varepsilon^4).$$

We note that

$$\frac{d\tilde{x}^{\varepsilon}}{dt} = A\tilde{x}^{\varepsilon} + B\tilde{y}^{\varepsilon} + G\tilde{u}^{\varepsilon} + \varepsilon(A(X_o(\tfrac{t}{\varepsilon}) + U_o(\tfrac{T-t}{\varepsilon})) + \tag{2.80}$$

$$+ B(Y_1(\tfrac{t}{\varepsilon}) + Z_1(\tfrac{T-t}{\varepsilon})) + G(\xi_1(\tfrac{t}{\varepsilon}) + \eta_1(\tfrac{T-t}{\varepsilon}))$$

$$\tilde{x}^{\varepsilon}(0) = - \varepsilon U_o(\tfrac{T}{\varepsilon})$$

$$\varepsilon\frac{d\tilde{y}^{\varepsilon}}{dt} = C\tilde{x}^{\varepsilon} + D\tilde{y}^{\varepsilon} + H\tilde{u}^{\varepsilon} - \varepsilon^2 \frac{dy_1}{dt} \tag{2.81}$$

$$\tilde{y}^{\varepsilon}(0) = - Z_o(\tfrac{T}{\varepsilon}) - \varepsilon Z_1(\tfrac{T}{\varepsilon}).$$

We compute the quantity

$$\int_0^T [\tilde{x}^{\varepsilon}(Qx_o + q + \varepsilon Qx_1) + \tilde{y}^{\varepsilon}(Ry_o + r + R(Y_o + Z_o) + \varepsilon R(Y_1 + Z_1)) + \tag{2.82}$$

$$+ \tilde{u}^{\varepsilon}(Nu_o + n + \varepsilon Nu_1 + N(\xi_o + \eta_o) + \varepsilon N(\xi_1 + \eta_1))]dt +$$

$$+ \tilde{x}^{\varepsilon}(T)(Mx_o(T) + \varepsilon M(x_1(T) + U_o(0)))$$

$$= \int_0^T [\tilde{x}^{\varepsilon}(- \frac{d}{dt}(p_o + \varepsilon p_1) - A^*(p_o + \varepsilon p_1) - C^*(q_o + \varepsilon q_1)) +$$

$$+ \tilde{y}^{\varepsilon}(- B^*(p_o + \varepsilon p_1) - D^*(q_o + \varepsilon q_1) - \varepsilon\frac{dq_o}{dt} - \varepsilon\frac{d}{dt}(L_o + \varepsilon L_1) -$$

$$- D^*(L_o + \varepsilon L_1) - \varepsilon B^* M_o - \varepsilon\frac{d}{dt}(Q_o + \varepsilon Q_1) - D^*(Q_o + \varepsilon Q_1) - \varepsilon B^* K_o) +$$

$$+ \tilde{u}^{\varepsilon}(- G^*(p_o + \varepsilon p_1) - H^*(q_o + \varepsilon q_1) - H^*(L_o + \varepsilon L_1) - H^*(Q_o + \varepsilon Q_1) -$$

$$- \varepsilon G^*(M_o + K_o))]dt + \tilde{x}^{\varepsilon}(T)M(x_o(T) + \varepsilon x_1(T) + \varepsilon U_o(0))$$

$$= \varepsilon K_o(0)\tilde{x}^\varepsilon(T) + \int_0^T [\varepsilon(p_o + \varepsilon p_1)(A(X_o + U_o) + B(Y_1 + Z_1) +$$

$$+ G(\xi_1 + \eta_1)) + (q_o + \varepsilon q_1)( - \varepsilon\frac{d\tilde{y}^\varepsilon}{dt} - \varepsilon^2 \frac{dy_1}{dt}) - \varepsilon\tilde{y}^\varepsilon \frac{dq_o}{dt} +$$

$$+ (L_o + \varepsilon L_1)(C\tilde{x}^\varepsilon - \varepsilon^2 \frac{dy_1}{dt}) - \varepsilon(Q_o(0) + \varepsilon Q_1(0))\tilde{y}^\varepsilon(T) +$$

$$+ (Q_o + \varepsilon Q_1)(C\tilde{x}^\varepsilon - \varepsilon^2\frac{dy_1}{dt}) - \varepsilon(B\tilde{y}^\varepsilon + G\tilde{u}^\varepsilon)(M_o + K_o)]dt + 0(\varepsilon^4)$$

and taking account of $||\tilde{x}^\varepsilon||_{L^\infty} \leq C\varepsilon^2$,

$$= \int_0^T [\varepsilon^2(M_o + K_o)(A(X_o + U_o) + B(Y_1 + Z_1) + G(\xi_1 + \eta_1)) +$$

$$+ \varepsilon(p_o + \varepsilon p_1)(A(X_o + U_o) + B(Y_1 + Z_1) + G(\xi_1 + \eta_1)) -$$

$$- \varepsilon^2(q_o + \varepsilon q_1) \frac{dy_1}{dt} - \varepsilon^2 L_o \frac{dy_1}{dt} - \varepsilon^2 Q_o \frac{dy_1}{dt}]dt + 0(\varepsilon^4).$$

Collecting results

$$J^\varepsilon(u^\varepsilon) = \frac{1}{2} \int_0^T [Q(x_o + \varepsilon x_1 + \varepsilon(X_o + U_o))^2 + R(y_o+\varepsilon y_1+Y_o+Z_o+\varepsilon(Y_1+Z_1))^2 +$$

$$+ N(u_o + \varepsilon u_1 + \xi_o + \eta_o + \varepsilon(\xi_1 + \eta_1))^2 + q(x_o + \varepsilon x_1 + \varepsilon(X_o + U_o)) +$$

$$+ r(y_o + \varepsilon y_1 + Y_o + Z_o + \varepsilon(Y_1 + Z_1)) +$$

$$+ n(u_o + \varepsilon u_1 + \xi_o + \eta_o + \varepsilon(\xi_1 + \eta_1))]dt +$$

$$+ \int_0^T [\varepsilon^2(M_o + K_o)(A(X_o + U_o) + B(Y_1 + Z_1) + G(\xi_1 + \eta_1)) +$$

$$+ \varepsilon(p_o + \varepsilon p_1)(A(X_o + U_o) + B(Y_1 + Z_1) + G(\xi_1 + \eta_1)) -$$

$$- \varepsilon^2(q_o + \varepsilon q_1) \frac{dy_1}{dt} - \varepsilon^2 L_o \frac{dy_1}{dt} - \varepsilon^2 Q_o \frac{dy_1}{dt}]dt +$$

$$+ \frac{1}{2} M(x_o(T) + \varepsilon x_1(T) + \varepsilon U_o(0))^2 + 0(\varepsilon^4)$$

Considering first the terms which may be of order $\varepsilon$, we get

$$\int_0^T [\varepsilon x_1(Qx_o + q) + (\varepsilon y_1 + Y_o + Z_o)(Ry_o + r) + (\varepsilon u_1 + \xi_o + \eta_o)(Nu_o + n) +$$

$$+ \frac{1}{2} R(Y_o^2 + Z_o^2) + \frac{1}{2} N(\xi_o^2 + \eta_o^2)]dt + \varepsilon Mx_o(T)(x_1(T) + U_o(0)).$$

But

$$\int_0^T [\varepsilon x_1(Qx_o + q) + (\varepsilon y_1 + Y_o + Z_o)(Ry_o + r) + (\varepsilon u_1 + \xi_o + \eta_o)(Nu_o + n)]dt$$

$$+ \varepsilon Mx_o(T)(x_1(T) + U_o(0)) = \varepsilon Mx_o(T)U_o(0) + \varepsilon p_o(0)x_1(0) -$$

$$- \varepsilon \int_0^T p_o(\frac{dX_o}{dt} + \frac{dU_o}{dt})dt - \varepsilon \int_0^T q_o \frac{dy_o}{dt} dt -$$

$$- \varepsilon \int^T q_o(\frac{dY_o}{dt} + \frac{dZ_o}{dt}) dt$$

Collecting results we obtain the term $\varepsilon X_1$ and a term of order 2,
$\varepsilon \int_0^T \frac{dp_o}{dt} (X_o + U_o)dt + \varepsilon \int_0^T \frac{dq_o}{dt}(Y_o + Z_o)dt.$

Considering next the terms of order 2, we get

$$\frac{1}{2} \varepsilon^2 \int_0^T (Qx_1^2 + Ry_1^2 + Nu_1^2)dt + \frac{\varepsilon^2}{2} M(x_1(T) + U_o(0))^2 - \varepsilon^2 \int_0^T q_o \frac{dy_1}{dt} dt +$$

$$+ \varepsilon \int_0^T [(\frac{dp_o}{dt} + Qx_o + q)(X_o + U_o) + R(Y_o + Z_o)(y_1 + Y_1 + Z_1) +$$

$$+(Ry_o + r)(Y_1 + Z_1) + N(\xi_o + \eta_o)(u_1 + \xi_1 + \eta_1) + (Nu_o + n)(\xi_1 + \eta_1) +$$

$$+ \frac{dq_o}{dt}(Y_o + Z_o) + p_o(A(X_o + U_o) + B(Y_1 + Z_1) + G(\xi_1 + \eta_1))]dt -$$

$$- \varepsilon^2 \int_0^T (L_o \frac{dy_1}{dt} + Q_o \frac{dy_1}{dt})dt.$$

Now we use

$$\varepsilon \int_0^T [(\frac{dp_o}{dt} + Qx_o + q)(X_o + U_o) + R(Y_o + Z_o)(y_1 + Y_1 + Z_1) +$$

$$+ (Ry_o + r)(Y_1 + Z_1) + N(\xi_o + \eta_o)(u_1 + \xi_1 + \eta_1) + (Nu_o + n)(\xi_1 + \eta_1) +$$

$$+ p_o(A(X_o + U_o) + B(Y_1 + Z_1) + G(\xi_1 + \eta_1)]dt =$$

$$= \varepsilon \int_0^T [-(A^* p_o + C^* q_o)(X_o + U_o) + (-\varepsilon \frac{dL_o}{dt} - D^* L_o - \varepsilon \frac{dQ_o}{dt} - D^* Q_o)(y_1 + Y_1 + Z_1) -$$

$$- (B^* p_o + D^* q_o)(Y_1 + Z_1) - H^*(L_o + Q_o)(u_1 + \xi_1 + \eta_1) -$$

$$- (G^* p_o + H^* q_o)(\xi_1 + \eta_1) + p(A(X_o + U_o) + B(Y_1 + Z_1) + G(\xi_1 + \eta_1)]dt =$$

$$= - \varepsilon^2 \int_0^T q_o(\frac{dY_1}{dt} + \frac{dZ_1}{dt})dt - \varepsilon^2 Q_o(0)(y_1(T) + Z_1(0)) +$$

$$+ \varepsilon^2 \int_0^T (L_o + Q_o) \frac{dy_1}{dt} dt + \varepsilon \int_0^T [L_o(Cx_1 - \frac{dy_o}{dt} + CX_o) +$$

$$+ Q_o(Cx_1 - \frac{dy_o}{dt} + CU_o)]dt + O(\varepsilon^4)$$

and

$$\varepsilon \int_0^T (L_o Cx_1 + Q_o Cx_1)dt = - \int_0^T \varepsilon^2 x_1(\frac{dM_o}{dt} + \frac{dK_o}{dt})dt$$

$$= \varepsilon^2(- CX_o(0) \int_0^\infty L_o d\tau - K_o(0)x_1(T))$$

$$+ \varepsilon^2 \int_0^T \frac{dx_1}{dt} (M_o + K_o)dt.$$

Collecting results we obtain

$$\varepsilon^2 X_2 + \varepsilon^2 \int_0^T [\frac{dq_o}{dt} (Y_1 + Z_1) + \frac{dx_1}{dt} (M_o + K_o)]dt$$

$$+ \varepsilon^3 [\frac{d^2 q_o}{dt^2}(0) \int_0^\infty \tau Y_o d\tau - \frac{d^2 q_o}{dt^2}(T) \int_0^\infty \tau Z_o d\tau - \frac{d^2 y_o}{dt^2}(0) \int_0^\infty \tau L_o d\tau$$

$$+ \frac{d^2 y_o}{dt^2}(T) \int_0^\infty \tau Q_o d\tau] + O(\varepsilon^4).$$

We now collect the terms of order 3. Letting aside those in the last bracket of the preceding expression, we obtain

$$\frac{1}{2}\int_0^T \varepsilon^2[Q(X_o^2 + U_o^2) + R(Y_1^2 + Z_1^2) + N(\xi_1^2 + \eta_1^2) + 2(M_oAX_o + K_oAU_o) +$$

$$+ 2(M_oBY_1 + K_oBZ_1) + 2(M_oG\xi_1 + K_oG\eta_1)]dt + \int_0^T \varepsilon^2[Qx_1(X_o + U_o) +$$

$$+ (\frac{dq_o}{dt} + Ry_1)(Y_1 + Z_1) + Nu_1(\xi_1 + \eta_1) + \frac{dx_1}{dt}(M_o + K_o) +$$

$$+ p_1(AX_o + AU_o + BY_1 + BZ_1 + G\xi_1 + G\eta_1)]dt - \varepsilon^3\int_0^T q_1 \frac{dy_1}{dt} dt$$

Nothing that

$$\varepsilon^2\int_0^T [Qx_1(X_o + U_o) + (\frac{dq_o}{dt} + Ry_1)(Y_1 + Z_1) + Nu_1(\xi_1 + \eta_1) +$$

$$+ p_1(AX_o + AU_o + BY_1 + BZ_1 + G\xi_1 + G\eta_1)]dt =$$

$$\varepsilon^2\int_0^T [(-\frac{dp_1}{dt} - A^*p_1 - C^*q_1)(X_o + U_o) + (-B^*p_1 - D^*q_1)(Y_1 + Z_1) +$$

$$+ (-G^*p_1 - H^*q_1)(\xi_1 + \eta_1) + p_1(A(X_o + U_o) + B(Y_1 + Z_1) + G(\xi_1 + \eta_1)]dt$$

$$= -\varepsilon^2\int_0^T \frac{dp_1}{dt}(X_o + U_o)dt - \varepsilon^3\int_0^T q_1 \frac{dY_1}{dt} dt - \varepsilon^3\int_0^T q_1 \frac{dZ_1}{dt} dt$$

$$= -\varepsilon^3 \frac{dp_1}{dt}(0)\int_0^\infty X_o d\tau + \varepsilon^3 \frac{dp_1}{dt}(T)\int_0^\infty U_o d\tau + \varepsilon^3 q_1(0)Y_1(0) -$$

$$- \varepsilon^3 q_1(T)Z_1(0) + O(\varepsilon^4).$$

Collecting results we indeed obtain $\varepsilon^3 X_3 + O(\varepsilon^4)$, which completes the proof of (2.76) □

## 3. EXPANSION WITH BOUNDARY LAYER TERMS IN THE GENERAL CASE : FORMULAS.

Let us go back to the general case (1.3), (1.4). We have proved in Theorem 1.1 a convergence result and in Theorem 1.2 we have given an estimate of the rate of convergence, provided some regularity properties were satisfied. This rate is of order $\varepsilon$. Our calculations in the linear quadratic case show that we cannot improve the accuracy of the convergence without boundary layer terms. Our objective in this section is to do a treatment analogous to that of section 2 for the general case. The approach however cannot be identical since in the L.Q. case we knew that a unique optimal control was available for any $\varepsilon$ problem. This is no longer true in the general case.

We shall derive the formulas by an asymptotic expansion from the set of necessary conditions. The convergence results will be given in the next section.

### 3.1. The form of the expansion.

Using the notation

$$w^{\varepsilon}(t) = (x^{\varepsilon}(t), y^{\varepsilon}(t), u^{\varepsilon}(t), p^{\varepsilon}(t), q^{\varepsilon}(t))$$

we can write the set of necessary conditions

$$\text{(3.1)} \qquad \frac{dx^{\varepsilon}}{dt} = f(w^{\varepsilon}) \qquad x^{\varepsilon}(0) = x_0$$

$$\varepsilon\frac{dy^{\varepsilon}}{dt} = g(w^{\varepsilon}) \qquad y^{\varepsilon}(0) = y_0$$

$$-\frac{dp^{\varepsilon}}{dt} = H_x(w^{\varepsilon}) \qquad p^{\varepsilon}(T) = h_x(x^{\varepsilon}(T))$$

$$-\varepsilon\frac{dq^{\varepsilon}}{dt} = H_y(w^{\varepsilon}) \qquad q^{\varepsilon}(T) = 0$$

$$H_v(w^{\varepsilon}) = 0.$$

As in the in L.Q. case we consider an expansion of the form

$$\text{(3.2)} \qquad x^{\varepsilon}(t) = x_0(t) + \varepsilon x_1(t) + \varepsilon(X_0(\frac{t}{\varepsilon}) + U_0(\frac{T-t}{\varepsilon})) + \tilde{x}^{\varepsilon}(t)$$

$$y^\varepsilon(t) = y_o(t) + Y_o(\tfrac{t}{\varepsilon}) + Z_o(\tfrac{T-t}{\varepsilon}) + \varepsilon y_1(t) + \varepsilon(Y_1(\tfrac{t}{\varepsilon}) + Z_1(\tfrac{T-t}{\varepsilon})) + \tilde{y}^\varepsilon$$

$$u^\varepsilon(t) = u_o(t) + \xi_o(\tfrac{t}{\varepsilon}) + \eta_o(\tfrac{T-t}{\varepsilon}) + \varepsilon u_1(t) + \varepsilon(\xi_1(\tfrac{t}{\varepsilon}) + \eta_1(\tfrac{T-t}{\varepsilon})) + \tilde{u}^\varepsilon(t)$$

$$p^\varepsilon(t) = p_o(t) + \varepsilon p_1(t) + \varepsilon(M_o(\tfrac{t}{\varepsilon}) + K_o(\tfrac{T-t}{\varepsilon})) + \tilde{p}^\varepsilon(t)$$

$$q^\varepsilon(t) = q_o(t) + L_o(\tfrac{t}{\varepsilon}) + Q_o(\tfrac{T-t}{\varepsilon}) + \varepsilon q_1(t) + \varepsilon(L_1(\tfrac{t}{\varepsilon}) + Q_1(\tfrac{T-t}{\varepsilon})) + \tilde{q}^\varepsilon(t).$$

We shall write (using a notation of HABETS [1])

$$W_o^o = (0,Y_o,\xi_o,0,L_o)$$

$$W_o^T = (0,Z_o,\eta_o,0,Q_o)$$

$$W_1^o = (X_o,Y_1,\xi_1,M_o,L_1)$$

$$W_1^T = (U_o,Z_1,\eta_1,K_o,Q_1)$$

These functions depend on an argument $\tau$ which plays the role of $\frac{t}{\varepsilon}$ or $\frac{T-t}{\varepsilon}$. The first case corresponds to the upper index 0 and the second to the upper index T. With this notation we have the asymptotics

$$w^\varepsilon(t) = w_o(t) + W_o^o(\tfrac{t}{\varepsilon}) + W_o^T(\tfrac{T-t}{\varepsilon}) + \varepsilon(w_1(t) + W_1^o(\tfrac{t}{\varepsilon}) + W_1^T(\tfrac{T-t}{\varepsilon})) + \tilde{w}^\varepsilon(t). \tag{3.3}$$

We shall also use the notation

$$\sigma^\varepsilon(t) = (x^\varepsilon(t),y^\varepsilon(t),u^\varepsilon(t))$$

$$\sigma_o(t) = (x_o(t),y_o(t),u_o(t))$$

$$\Sigma_o^o(\tau) = (0,Y_o(\tau),\xi_o(\tau))$$

$$\Sigma_1^o(\tau) = (X_o,Y_1,\xi_1)$$

and similar definitions for $\sigma_1(t)$, $\Sigma_o^T(\tau)$, $\Sigma_1^T(\tau)$. The key to write the expansion lies in the following formal consideration. Let be an expansion $\Phi(w^\varepsilon(t))$.

We have

$$\Phi(w^{\varepsilon}(t)) \simeq \Phi(w_o + W_o^o + \varepsilon(w_1 + W_1^o)) + \Phi(w_o + W_o^T + \varepsilon(w_1 + W_1^T)) - \tag{3.4}$$

$$- \Phi(w_o + \varepsilon w_1) + \text{negligeable terms}$$

$$\simeq \Phi(w_o(0) + W_o^o(\tfrac{t}{\varepsilon}) + \varepsilon(\tfrac{t}{\varepsilon}\dot{w}_o(0) + w_1(0) + W_1^o(\tfrac{t}{\varepsilon})) +$$

$$+ \varepsilon^2(\tfrac{t^2}{2\varepsilon^2}\ddot{w}_o(0) + \tfrac{t}{\varepsilon}\dot{w}_1(0)) - \Phi(w_o(0) + \varepsilon(\tfrac{t}{\varepsilon}\dot{w}(0) + w_1(0)) +$$

$$+ \varepsilon^2(\tfrac{t^2}{2\varepsilon^2}\ddot{w}_o(0) + \tfrac{t}{\varepsilon}\dot{w}_1(0))) + \Phi(w_o(T) + W_o^T(\tfrac{T-t}{\varepsilon}) +$$

$$+ \varepsilon(-\tfrac{T-t}{\varepsilon}\dot{w}_o(T) + w_1(T) + W_1^T(\tfrac{T-t}{\varepsilon})) + \varepsilon^2(\tfrac{(t-T)^2}{2\varepsilon^2}\ddot{w}_o(T) + \tfrac{t-T}{\varepsilon}\dot{w}_1(T))) -$$

$$- \Phi(w_o(T) + \varepsilon(-\tfrac{T-t}{\varepsilon}\dot{w}_o(T) + w_1(T)) + \varepsilon^2(\tfrac{(t-T)^2}{2\varepsilon^2}\ddot{w}_o(T) + \tfrac{t-T}{\varepsilon}\dot{w}_1(T)))$$

$$+ \Phi(w_o + \varepsilon w_1)$$

$$\simeq \Phi(w_o(0) + W_o^o(\tfrac{t}{\varepsilon})) - \Phi(w_o(0)) + \varepsilon\,\Phi_w(w_o(0) + W_o^o)W_1^o +$$

$$+ \varepsilon(\Phi_w(w_o(0) + W_o^o) - \Phi_w(w_o(0)))(\tfrac{t}{\varepsilon}\dot{w}_o(0) + w_1(0)) +$$

$$+ \Phi(w_o(T) + W_o^T) - \Phi(w_o(T)) + \varepsilon\,\Phi_w(w_o(T) + W_o^T)\,W_1^T +$$

$$+ \varepsilon(\Phi_w(w_o(T) + W_o^T) - \Phi_w(w_o(T)))(-\tfrac{T-t}{\varepsilon}\dot{w}_o(T) + w_1(T)) +$$

$$+ \varepsilon^2[(\Phi_w(w_o(0) + W_o^o) - \Phi_w(w_o(0)))(\tfrac{1}{2}\tfrac{t^2}{\varepsilon^2}\ddot{w}_o(0) + \tfrac{t}{\varepsilon}\dot{w}_1(0)) +$$

$$+ \tfrac{1}{2}(\Phi_{ww}(w_o(0) + W_o^o) - \Phi_{ww}(w_o(0)))(\tfrac{t}{\varepsilon}\dot{w}_o(0) + w_1(0))^2 +$$

$$+ \tfrac{1}{2}\Phi_{ww}(w_o(0) + W_o^o)W_1^o(W_1^o + 2(\tfrac{t}{\varepsilon}\dot{w}_o(0) + w_1(0)))] +$$

$$+ \varepsilon^2[(\Phi_w(w_o(T) + W_o^T) - \Phi_w(w_o(T)))(\tfrac{1}{2}\tfrac{(t-T)^2}{\varepsilon^2}\ddot{w}_o(T) + \tfrac{t-T}{\varepsilon}\dot{w}_1(T)) +$$

$$+ \frac{1}{2}(\Phi_{ww}(w_o(T) + W_o^T) - \Phi_{ww}(w_o(T)))(\frac{t-T}{\varepsilon} \dot{w}_o(T) + w_1(T))^2$$

$$+ \frac{1}{2} \Phi_{ww}(w_o(T) + W_o^T) W_1^T(W_1^T + 2(\frac{t-T}{\varepsilon} \dot{w}_o(T) + w_1(T)))]$$

$$+ \Phi(w_o) + \varepsilon\Phi_w(w_o)w_1 + \frac{1}{2} \varepsilon^2 \Phi_{ww}(w_o)w_1^2 + \frac{1}{6} \varepsilon^3\Phi_{www}(w_o)w_1^3 + \ldots$$

Using this expansion in (3.1) we obtain

(3.5) $$\frac{dx_o}{dt} = f(w_o) \qquad x_o(0) = x_o$$

$$g(w_o) = 0$$

$$-\frac{dp_o}{dt} = H_x(w_o) \qquad p_o(T) = h_x(x_o(T))$$

$$H_y(w_o) = 0 \qquad H_v(w_o) = 0$$

We then have

(3.6) $$\frac{dY_o}{d\tau} = g(\sigma_o(0) + \Sigma_o^0(\tau)) \ , \ Y_o(0) = y_o - y_o(0)$$

$$-\frac{dL_o}{d\tau} = H_y(w_o(0) + W_o^0(\tau))$$

(3.7) $$-\frac{dZ_o}{d\tau} = g(\sigma_o(T) + \Sigma_o^T(\tau))$$

$$\frac{dQ_o}{d\tau} = H_y(w_o(T) + W_o^T(\tau))$$

$$Q_o(0) = - q_o(T)$$

(3.8) $$\frac{dX_o}{d\tau} = f(\sigma_o(0) + \Sigma_o^0(\tau)) - f(\sigma_o(0))$$

(3.9) $$-\frac{dU_o}{d\tau} = f(\sigma_o(T) + \Sigma_o^T(\tau)) - f(\sigma_o(T))$$

(3.10) $$-\frac{dM_o}{d\tau} = H_x(w_o(0) + W_o^0(\tau)) - H_x(w_o(0))$$

(3.11) $$\frac{dK_o}{d\tau} = H_x(w_o(T) + W_o^T(\tau)) - H_x(w_o(T))$$

(3.12) $$\frac{dx_1}{dt} = f_\sigma(\sigma_o)\sigma_1$$

$$x_1(0) = - X_o(0)$$

$$\frac{dy_o}{dt} = g_\sigma(\sigma_o)\sigma_1$$

(3.13) $$-\frac{dp_1}{dt} = H_{xw}(w_o)w_1$$

$$p_1(T) = h_{xx}(x_o(T))(x_1(T) + U_o(0)) - K_o(0)$$

$$-\frac{dq_o}{dt} = H_{yw}(w_o)w_1$$

$$H_{vw}(w_o)w_1 = 0$$

(3.14) $$\frac{dY_1}{d\tau} = g_\sigma(\sigma_o(0) + \Sigma_o^o(\tau))\Sigma_1^o + (g_\sigma(\sigma_o(0) + \Sigma_o^o(\tau)) - g_\sigma(\sigma_o(0)))(\tau\dot{\sigma}_o(0)+\sigma_1(0))$$

$$Y_1(0) = - y_1(0)$$

$$-\frac{dL_1}{d\tau} = H_{yw}(w_o(0) + W_o^o(\tau))W_1^o(\tau) + (H_{yw}(w_o(0) + W_o^o(\tau)) -$$

$$- H_{yw}(w_o(0)))(\tau\dot{w}_o(0) + w_1(0))$$

(3.15) $$-\frac{dZ_1}{d\tau} = g_\sigma(\sigma_o(T) + \Sigma_o^T(\tau))\Sigma_1^T + (g_\sigma(\sigma_o(T) + \Sigma_o^T(\tau)) -$$

$$- g_\sigma(\sigma_o(T)))(- \tau\dot{\sigma}_o(T) + \sigma_1(T))$$

$$\frac{dQ_1}{d\tau} = H_{yw}(w_o(T) + W_o^T(\tau))W_1^T(\tau) + (H_{yw}(w_o(T) + W_o^T(\tau)) -$$

$$- H_{yw}(w_o(T)))(-\tau\dot{w}_o(T) + w_1(T))$$

$$Q_1(0) = - q_1(T)$$

We note also the relations, obtained by expanding the last relation (3.1)

(3.16) $H_v(w_o(0) + W_o^o(\tau)) = 0$

(3.17) $H_v(w_o(T) + W_o^T(\tau)) = 0$

(3.18) $H_{vw}(w_o(0) + W_o^o(\tau))W_1^o(\tau) + (H_{vw}(w_o(0) + W_o^o(\tau)) -$

$- H_{vw}(w_o(0)))(\tau\dot{w}_o(0) + w_1(0)) = 0$

(3.19) $H_{vw}(w_o(T) + W_o^T(\tau))W_1^T(\tau) + (H_{vw}(w_o(T) + W_o^T(\tau)) -$

$- H_{vw}(w_o(T)))(- \tau\dot{w}_o(T) + w_1(T)) = 0$

Note also that in fact

(3.20) $H_{yw}(w(t))\dot{w}(t) = 0 \qquad \forall\ t$

$H_{vw}(w(t))\dot{w}(t) = 0$

3.2. Optimization problems

We shall now relate the relations (3.5) to (3.20) to some optimization problems which will allow us to prove the existence of solutions.

The system (3.5) relates to the limit problem

(3.21) $\frac{dx_o}{dt} = f(x_o,y_o,u_o) \qquad x_o(0) = x_o$

$g(x_o,y_o,u_o) = 0$

$J(u_o) = \int_0^T \ell(x_o,y_o,u_o)dt + h(x_o(T)).$

The relations (3.7) together with (3.16) relate to the control problem

(3.22) $\frac{dY_o}{d\tau} = g(x_o(0),y_o(0) + Y_o,u_o(0) + \xi_o)$

$$Y_o(0) = y_o - y_o(0)$$

$$\mathcal{G}_o(\xi_o) = \int_0^\infty [H(x_o(0), y_o(0) + Y_o, u_o(0) + \xi_o, p_o(0), q_o(0) - H(w_o(0))]d\tau$$

Similarly (3.7) and (3.17) are related to

$$\text{(3.23)} \qquad -\frac{dZ_o}{d\tau} = g(x_o(T), y_o(T) + Z_o, u_o(T) + \eta_o)$$

$$\mathcal{K}_o(\eta_o) = \int_0^\infty [H(x_o(T), y_o(T) + Z_o, u_o(T) + \eta, p_o(T), q_o(T)) - H(w_o(T))]d\tau$$

$$- Z_o(0)q_o(T).$$

The conditions (3.12), (3.14) express the Pontryagin principle for the problem (written with the abbreviated symbols $\sigma_o, \sigma_1$)

$$\text{(3.24)} \qquad \frac{dx_1}{dt} = f_\sigma(\sigma_o)\sigma_1 \qquad\qquad x_1(0) = - X_o(0)$$

$$\frac{dy_o}{dt} = g_\sigma(\sigma_o)\sigma_1$$

$$J_1(u_1) = \frac{1}{2}\int_0^T [H_{\sigma\sigma}(w_o)\sigma_1^2 + 2\,\frac{dq_o}{dt}\,y_1]dt + \frac{1}{2}\,h_{xx}(x_o(T))x_1(T)^2 +$$

$$+ (h_{xx}(x_o(T))U_o(0) - K_o(0))x_1(T)$$

Consider next the problem related to (3.14), (3.18). It is given by the state equation (3.14) (in $Y_1$) and the cost function

$$\text{(3.25)} \qquad \mathcal{G}_1(\xi_1) = \frac{1}{2}\int_0^\infty [H_{yy}(w_o(0) + W_o^o)Y_1^2 + H_{vv}\,\xi_1^2 + 2\,H_{yv}\,\xi_1 Y_1]d\tau +$$

$$+ \int_0^\infty [(H_{yx}(w_o(0) + W_o^o)X_o + M_o f_y)Y_1 + (H_{vx}X_o + M_o f_v)\xi_1 +$$

$$+ (H_{yw}(w_o(0) + W_o^o)(\tau\,\dot{w}_o(0) + w_1(0)) + \frac{dq_o}{dt})Y_1 +$$

$$+ (H_{vw}(w_o(0) + W_o^o)(\tau\,\dot{w}_o(0) + w_1(0))\xi_1]d\tau$$

where we have used the two last relations (3.13) as well as (3.20).

Similarly the conditions (3.15), (3.19) are related to the control problem in which the state equation is given by the 1st equation (3.15) and the cost functional is given by

$$(3.26)\qquad \mathcal{K}_1(\eta_1) = \frac{1}{2}\int_0^\infty [H_{yy}(w_o(T) + W_o^T)Z_1^2 + H_{vv}\eta_1^2 + 2H_{yv}\eta_1 Z_1]d\tau +$$

$$+ \int_0^\infty [(H_{yx}(w_o(T) + W_o^T)U_o + f_y K_o)Z_1 + (H_{yx}U_o + f_v K_o)\eta_1 +$$

$$+ ((H_{yw}(w_o(T) + W_o^T)(-\tau\,\dot{w}_o(T) + w_1(T)) + \frac{dq_o}{dt}(T))\,Z_1 +$$

$$+ (H_{vw}(w_o(T) + W_o^T)(-\tau\,\dot{w}_o(T) + w_1(T))\eta_1]d\tau - q_1(T)Z_1(0)$$

3.3. Solution of the optimization problems.

Under the conditions of Theorem 1.1 we know that the limit problem (3.21) posesses a unique solution $u_o$. We shall then concentrate on (3.22), (3.23). As we have introduced Riccati equations in the L.Q. case (cf (2.26), (2.47)) we shall consider analogously Bellman equations in duality.

3.2.1. Bellman equations in duality.

Let us introduce some notation

$$(3.27)\qquad F(Y,\xi) = H(x_o(0),y_o(0) + Y,u_o(0) + \xi,p_o(0),q_o(0)) - H(w_o(0))$$

$$(3.28)\qquad G(Y,\xi) = g(x_o(0),y_o(0) + Y,u_o(0) + \xi).$$

We note the properties

$$(3.29)\qquad F(0,0) = 0,\ F_Y(0,0) = 0,\ F_\xi(0,0) = 0$$

$$\begin{pmatrix} F_{YY} & F_{Y\xi} \\ F_{\xi Y} & F_{\xi\xi} \end{pmatrix} \geq \beta I$$

$F_{YY}, F_{Y\xi}, F_{\xi\xi}$ bounded

(3.30) $G(0,0) = 0 \ , \ G_Y(Y,\xi) \leq -\mu I$

$G_Y, G_\xi$ bounded.

Let us consider the Bellman equation (generalizing the Riccati equation (2.26))

(3.31) $$\operatorname*{Inf}_{\xi} [F(h,\xi) + D\Phi . G(h,\xi)] = 0 \qquad \text{a.e.h.}$$

Let us consider the control problem

(3.32) $$\frac{dY}{d\tau} = G(Y,\xi) \qquad Y(0) = h \ , \qquad \xi \in L^2(0,\infty;R^k)$$

$$\mathcal{J}_h(\xi) = \int_0^\infty F(Y,\xi)d\tau$$

then we have

Proposition 3.1. *If* (3.29), (3.30) *hold, the function*

(3.33) $$\Phi(h) = \inf_{\xi} \mathcal{J}_h(\xi)$$

*satisfies*

(3.34) $$0 \leq \Phi(h) \leq C|h|^2$$

$$|\Phi(h_1) - \Phi(h_2)| \leq C(|h_1| + |h_2|)\ |h_1 - h_2|$$

*and is the maximum solution of* (3.31), *belonging to the class* (3.34). *The function* $\Phi$ *defined by* (3.33) *satisfies also*

(3.35) $$\Phi(h) \geq a|h|^2, \quad a > 0.$$

Proof.

The fact that $\Phi$ defined by (3.33) satisfies (3.34) and is the maximum solution of (3.31) in the class (3.34) is a well known result on Hamilton-Jacobi Bellman equations.

Let us prove (3.35). Set

$$\psi(h) = \Phi(h) - a|h|^2$$

then from (3.33) we have

$$\psi(h) = \inf \int_0^\infty [F(Y,\xi) + 2aY.G(Y,\xi)]d\tau$$

and

$$F(Y,\xi) + 2aY.G(Y,\xi) \geq \beta(|Y|^2 + |\xi|^2) - 2aC|Y|(|Y| + |\xi|) \geq 0$$

for a sufficiently small, hence (3.35)

□

Remark 3.1. As in the L.Q. case it is possible to introduce the problem

$$-\frac{dY}{d\tau} = G(Y,\xi) \quad , \quad \xi \in L^2(0,\infty;R^k)$$

and to minimize

$$\mathcal{Y}(\xi) = \int_0^\infty F(Y,\xi)d\tau$$

on the set

$$E(h) = \{\xi |\ Y(0) = h\}$$

assuming this set not empty for any h (controllability condition). Let us term

$$(3.36) \qquad \underline{\Phi}(h) = - \operatorname*{Inf}_{\xi \in E(h)} \mathcal{Y}(\xi) \leq 0.$$

Then we may wonder wether $\underline{\Phi}(h)$ is the minimum solution of (3.31). One of the main difficulties lies in the fact that it is not possible easily to establish that $\underline{\Phi}$ belongs to the class (3.34). Note however the optimality principle

$$-\underline{\Phi}(h) = \inf_{\xi(.)} \{ \int_0^\delta F(Y,\xi)d\tau - \underline{\Phi}(Y(\delta))\}$$

from which (3.31) can be derived in any point of differentiability of $\underline{\Phi}$.

Note also that (3.31) will have at most one positive smooth function, and one negative smooth function. They coïncide with those identified by (3.33) and (3.36). □

We shall introduce duality as follows. Let us consider the equation (Dual Bellman equation)

$$(3.37) \qquad \inf_{\eta} [F(D\psi,\eta) - h\, G(D\psi,\eta)] = 0.$$

In the L.Q. case, it has a solution of the form $\frac{1}{2}\Sigma h^2$ with $\Sigma$ satisfying (2.47) (the dual Riccati equation).

Unfortunately, unlike the L.Q. case, (3.37) is not apparently of the same type as (3.31).

Nethertheless, it is possible to identify a non negative solution of (3.37), as follows : Consider the control problem

$$(3.38) \qquad -\frac{dZ}{d\tau} = G(Z,\eta) \quad , \quad \eta \in L^2(0,\infty;R^k)$$

$$\mathscr{K}_h(\eta) = \int_0^\infty F(Z,\eta)d\tau - hZ(0)$$

and

$$(3.39) \qquad \psi(h) = -\inf \mathscr{K}_\eta(\eta).$$

One has

<u>Proposition 3.2.</u> <u>The function</u> $\psi$ <u>defined by</u> (3.39) <u>belongs to the class</u> (3.34) <u>and is a solution of</u> (3.37).

<u>Proof.</u>

Note that $\mathscr{K}_\eta(0) = 0$, hence $\psi(h) \geq 0$. It is also easy to check that the set of admissible controls can be reduced to

$$(3.40) \qquad |\eta|_{L^2} \leq C|h|.$$

This gives the 2nd estimate (3.34). The Lipschitz property (3.34) is almost obvious from the definition of $\mathcal{K}_h(\eta)$. Pick a point h where $\psi$ is differentiable, and for any $\varepsilon$ a control $\eta_\varepsilon$ such that

$$\mathcal{K}_h(\eta_\varepsilon) \leq - \psi(h) + \varepsilon^2.$$

Let $Z_\varepsilon$ be the corresponding trajectory. One can assert that

(3.41) $\qquad Z_\varepsilon(0) \to D\psi(h) \quad \text{as } \varepsilon \to 0.$

Indeed $Z_\varepsilon(0)$ is bounded, since $|\eta_\varepsilon| \leq C|h|$, and

$$\psi(h + \varepsilon k) - \psi(h) \geq - \mathcal{K}_{h+\varepsilon k}(\eta_\varepsilon) + \mathcal{K}_h(\eta_\varepsilon) - \varepsilon^2$$

$$= + \varepsilon k \, Z_\varepsilon(0) - \varepsilon^2.$$

Pick a cluster point of $Z_\varepsilon(0)$, $Z^*$, using the differentiability in h, we get after dividing by $\varepsilon$ and letting $\varepsilon$ tend to 0

$$D\psi(h)k \geq kZ^*$$

and since k is arbitrary $Z^* = D\psi(h)$. By the uniqueness of the limit (3.41) follows.

Note that

$$\mathcal{K}_h(\eta_\varepsilon) = \int_0^\varepsilon [F(Z_\varepsilon,\eta_\varepsilon) - hG(Z_\varepsilon,\eta_\varepsilon)]d\tau + \int_\varepsilon^\infty F(Z_\varepsilon,\eta_\varepsilon)d\tau - hZ_\varepsilon(\varepsilon).$$

But setting

$$\tilde{\eta}_\varepsilon(\tau) = \eta_\varepsilon(\varepsilon + \tau)$$

we have

$$\tilde{Z}_\varepsilon(\tau) = Z_\varepsilon(\varepsilon + \tau), \qquad \tilde{Z}_\varepsilon(0) = Z_\varepsilon(\varepsilon)$$

hence

$$\int_\varepsilon^\infty F(Z_\varepsilon,\eta_\varepsilon)d\tau - hZ_\varepsilon(\varepsilon) = \mathcal{K}_h(\tilde{\eta}_\varepsilon)$$

Therefore we deduce

$$(3.42) \qquad \int_0^{\varepsilon} [F(Z_\varepsilon,\eta_\varepsilon) - h\, G(Z_\varepsilon,\eta_\varepsilon)]d\tau \le \varepsilon^2.$$

Define

$$\mathcal{H}(\lambda) = \inf_\eta [F(\lambda,\eta) - h\, G(\lambda,\eta)]$$

which is a continuous function. From (3.42) we deduce

$$\varepsilon^2 \ge \int_0^{\varepsilon} \mathcal{H}(Z_\varepsilon(\varepsilon))d\tau$$

hence

$$\varepsilon \ge \int_0^1 \mathcal{H}(Z_\varepsilon(\varepsilon s))ds.$$

But for $s \in [0,1]$

$$|Z_\varepsilon(\varepsilon s) - Z_\varepsilon(0)| \le C\sqrt{\varepsilon}.$$

Therefore from (3.41) and Lebesgue Theorem, we deduce

$$\mathcal{H}(D\psi(h)) \le 0$$

i.e

$$(3.43) \qquad \inf_\eta [F(D\psi,\eta) - h\, G(D\psi,\eta)] \le 0.$$

To obtain the reverse inequality, let

$$\tilde{\eta}_\varepsilon(s) = \begin{cases} \eta_\varepsilon(s-\varepsilon) & \text{for } s \ge \varepsilon \\ \eta_0 & \text{for } s < \varepsilon,\ \eta_0 \text{ fixed} \end{cases}$$

Let $\tilde{Z}_\varepsilon(s)$ be the trajectory corresponding to $\tilde{\eta}_\varepsilon$ , we have

$$Z_\varepsilon(s) = Z_\varepsilon(s-\varepsilon) \quad \text{for } s \ge \varepsilon$$

hence

$$\tilde{Z}_\varepsilon(\varepsilon) = Z_\varepsilon(0).$$

Therefore

$$\mathscr{K}_h(\tilde{\eta}_\varepsilon) = \int_0^\varepsilon [F(\tilde{Z}_\varepsilon(s),\eta_o) - h\, G(\tilde{Z}_\varepsilon(s),\eta_o)]ds + \mathscr{K}_h(\eta_\varepsilon).$$

It is easy to check that

$$|\tilde{Z}_\varepsilon(s) - Z_\varepsilon(0)| \le C\varepsilon \qquad \text{for } s \in [0,\varepsilon]$$

hence

$$\mathscr{K}_h(\tilde{\eta}_\varepsilon) \le \varepsilon\, [F(Z_\varepsilon(0),\eta_o) - h\, G(Z_\varepsilon(0),\eta_o)] + C\varepsilon^2 + \mathscr{K}_h(\eta_\varepsilon)$$

and from the definition of $\eta_\varepsilon$

$$-\psi(h) \le \varepsilon\, [F(Z_\varepsilon(0),\eta_o) - h\, G(Z_\varepsilon(0),\eta_o)] + C\varepsilon^2 - \psi(h) + \varepsilon^2$$

or

$$0 \le F(Z_\varepsilon(0),\eta_o) - h\, G(Z_\varepsilon(0),\eta_o)$$

and letting $\varepsilon$ tend to 0, we deduce

$$0 \le F(D\psi,\eta_o) - h\, G(D\psi,\eta_o).$$

Since $\eta_o$ is arbitrary, we have proved the reverse inequality of (3.43), hence the result desired □

Consider next the control problem

$$(3.44) \qquad \frac{dZ}{d\tau} = G(Z,\eta) \qquad\qquad Z(0) = \zeta$$

$$\mathscr{L}_h(\eta,\zeta) = \int_0^\infty F(Z,\eta)d\tau + h\zeta$$

and set

$$(3.45) \qquad \psi(h) = \inf_{\eta,\zeta} \mathcal{L}_h(\eta,\zeta)$$

then one has

<u>Proposition 3.3. The function defined by</u> (3.45) <u>satisfies</u>

$$(3.46) \qquad 0 \geq \psi(h) \geq -C|h|^2$$

$$|\psi(h_1) - \psi(h_2)| \leq C|h_1 - h_2|(|h_1| + |h_2|)$$

<u>and is a solution of</u> (3.37).

<u>Proof</u>.

We note that

$$\psi(h) = \operatorname{Inf}_{\zeta} [\Phi(\zeta) + h\zeta] .$$

From the estimates on $\Phi$, one deduces easily (3.46). By arguments similar to those of Proposition 3.2, one checks that $\psi$ is a solution of (3.37). □

<u>Remark 3.2</u>. If $\Phi$ is convex then

$$\psi(h) = -\Phi^*(-h)$$

where $\Phi^*$ is the conjugate of $\Phi$. The transformation $\Phi \to -\Phi^*(-h)$ is the duality transformation (Legendre transformation) which plays the role of $\pi \to -\pi^{-1}$. In fact convexity of $\Phi$ is not necessary to define $\psi$, at least in our framework

<u>Remark 3.3</u>. The structure of the set of solutions of the dual Bellman equation is an open problem □

### 3.3.2. <u>Additional properties</u>

In the following we are interested only by the functions $\Phi$ and $\psi$, and the corresponding control problems (3.32) and (3.38).

Let us set

$$\mathcal{H}(Y,\xi,L) = F(Y,\xi) + L\ G(Y,\xi)$$

and assume that there exists a triple $(Y_0,\xi_0,L_0)$ such that

$$(3.47)\qquad \frac{dY_0}{d\tau} = G(Y_0,\xi_0) \qquad\qquad Y_0(0) = h$$

$$-\frac{dL_0}{d\tau} = \mathcal{H}_Y(Y_0,\xi_0,L_0)$$

$$\mathcal{H}_\xi(Y_0,\xi_0,L_0) = 0$$

$$(3.48)\qquad \begin{pmatrix} \mathcal{H}_{YY} & \mathcal{H}_{Y\xi} \\ \mathcal{H}_{\xi Y} & \mathcal{H}_{\xi\xi} \end{pmatrix}(Y,\xi,L_0) \geq \beta I, \quad \forall\ Y,\xi$$

Remark 3.4. For $h = 0$, the relations (3.47), (3.48) are satisfied with $Y_0,\xi_0,\ L_0$ equal to 0. From the assumptions (3.29), (3.30) it can be established that (3.47), (3.48) hold for h small.

□

When (3.47), (3.48) hold it follows from classical results on Bellman equations that :

(3.49) $\quad\Phi$ is differentiable at any point $Y_0(\tau)$ of the trajectory and

$$D\Phi(Y_0(\tau)) = L_0(\tau).$$

We can then prove the exponential decay

Proposition 3.4. Assume (3.47), (3.48) then one has

$$(3.50)\qquad |Y_0(\tau)|,\ |L_0(\tau)|,\ |\xi_0(\tau)| \leq C|h|e^{-\gamma\tau},\ \gamma > 0$$

Proof.

From (3.49) we can assert that $\Phi(Y_0(\tau))$ is a.e. differentiable and

$$\frac{d}{d\tau}\Phi(Y_0(\tau)) = D\Phi(Y_0(\tau))\ G(Y_0(\tau),\xi_0(\tau)).$$

We also have

$$D\Phi(Y_0(\tau))G(Y_0(\tau),\xi_0(\tau)) + F(Y_0(\tau),\xi_0(\tau)) = 0 \qquad \text{a.e} \tag{3.51}$$

Indeed call $\rho(\tau)$ the function (3.51). It is $\geq 0$ by (3.31). From the estimates (3.34) follows

$$- \Phi(h) = \int_0^\infty D\Phi\ G\ d\tau = - \int_0^\infty F(Y_0,\xi_0)d\tau + \int_0^\infty \rho(\tau)d\tau$$

$$= - \int_0^\infty F(Y_0,\xi_0)d\tau,$$

by the optimality of $\xi_0$. Hence (3.51).

But then

$$\frac{d}{d\tau}\Phi(Y_0(\tau)) = - F(Y_0(\tau),\xi_0(\tau))$$

$$\leq - \beta|Y_0(\tau)|^2$$

and from (3.34)

$$\leq - k\ \Phi(Y_0(\tau)).$$

Integrating we deduce

$$\Phi(Y_0(\tau)) \leq \Phi(h)e^{-k\tau} \leq C|h|^2e^{-k\tau}$$

and from (3.35) we deduce

$$|Y_0(\tau)| \leq C|h|e^{-\gamma\tau}$$

From (3.49) and the $2^{nd}$ estimate (3.34) it follows

$$|L_0(\tau)| \leq C|h|e^{-\gamma\tau}$$

The $3^{rd}$ equation (3.47) implies the same estimate for $\xi_o(\tau)$, which completes the proof

□

We next make similar consideration for $\psi(h)$. We assume that there exists a triple $Z_o, Q_o, \eta_o$ such that

$$(3.52) \qquad - \frac{dZ_o}{d\tau} = G(Z_o, \eta_o)$$

$$\frac{dQ_o}{d\tau} = \mathcal{H}_Z(Z_o, \eta_o, Q_o) \qquad\qquad Q_o(0) = - h$$

$$\mathcal{H}_\eta(Z_o, \eta_o, Q_o) = 0 \ , \quad Z_o, \eta_o, Q_o \in L^2$$

$$(3.53) \qquad \begin{pmatrix} \mathcal{H}_{ZZ} & \mathcal{H}_{Z\eta} \\ \mathcal{H}_{\eta Z} & \mathcal{H}_{\eta\eta} \end{pmatrix} (Z, \eta, Q_o) \geq \beta I \ , \quad \forall \ Z, \eta .$$

We then have

<u>Proposition 3.5</u>. <u>When</u> (3.52), (3.53) <u>hold</u>, $\eta_o$ <u>is the unique optimal control for</u> (3.38). <u>Moreover</u> $\psi$ <u>is differentiable along the trajectory</u> $Q_o(\tau)$ <u>and</u>

$$(3.54) \qquad D\psi(- Q_o(\tau)) = + Z_o(\tau).$$

<u>Proof</u>.

One easily checks the formula

$$(3.55) \qquad \mathcal{K}_{\tilde{h}}(\eta) - \mathcal{K}_h(\eta_o) = \int_0^\infty \int_0^1 \int_0^1 \lambda[\mathcal{H}_{ZZ}(Z_o + \lambda\mu(Z - Z_o), \eta_o + \lambda\mu(\eta - \eta_o))$$

$$(Z - Z_o)^2 + 2\mathcal{H}_{Z\eta}(\eta - \eta_o)(Z - Z_o) + \mathcal{H}_{\eta\eta}(\eta - \eta_o)^2] d\mu d\lambda d\tau - (\tilde{h} - h) Z(0).$$

Therefore if $\tilde{h} = h$, (3.53) implies

$$\mathcal{K}_h(\eta) > \mathcal{K}_h(\eta_o) \qquad \text{if } \eta \neq \eta_o$$

which proves that $\eta_o$ is the unique optimal control.

It also follows that

$$\mathcal{K}_{\tilde{h}}(\eta) - \mathcal{K}_h(\eta_o) \geq - (\tilde{h} - h)Z_o(0) + (\tilde{h} - h)(Z_o(0) - Z(0)) +$$

$$+ \frac{\beta}{2} |\eta - \eta_o|^2_{L^2} \geq - (\tilde{h} - h)Z_o(0) - C|\tilde{h} - h|^2$$

since

$$|Z(0) - Z_o(0)| \leq C|\eta - \eta_o|_{L^2}.$$

Therefore

$$- \psi(\tilde{h}) + \psi(h) \geq - (\tilde{h} - h)Z_o(0) - c|\tilde{h} - h|^2.$$

On the other hand

$$- \psi(\tilde{h}) + \psi(h) \leq \mathcal{K}_{\tilde{h}}(\eta_o) - \mathcal{K}_h(\eta_o) = - (\tilde{h} - h)Z_o(0)$$

hence

$$D\psi(h) = + Z_o(0)$$

or

$$D\psi(- Q_o(0)) = + Z_o(0)$$

i.e (3.53) with $\tau = 0$. A similar argument holds for any $\tau$.

□

To get a result similar to that of Proposition 3.4, we shall need the additional assumption

(3.56) $G^*_\eta(Z,\eta)$ has a left inverse, bounded in $Z,\eta$.

Note that in the linear case (cf.(2.2)) one has

$$G^*_\eta = H^*$$

and thus (3.56) is satisfied whenever (2.60) holds.

We then have

<u>Proposition 3.6</u>. <u>Assume</u> (3.52), (3.53) <u>and</u> (3.56). <u>Then one has</u>

(3.57) $$|Z_0(\tau)|\ ,\ |Q_0(\tau)|\ ,\ |\eta_0(\tau)| \le C|h|e^{-\gamma\tau},\quad \gamma > 0.$$

<u>Proof</u>.

We first prove that

(3.58) $$\psi(h) \ge a|h|^2$$

Indeed, notice that

(3.59) $$\psi(h) = \int_0^\infty (-\mathcal{H} + \mathcal{H}_Z Z_0)(Z_0,\eta_0,Q_0)d\tau$$

and from the last relation (3.52) and (3.53)

$$\ge \beta \int_0^\infty (Z_0^2 + \eta_0^2)d\tau.$$

On the other hand, thanks to (3.56) and the 1st and 3$^{rd}$ relations (3.52) one has

$$\int_0^\infty Q_0^2 d\tau,\ \int_0^\infty Z_0^2 d\tau \le C \int_0^\infty \eta_0^2 d\tau$$

and from the 2$^{nd}$ relation (3.52)

$$-\frac{1}{2}|h|^2 = \int_0^\infty \mathcal{H}_Z Q_0 d\tau$$

or

$$|h|^2 \le C \int_0^\infty [Q_0^2 + |Q_0|\ (|Z_0| + |\eta_0|)]d\tau.$$

Collecting results (3.58) obtains.

Let us prove (3.57). From the 3[rd] relation (3.52) and (3.56) we deduce

$$(3.60) \qquad |Q_o(\tau)| \leq C(|Z_o(\tau)| + |\eta_o(\tau)|)$$

On the other hand, one can check that

$$\frac{d}{d\tau}\mathcal{H} = 0 \quad , \quad \text{hence } \mathcal{H}(Z_o(\tau),\eta_o(\tau),Q_o(\tau)) = 0.$$

Therefore also taking account of the 3[rd] relation (3.52) and (3.53),

$$\mathcal{H}_Z\, Z_o \geq \beta|\eta_o(\tau)|^2$$

whence

$$|\eta_o(\tau)|^2 \leq C(|Z_o(\tau)|^2 + (|Q_o(\tau)| + |\eta_o(\tau)|)|Z_o(\tau)|)$$

which with (3.60) implies

$$(3.61) \qquad |Q_o(\tau)| \leq C|Z_o(\tau)| \quad , \quad |\eta_o(\tau)| \leq C|Z_o(\tau)|.$$

Now we compute

$$\frac{d}{d\tau}\,\psi(-\,Q_o(\tau)) = -\,Z_o(\tau)\,\mathcal{H}_Z - \beta\,(|Z_o(\tau)|^2 + |\eta_o(\tau)|^2)$$

$$\leq -\,\gamma_o|Q_o(\tau)|^2$$

$$\leq -\,\gamma_1\,\psi(-\,Q_o(\tau))$$

hence

$$\psi(-\,Q_o(\tau)) \leq \psi(h)e^{-\gamma_1\tau}$$

and from (3.58) one obtains the estimate (3.57) related to $Q_o$. Using (3.54) and the 2[nd] estimate (3.61) we complete the proof of the results desired

□

### 3.3.3. Application to the boundary layer problems

We assume that there exist solutions of the systems (3.5), (3.6) and (3.16), (3.7) and (3.17). We need also $2^{nd}$ order conditions like (3.48), (3.53). In fact for the convergence we shall need more stringent assumptions, which we make right now for the sake of brevity.

$$(3.62)\qquad \begin{pmatrix} H_{yy} & H_{yv} \\ H_{vy} & H_{vv} \end{pmatrix} (x,y,v,p_0(t),\ q_0(t) + L_0(\tau)) \geq \beta I$$

$$, \qquad \begin{pmatrix} H_{yy} & H_{yv} \\ H_{vy} & H_{vv} \end{pmatrix} (x,y,v,p_0(t),\ q_0(t) + Q_0(\tau)) \geq \beta I$$

$$(3.63)\qquad H_{xx} - (H_{xy}\ H_{xv}) \begin{pmatrix} H_{yy} & H_{yv} \\ H_{vy} & H_{vv} \end{pmatrix}^{-1} \begin{pmatrix} H_{yx} \\ H_{vx} \end{pmatrix} \geq 0$$

with the same arguments as in (3.62), $\forall\ x,y,v,t,\tau$.

Note that (3.62), (3.66) contain (1.13), since $L_0$, $Q_0$ vanish at $\infty$. Using them with either $x = x_0(0)$, $t = 0$, or $x = x_0(T)$, $t = T$, we recover (3.48) and (3.53) correctly interpreted with F, G given by (3.27), (3.28) or the analogous ones with $w_0(T)$ instead of $w_0(0)$.

We also need (3.56) only for the boundary layer at T. It reads

(3.64) $\quad g_v^*(x_0(T),Z,\eta)$ has a left inverse which is bounded in $Z,\eta$.

Under these assumptions we can assert the following exponential decay properties

$$(3.65)\qquad |W_0^0(\tau)|\ ,\ \ |W_0^T(\tau)| \leq Ce^{-\gamma\tau},\ \gamma > 0.$$

From this and (3.8)...,(3.11) we deduce

$$(3.66)\qquad |X_0(\tau)|\ ,\ |U_0(\tau)|\ ,\ |M_0(\tau)|\ ,\ |K_0(\tau)| \leq Ce^{-\gamma\tau}.$$

The problems (3.24), (3.25), (3.26) are linear quadratic. It is not difficult to check that they have unique optimal controls, that (3.12), (3.18) ; (3.14), (3.18); (3.15), (3.19) are satisfied and that the exponential decay property

$$(3.67) \qquad |W_1^0(\tau)| \ , \ |W_1^T(\tau)| \le Ce^{-\gamma\tau}$$

holds.

## 4. CONVERGENCE IN THE GENERAL CASE

### Preliminary remark.

Our objective in this section is to derive the analogue of Theorem 2.2, in the general case. Besides the technicalities, there is an essential difference, due to the fact that the existence of an optimal control for the $\varepsilon$ problem is not assumed, whereas it is automatic in the L.Q. case. For this reason the proof will be different, since the direct method used in Theorem 2.2 does not extend.

### 4.1. Assumptions and preliminary properties.

We shall assume that

(4.1) $f,g,\ell,h$ are $C^4$ in $x,y,v$ ; all derivatives of $f,g$ are bounded ; all derivatives of $\ell$, $h$ starting from the $2^{nd}$ order are bounded.

We of course assume (1.2), the existence of solutions of the systems (3.5) ; (3.6) and (3.16) ; (3.7) and (3.17), as well as (3.62), (3.63), (3.64).

We deduce from the definitions (3.2) the following relations for $\tilde{x}^\varepsilon, \tilde{y}^\varepsilon$

$$(4.2) \qquad \frac{d\tilde{x}^\varepsilon}{dt} = f(\sigma^\varepsilon) - f(\sigma_0) - \varepsilon f_\sigma(\sigma_0)\sigma_1 - (f(\sigma_0(0) + \Sigma_0^0) - f(\sigma_0(0)) -$$

$$- (f(\sigma_0(T) + \Sigma_0^T) - f(\sigma_0(T)))$$

$$\tilde{x}^\varepsilon(0) = - U_0(\frac{T}{\varepsilon})$$

$$(4.3)\qquad \varepsilon \frac{d\tilde{y}^\varepsilon}{dt} + \varepsilon^2 \frac{dy_1}{dt} = g(\sigma_\varepsilon) - \varepsilon g_\sigma(\sigma_0)\sigma_1 - g(\sigma_0(0) + \Sigma_0^0) - g(\sigma_0(T) + \Sigma_0^T) -$$

$$- \varepsilon g_\sigma(\sigma_0(0) + \Sigma_0^0)\Sigma_1^0 - \varepsilon g_\sigma(\sigma_0(T) + \Sigma_0^T)\Sigma_1^T - \varepsilon(g_\sigma(\sigma_0(0)+\Sigma_0^0)-$$

$$- g_\sigma(\sigma_0(0)))(\sigma_1(0) + \frac{t}{\varepsilon}\dot{\sigma}_0(0)) - \varepsilon\,(g_\sigma(\sigma_0(T) + \Sigma_0^T) -$$

$$- g_\sigma(\sigma_0(T)))(\sigma_1(T) + \frac{t-T}{\varepsilon}\dot{\sigma}_0(T)))$$

$$\tilde{y}^\varepsilon(0) = -(Z_0 + \varepsilon Z_1)(\frac{T}{\varepsilon}).$$

Moreover we can express $J^\varepsilon(u^\varepsilon)$ as

$$(4.4)\qquad J^\varepsilon(u^\varepsilon) = J(u_0) - \varepsilon\int_0^T q_0 \frac{dy^\varepsilon}{dt}\,dt + \int_0^T [H(\sigma^\varepsilon,p_0,q_0) - H(w_0) -$$

$$- H_x(w_0)(x^\varepsilon - x_0)]dt + h(x^\varepsilon(T)) - h(x_0(T)) - h_x(x_0(T))(x^\varepsilon(T)-x_0(T))$$

We shall make use of the following properties

$$(4.5)\qquad \left|\int_0^T \phi_\varepsilon(t)X_0(\frac{t}{\varepsilon})U_0(\frac{T-t}{\varepsilon})dt\right| \le e^{-K\frac{T}{\varepsilon}}\int_0^T |\phi_\varepsilon(t)|dt$$

$$(4.6)\qquad \begin{aligned} &|t^m X_0(\tfrac{t}{\varepsilon})|_{L^1} \le C\varepsilon^{m+1} \\ &|t^m X_0(\tfrac{t}{\varepsilon})|_{L^2} \le C\varepsilon^{m+1/2},\quad m \ge 0 \end{aligned}$$

and analogue properties for $U_0(\frac{T-t}{\varepsilon})$.

In (4.5), (4.6) $X_0$, $u_0$ are generic functions with exponential decay.

$$(4.7)\qquad \left|\int_0^T F_\varepsilon(t;X_0(\tfrac{t}{\varepsilon}) + U_0(\tfrac{T-t}{\varepsilon}))dt - \int_0^T F_\varepsilon(t;X_0(\tfrac{t}{\varepsilon}))dt -\right.$$

$$\left.-\int_0^T F_\varepsilon(t;U_0(\tfrac{T-t}{\varepsilon}))dt + \int_0^T F_\varepsilon(t;0)dt\right| \le Ce^{-k\frac{T}{\varepsilon}}$$

provided all integrals of the form $\int_0^T\int_0^1 |F_{\varepsilon,\xi}(t,X_0 + \lambda U_0)|dtd\lambda$ are bounded and $\xi$ refers to the $2^{nd}$ argument.

Lemma 4.1. We have the estimate

$$(4.8) \qquad |\tilde{x}^{\varepsilon}|_{C(0,T)} \le C(|\tilde{y}^{\varepsilon}|_{L^2} + |\tilde{u}^{\varepsilon}|_{L^2} + \varepsilon^2)$$

Proof.

We have

$$(4.9) \qquad f(\sigma^{\varepsilon}) = f(\sigma_0 + \Sigma_0^0 + \Sigma_0^T + \varepsilon\sigma_1 + \varepsilon\Sigma_1^0 + \varepsilon\Sigma_1^T + \tilde{\sigma}^{\varepsilon})$$

$$= f(\sigma_0 + \Sigma_0^0 + \Sigma_0^T + \varepsilon\sigma_1 + \varepsilon\Sigma_1^0 + \varepsilon\Sigma_1^T) +$$

$$+ \int_0^1 f_{\sigma}(\sigma^{\varepsilon} - (1 - \lambda)\tilde{\sigma}^{\varepsilon})\tilde{\sigma}^{\varepsilon} d\lambda .$$

Therefore the differential equation (4.2) can be written as

$$(4.10) \qquad \frac{d\tilde{x}^{\varepsilon}}{dt} = A\tilde{x}^{\varepsilon} + B\tilde{y}^{\varepsilon} + C\tilde{u}^{\varepsilon} + \phi^{\varepsilon}$$

$$\tilde{x}^{\varepsilon}(0) = - \varepsilon U_0(\frac{T}{\varepsilon})$$

where

$$(4.11) \qquad \phi^{\varepsilon} = f(\sigma_0 + \Sigma_0^0 + \Sigma_0^T + \varepsilon\sigma_1 + \varepsilon\Sigma_1^0 + \varepsilon\Sigma_1^T) - f(\sigma_0) - \varepsilon f_{\sigma}(\sigma_0)\sigma_1 -$$

$$- (f(\sigma_0(0) + \Sigma_0^0) - f(\sigma_0(0)) - (f(\sigma_0(T) + \Sigma_0^T) - f(\sigma_0(T))$$

Introducing $\psi^{\varepsilon}$ which is the solution of

$$\frac{d\psi^{\varepsilon}}{dt} = A\psi^{\varepsilon} + \phi^{\varepsilon} \qquad , \quad \psi^{\varepsilon}(0) = 0$$

then

$$|\psi^{\varepsilon}(t)| \le C \int_0^T |\phi^{\varepsilon}(s)| ds$$

and the result desired will be a consequence of

$$(4.12)\qquad \int_0^T |\phi^\varepsilon(t)|dt \le C\varepsilon^2.$$

which easily follows from (4.5), (4.6), (4.7)

□

4.2. Expansions

We write the 3rd integral at the right side of (4.4) as follows

$$(4.13)\qquad \int_0^T [H(w_o + \Sigma_o^o + \Sigma_o^T + \varepsilon\sigma_1 + \varepsilon\Sigma_1^o + \varepsilon\Sigma_1^T + \tilde{\sigma}^\varepsilon) - H(w_o) -$$

$$- H_x(w_o)(\varepsilon x_1 + \varepsilon X_o + \varepsilon U_o + \tilde{x}^\varepsilon)]dt.$$

This notation deserves some explanation. Remember that the variables w have 5 components, and $\sigma$ (or $\Sigma$) have 3, which are the first three of w. When we add $w + \sigma$ we implicitly imbed $\sigma$ in a 5 component vector, by equating to 0 the last two. So, for instance

$$w_o + \Sigma_o^o + \Sigma_o^T + \varepsilon\sigma_1 + \varepsilon\Sigma_1^o + \varepsilon\Sigma_1^T + \tilde{\sigma}^\varepsilon = (x_o + \varepsilon x_1 + \varepsilon X_o + \varepsilon U_o + \tilde{x}^\varepsilon,$$

$$y_o + Y_o + Z_o + \varepsilon y_1 + \varepsilon Y_1 + \varepsilon Z_1 + \tilde{y}^\varepsilon, u_o + \xi_o + \eta_o + \varepsilon u_1 + \varepsilon\xi_1 + \varepsilon\eta_1 + \tilde{u}^\varepsilon, p_o, q_o)$$

The expression (4.13) is written as

$$(4.14)\qquad \int_0^T [H(w_o + \Sigma_o^o + \Sigma_o^T + \varepsilon\sigma_1 + \varepsilon\Sigma_1^o + \varepsilon\Sigma_1^T) - H(w_o) - H_x(w_o)(\varepsilon x_1 + \varepsilon X_o + \varepsilon U_o) +$$

$$+ H_\sigma(w_o + \Sigma_o^o + \Sigma_o^T + \varepsilon\sigma_1 + \varepsilon\Sigma_1^o + \varepsilon\Sigma_1^T)\tilde{\sigma}^\varepsilon - H_x(w_o)\tilde{x}^\varepsilon]dt +$$

$$+ \int_0^T \int_0^1 \int_0^1 \lambda\, H_{\sigma\sigma}(w_{\lambda\mu}^\varepsilon)(\tilde{\sigma}^\varepsilon)^2 dt d\lambda d\mu$$

in which we have set

$$(4.15)\qquad w_{\lambda\mu}^\varepsilon = w_o + \Sigma_o^o + \Sigma_o^T + \varepsilon(\sigma_1 + \Sigma_1^o + \Sigma_1^T) + \lambda\mu\tilde{\sigma}^\varepsilon .$$

We next concentrate on the terms of order 2 in $\tilde{\sigma}^\varepsilon$ ,namely

$$\tilde{X}^\varepsilon = \int_0^T [H_\sigma(w_o + \Sigma_o^o + \Sigma_o^T + \varepsilon\sigma_1 + \varepsilon\Sigma_1^o + \varepsilon\Sigma_1^T)\tilde{\sigma}^\varepsilon - H_x(w_o)\tilde{x}^\varepsilon]dt$$

making use of (4.5), (4.6), (4.7)

$$\begin{aligned}\tilde{X}^\varepsilon = \int_0^T & [H_\sigma(w_o(0) + \Sigma_o^o) - H_\sigma(w_o(0)) + H_\sigma(w_o(T) + \Sigma_o^T) - H_\sigma(w_o(T)) + \\ & + H_{\sigma\sigma}(w_o)\varepsilon\sigma_1 + H_{\sigma\sigma}(w_o(0) + \Sigma_o^o)\varepsilon\Sigma_1^o + H_{\sigma\sigma}(w_o(T) + \Sigma_o^T)\varepsilon\Sigma_1^T + \\ & + (H_{\sigma w}(w_o(0) + \Sigma_o^o) - H_{\sigma w}(w_o(0)))t\dot{w}_o(0) + (H_{\sigma w}(w_o(T) + \Sigma_o^T) - \\ & - H_{\sigma w}(w_o(T)))(t - T)\dot{w}_o(T) + (H_{\sigma\sigma}(w_o(0) + \Sigma_o^o) - \\ & - H_{\sigma\sigma}(w_o(0)))\varepsilon\sigma_1(0) + (H_{\sigma\sigma}(w_o(T) + \Sigma_o^T) - H_{\sigma\sigma}(w_o(T)))\varepsilon\sigma_1(T)]\tilde{\sigma}^\varepsilon dt + \\ & + \varepsilon^2 O(|\tilde{\sigma}^\varepsilon|_{L^2}).^{(1)}\end{aligned}$$

From Lemma 4.1, we also deduce

$$\begin{aligned}\tilde{X}^\varepsilon = \int_0^T & [H_\sigma(w_o(0) + \Sigma_o^o) - H_\sigma(w_o(0)) + H_\sigma(w_o(T) + \Sigma_o^T) - H_\sigma(w_o(T)) + \\ & + H_{\sigma\sigma}(w_o)\varepsilon\sigma_1]\tilde{\sigma}^\varepsilon dt + \int_0^T [H_{y\sigma}(w_o(0) + \Sigma_o^o)\varepsilon\Sigma_1^o + H_{y\sigma}(w_o(T) + \Sigma_o^T)\varepsilon\Sigma_1^T + \\ & + (H_{yw}(w_o(0) + \Sigma_o^o) - H_{yw}(w_o(0)))t\dot{w}_o(0) + (H_{yw}(w_o(T) + \Sigma_o^T) - \\ & - H_{yw}(w_o(T)))(t - T)\dot{w}_o(T) + (H_{y\sigma}(w_o(0) + \Sigma_o^o) - H_{y\sigma}(w_o(0)))\varepsilon\sigma_1(0) + \\ & + (H_{y\sigma}(w_o(T) + \Sigma_o^T) - H_{y\sigma}(w_o(T)))\varepsilon\sigma_1(T)]\tilde{y}^\varepsilon dt + \\ & + \int_0^T [H_{v\sigma}(w_o(0) + \Sigma_o^o)\varepsilon\Sigma_1^o + H_{v\sigma}(w_o(T) + \Sigma_o^T)\varepsilon\Sigma_1^T + (H_{vw}(w_o(0) + \Sigma_o^o) - \\ & - H_{vw}(w_o(0)))t\dot{w}_o(0) + (H_{vw}(w_o(T) + \Sigma_o^T) - H_{vw}(w_o(T)))(t - T)\dot{w}_o(T) +\end{aligned} \tag{4.16}$$

(1) $|O(X)| \le C|X|$

$$+ (H_{v\sigma}(w_o(0) + \Sigma_o^0) - H_{v\sigma}(w_o(0)))\varepsilon\sigma_1(0) + (H_{v\sigma}(w_o(T) + \Sigma_o^T) -$$

$$- H_{v\sigma}(w_o(T)))\varepsilon\sigma_1(T)]\tilde{u}^\varepsilon dt + \varepsilon^2\, O(|\tilde{y}^\varepsilon|_{L^2} + |\tilde{u}^\varepsilon|_{L^2} + \varepsilon^2).$$

But

$$\int_0^T [H_\sigma(w_o(0) + \Sigma_o^0) - H_\sigma(w_o(0))]\tilde{\sigma}^\varepsilon dt = \varepsilon \int_0^T M_o \frac{d\tilde{x}^\varepsilon}{dt}\, dt + \tag{4.17}$$

$$+ \int_0^T L_o(\varepsilon \frac{d\tilde{y}^\varepsilon}{dt} - g_\sigma(\sigma_o(0) + \Sigma_o^0)\tilde{\sigma}^\varepsilon)dt + O(e^{-k\frac{T}{\varepsilon}}).$$

From (4.2) we derive, taking account of (4.5)

$$\varepsilon\int_0^T M_o \frac{d\tilde{x}^\varepsilon}{dt}\, dt = \varepsilon \int_0^T M_o[f(\sigma_o + \Sigma_o^0 + \varepsilon\sigma_1 + \varepsilon\Sigma_1^0) - f(\sigma_o) - \tag{4.18}$$

$$- \varepsilon f_\sigma(\sigma_o)\sigma_1 - (f(\sigma_o(0) + \Sigma_o^0) - f(\sigma_o(0))]dt +$$

$$+ \varepsilon \int_0^T M_o(f_y(\sigma_o(0) + \Sigma_o^0)\tilde{y}^\varepsilon + f_v(\sigma_o(0) + \Sigma_o^0)\tilde{u}^\varepsilon)dt +$$

$$+ \varepsilon^2\, O(|\tilde{y}^\varepsilon| + |\tilde{u}^\varepsilon| + \varepsilon^2) + \varepsilon\, O(|\tilde{y}^\varepsilon|^2 + |\tilde{u}^\varepsilon|^2)$$

and

$$\int_0^T L_o(\varepsilon \frac{d\tilde{y}^\varepsilon}{dt} - g_\sigma(\sigma_o(0) + \Sigma_o^0)\tilde{\sigma}^\varepsilon)dt = \int_0^T L_o[-\varepsilon^2 \frac{dy_1}{dt} + g(\sigma_o + \Sigma_o^0 + \varepsilon\sigma_1 + \varepsilon\Sigma_1^0) - \tag{4.19}$$

$$- g(\sigma_o(0) + \Sigma_o^0) - \varepsilon g_\sigma(\sigma_o)\sigma_1 - \varepsilon g_\sigma(\sigma_o(0) + \Sigma_o^0)\Sigma_1^0 - \varepsilon(g_\sigma(\sigma_o(0) + \Sigma_o^0) -$$

$$- g_\sigma(\sigma_o(0)))(\sigma_1(0) + \frac{t}{\varepsilon}\, \dot{\sigma}_o(0))]dt +$$

$$+ \varepsilon \int_0^T L_o[g_{y\sigma}(\sigma_o(0) + \Sigma_o^0)(\frac{t}{\varepsilon}\, \dot{\sigma}_o(0) + \sigma_1(0) + \Sigma_1^0)\tilde{y}^\varepsilon +$$

$$+ g_{v\sigma}(\sigma_o(0) + \Sigma_o^0)(\frac{t}{\varepsilon}\, \dot{\sigma}_o(0) + \sigma_1(0) + \Sigma_1^0)\tilde{u}^\varepsilon]dt +$$

$$+ \int_0^T \int_0^1 \int_0^1 \lambda\, L_o g_{\sigma\sigma}(w_{\lambda\mu}^\varepsilon)(\tilde{\sigma}^\varepsilon)^2 dt d\lambda d\mu + \varepsilon^2\, O(|\tilde{y}^\varepsilon| + |\tilde{u}^\varepsilon| + \varepsilon^2).$$

Similarly one has

$$\int_0^T [H_\sigma(w_o(T) + \Sigma_o^T) - H_\sigma(w_o(T))]\tilde{\sigma}^\varepsilon dt = -\varepsilon K_o(0)\tilde{x}^\varepsilon(T) - \varepsilon Q_o(0)\tilde{y}^\varepsilon(T) + \tag{4.20}$$

$$+ \varepsilon \int_0^T K_o \frac{d\tilde{x}^\varepsilon}{dt} + \int_0^T Q_o(\varepsilon \frac{d\tilde{y}^\varepsilon}{dt} - g_\sigma(\sigma_o(0) + \Sigma_o^T)\tilde{\sigma}^\varepsilon)dt + O(e^{-k\frac{T}{\varepsilon}})$$

which can be expressed by formulas similar to (4.18), (4.19).

We next compute, using (3.13)

$$\varepsilon \int_0^T H_{\sigma\sigma}(w_o)\sigma_1\tilde{\sigma}^\varepsilon dt = \varepsilon \int_0^T [(-\frac{dp_1}{dt} - f_x^* p_1 - g_x^* q_1)\tilde{x}^\varepsilon + \tag{4.21}$$

$$+ (-\frac{dq_o}{dt} - f_y^* p_1 - g_y^* q_1)\tilde{y}^\varepsilon + (-f_v^* p_1 - g_v^* q_1)\tilde{u}^\varepsilon]dt$$

$$= -\varepsilon p_1(T)\tilde{x}^\varepsilon(T) - \varepsilon q_o(T)\tilde{y}^\varepsilon(T) + \varepsilon \int_0^T q_o \frac{d\tilde{x}^\varepsilon}{dt} dt +$$

$$+ \varepsilon \int_0^T p_1 [\frac{d\tilde{x}^\varepsilon}{dt} - f_\sigma(\sigma_o)\tilde{\sigma}^\varepsilon]dt + \varepsilon \int_0^T q_1(-g_\sigma(\sigma_o)\tilde{\sigma}^\varepsilon)dt =$$

$$= -\varepsilon p_1(T)\tilde{x}^\varepsilon(T) - \varepsilon(q_o(T) + \varepsilon q_1(T))\tilde{y}^\varepsilon(T) + \int_0^T q_o \frac{d\tilde{y}^\varepsilon}{dt} dt +$$

$$+ \varepsilon \int_0^T [p_1(f_y(\sigma_o(0) + \Sigma_o^0) - f_y(\sigma_o(0)) + f_y(\sigma_o(T) + \Sigma_o^T) - f_y(\sigma_o(T))) +$$

$$+ q_1(g_y(\sigma_o(0) + \Sigma_o^0) - g_y(\sigma_o(0)) + g_y(\sigma_o(T) + \Sigma_o^T) - g_y(\sigma_o(T)))]\tilde{y}^\varepsilon dt +$$

$$+ \varepsilon \int_0^T [p_1(f_v(\sigma_o(0) + \Sigma_o^0) - f_v(\sigma_o(0)) + f_v(\sigma_o(T) + \Sigma_o^T) - f_v(\sigma_o(T))) +$$

$$+ q_1(g_v(\sigma_o(0) + \Sigma_o^0) - g_v(\sigma_o(0)) + g_v(\sigma_o(T) + \Sigma_o^T) - g_v(\sigma_o(T)))]\tilde{u}^\varepsilon dt +$$

$$+ \varepsilon \int_0^T p_1[f(\sigma_o + \Sigma_o^0 + \Sigma_o^T + \varepsilon\sigma_1 + \varepsilon\Sigma_1^0 + \varepsilon\Sigma_1^T) - f(\sigma_o) - \varepsilon f_\sigma(\sigma_o)\sigma_1 -$$

$$-(f(\sigma_o(0) + \Sigma_o^0) - f(\sigma_o(0)) - (f(\sigma_o(T) + \Sigma_o^T) - f(\sigma_o(T)))]dt +$$

$$+ \varepsilon \int_0^T q_1[g(\sigma_o + \Sigma_o^0 + \Sigma_o^T + \varepsilon\sigma_1 + \varepsilon\Sigma_1^0 + \varepsilon\Sigma_1^T) - g(\sigma_o(0) + \Sigma_o^0) -$$

$$- g(\sigma_o(T) + \Sigma_o^T) - \varepsilon g_\sigma(\sigma_o)\sigma_1 - \varepsilon g_\sigma(\sigma_o(0) + \Sigma_o^o)\Sigma_1^o - \varepsilon(g_\sigma(\sigma_o(0) + \Sigma_o^o) -$$

$$- g_\sigma(\sigma_o(0)))(\sigma_1(0) + \frac{t}{\varepsilon}\dot{\sigma}_o(0)) - \varepsilon g_\sigma(\sigma_o(T) + \Sigma_o^T)\Sigma_1^T - \varepsilon(g_\sigma(\sigma_o(T) + \Sigma_o^T) -$$

$$- g_\sigma(\sigma_o(T)))(\sigma_1(T) + \frac{t-T}{\varepsilon}\dot{\sigma}_o(T))]dt - \varepsilon^3 \int_0^T q_1 \frac{dy_1}{dt} dt +$$

$$+ \varepsilon\, 0(|\tilde{y}^\varepsilon|^2 + |\tilde{u}^\varepsilon|^2) + \varepsilon^2 0(|\tilde{y}^\varepsilon| + |\tilde{u}^\varepsilon|)$$

the expressions (4.19) and (4.20) being a contribution to the quadratic form in $\tilde{\sigma}^\varepsilon$ entering in (4.14). Adding up we get the quadratic form

$\int_0^T \int_0^1 \int_0^1 \lambda\, H_{\sigma\sigma}(\bar{w}_{\lambda\mu}^\varepsilon)(\tilde{\sigma}^\varepsilon)^2 dt d\lambda d\mu$ where we have set

$$(4.22) \qquad \bar{w}_{\lambda\mu}^\varepsilon = w_o + W_o^o + W_o^T + \varepsilon(\sigma_1 + \Sigma_1^o + \Sigma_o^T) + \lambda\mu\tilde{\sigma}^\varepsilon .$$

Collecting the terms in $\tilde{y}^\varepsilon$ and $\tilde{u}^\varepsilon$ in the expression of $\tilde{X}^\varepsilon$, by (4.16), (4.18), (4.19), (4.20), (4.21) we obtain, considering the equations defining $L_1$, $Q_1$ (3.14), (3.15) as well as the relations (3.18), (3.19),

$$(4.23) \qquad - \varepsilon(p_1(T) + K_o(0))\tilde{x}^\varepsilon(T) - \varepsilon^2 q_1(T)\tilde{y}^\varepsilon(T) + \int_0^T q_o \frac{d\tilde{y}^\varepsilon}{dt} dt +$$

$$+ \int_0^T \tilde{y}^\varepsilon[- \varepsilon^2 \frac{dL_1}{dt} - \varepsilon L_1 g_y(\sigma_o(0) + \Sigma_o^o) - \varepsilon^2 \frac{dQ_1}{dt} -$$

$$- \varepsilon Q_1 g_y(\sigma_o(T) + \Sigma_o^T)]dt + \int_0^T \tilde{u}^\varepsilon[- \varepsilon L_1 g_v(\sigma_o(0) + \Sigma_o^o) -$$

$$- \varepsilon Q_1 g_v(\sigma_o(T) + \Sigma_o^T)]dt.$$

The expression (4.13) contains also a term which does not depend on the quantities $\tilde{\sigma}^\varepsilon$. Collecting terms arising from (4.14), (4.18), (4.19), (4.20), (4.21), we obtain

$$(4.24) \qquad \int_0^T \{H(w_o + W_o^o + W_o^T + \varepsilon w_1 + \varepsilon W_1^o + \varepsilon W_1^T) - \varepsilon L_1 g(\sigma_o + \Sigma_o^o + \varepsilon\sigma_1 + \varepsilon\Sigma_1^o) -$$

$$- \varepsilon Q_1 g(\sigma_o + \Sigma_o^T + \varepsilon\sigma_1 + \varepsilon\Sigma_1^T) - H(w_o) - H_w(w_o)(\varepsilon w_1 + \varepsilon W_1^o + \varepsilon W_1^T) +$$

$$+ \varepsilon M_o[-\varepsilon f_\sigma(\sigma_o)\sigma_1 - (f(\sigma_o(0) + \Sigma_o^0) - f(\sigma_o(0)))] - \varepsilon^2 L_o \frac{dy_1}{dt} -$$

$$- \varepsilon^3 q_1 \frac{dy_1}{dt} + L_o[- g(\sigma_o(0) + \Sigma_o^0) - \varepsilon g_\sigma(\sigma_o)\sigma_1 - \varepsilon g_\sigma(\sigma_o(0) + \Sigma_o^0)\Sigma_1^0 -$$

$$- \varepsilon(g_\sigma(\sigma_o(0) + \Sigma_o^0) - g_\sigma(\sigma_o(0)))(\sigma_1(0) + \frac{t}{\varepsilon}\dot{\sigma}_o(0))] +$$

$$+ \varepsilon K_o[- \varepsilon f_\sigma(\sigma_o)\sigma_1 - (f(\sigma_o(T) + \Sigma_o^T) - f(\sigma_o(T)))] - \varepsilon^2 Q_o \frac{dy_1}{dt} +$$

$$+ Q_o [- g(\sigma_o(T) + \Sigma_o^T) - \varepsilon g_\sigma(\sigma_o)\sigma_1 - \varepsilon g_\sigma(\sigma_o(T) + \Sigma_o^T)\Sigma_1^T -$$

$$- \varepsilon(g_\sigma(\sigma_o(T) + \Sigma_o^T) - g_\sigma(\sigma_o(T)))(\sigma_1(T) + \frac{t-T}{\varepsilon}\dot{\sigma}_o(T))] +$$

$$+ \varepsilon p_1 [- \varepsilon f_\sigma(\sigma_o)\sigma_1 - (f_\sigma(\sigma_o(0) + \Sigma_o^0) - f(\sigma_o(0))) - (f_\sigma(\sigma_o(T) + \Sigma_o^T) -$$

$$- f_\sigma(\sigma_o(T)))] + \varepsilon q_1 [- g(\sigma_o(0) + \Sigma_o^0) - g(\sigma_o(T) + \Sigma_o^T) -$$

$$- \varepsilon g_\sigma(\sigma_o)\sigma_1 - \varepsilon g_\sigma(\sigma_o(0) + \Sigma_o^0)\Sigma_1^0 - \varepsilon g_\sigma(\sigma_o(T) + \Sigma_o^T)\Sigma_1^T -$$

$$- \varepsilon(g_\sigma(\sigma_o(0) + \Sigma_o^0) - g_\sigma(\sigma_o(0)))(\sigma_1(0) + \frac{t}{\varepsilon}\dot{\sigma}_o(0)) - \varepsilon(g_\sigma(\sigma_o(T) + \Sigma_o^T) -$$

$$- g_\sigma(\sigma_o(T)))(\sigma_1(T) + \frac{t-T}{\varepsilon}\dot{\sigma}_o(T))]\}dt + O(\varepsilon^4).$$

Integrating by parts in (4.23) yields

(4.25)
$$-\varepsilon(p_1(T) + K_o(0))\tilde{x}^\varepsilon(T) + \int_0^T q_o \frac{d\tilde{y}^\varepsilon}{dt}\,dt + \varepsilon\int_0^T L_1(\varepsilon\frac{d\tilde{y}^\varepsilon}{dt} - g_\sigma(\sigma_o(0)+\Sigma_o^0)\tilde{\sigma}^\varepsilon)dt +$$

$$+ \varepsilon\int_0^T Q_1(\varepsilon\frac{d\tilde{y}^\varepsilon}{dt} - g_\sigma(\sigma_o(T) + \Sigma_o^T)\tilde{\sigma}^\varepsilon)dt + O(e^{-k\frac{T}{\varepsilon}})$$

which can been evaluated further as in (4.19).

We now come back to (4.4). We notice that we can evaluate

(4.26)
$$h(x^\varepsilon(T)) - h(x_o(T)) - h_x(x_o(T))(x^\varepsilon(T) - x_o(T)) = h(x_o(T) + \varepsilon(x_1(T)+U_o(0)))-$$

$$- h(x_o(T)) - \varepsilon h_x(x_o(T))(x_1(T) + U_o(0)) + \int_0^1\int_0^1 \lambda\, h_{xx}(\bar{x}^\varepsilon_{\lambda\mu}(T))(\tilde{x}^\varepsilon(T))^2 d\lambda d\mu +$$

$$+ \varepsilon\, h_{xx}(x_o(T))(x_1(T) + U_o(0))\tilde{x}^{\varepsilon}(T) + 0\,(e^{-k\frac{T}{\varepsilon}})$$

where we have set

$$\bar{x}^{\varepsilon}_{\lambda\mu}(T) = x_o(T) + \varepsilon(x_1(T) + U_o(0)) + \lambda\mu\, \tilde{x}^{\varepsilon}(T). \tag{4.27}$$

Collecting terms we deduce the expression

$$\begin{aligned}
J^{\varepsilon}(u^{\varepsilon}) = {} & J(u_o) + \varepsilon q_o(0)y_o - \varepsilon q_o(T)(y_o(T) + Z_o(0) + \varepsilon y_1(T) + \varepsilon Z_1(0)) \\
& + \varepsilon \int_0^T \frac{dq_o}{dt}(y_o + Y_o + Z_o + \varepsilon y_1 + \varepsilon Y_1 + \varepsilon Z_1)dt \\
& + h(x_o(T) + \varepsilon(x_1(T) + U_o(0))) - h(x_o(T)) - \varepsilon h_x(x_o(T))(x_1(T)+U_o(0))+ \\
& + \int_0^T (L_o + \varepsilon L_1)\,[- \varepsilon^2 \frac{dy_1}{dt} - g(\sigma_o(0) + \Sigma^0_o) - \varepsilon g_\sigma(\sigma_o)\sigma_1 - \\
& - \varepsilon g_\sigma(\sigma_o(0) + \Sigma^0_o)\Sigma^0_1 - \varepsilon(g_\sigma(\sigma_o(0)+\Sigma^0_o)-g_\sigma(\sigma_o(0)))(\sigma_1(0)+\frac{t}{\varepsilon}\dot{\sigma}_o(0))]dt+ \\
& + \int_0^T (Q_o + \varepsilon Q_1)\,[- \varepsilon^2 \frac{dy_1}{dt} - g(\sigma_o(T) + \Sigma^T_o) - \varepsilon g_\sigma(\sigma_o)\sigma_1 - \\
& - \varepsilon g_\sigma(\sigma_o(T) + \Sigma^T_o)\Sigma^T_1 - \varepsilon(g_\sigma(\sigma_o(T) + \Sigma^T_o) - \\
& - g_\sigma(\sigma_o(T)))(\sigma_1(T) + \frac{t-T}{\varepsilon}\dot{\sigma}_o(T))]dt - \varepsilon^3 \int_0^T q_1 \frac{dy_1}{dt}\,dt + \\
& + \int_0^T [H(w_o + W^0_o + W^T_o + \varepsilon w_1 + \varepsilon W^0_1 + \varepsilon W^T_1) - H(w_o) - \\
& - H_w(w_o)(\varepsilon w_1 + \varepsilon W^0_1 + \varepsilon W^T_1)]dt - \varepsilon^2 \int_0^T (p_1 + M_o + K_o)f_\sigma(\sigma_o)\sigma_1 dt - \\
& - \varepsilon \int_0^T [(p_1 + M_o)(f(\sigma_o(0) + \Sigma^0_o) - f(\sigma_o(0))) + \\
& + (p_1 + K_o)(f(\sigma_o(T) + \Sigma^T_o) - f(\sigma_o(T)))]dt - \varepsilon \int_0^T q_1[g(\sigma_o(0) + \Sigma^0_o) + \\
& + g(\sigma_o(T) + \Sigma^T_o) + \varepsilon g_\sigma(\sigma_o)\sigma_1 + \varepsilon g_\sigma(\sigma_o(0) + \Sigma^0_o)\Sigma^0_1 +
\end{aligned} \tag{4.28}$$

$$+ \varepsilon g_\sigma(\sigma_o(T) + \Sigma_o^T)\Sigma_1^T + \varepsilon(g_\sigma(\sigma_o(0) + \Sigma_o^0) -$$

$$- g_\sigma(\sigma_o(0)))(\sigma_1(0) + \frac{t}{\varepsilon}\dot{\sigma}_o(0)) + \varepsilon(g_\sigma(\sigma_o(T) + \Sigma_o^T) -$$

$$- g_\sigma(\sigma_o(T)))(\sigma_1(T) + \frac{t-T}{\varepsilon}\dot{\sigma}_o(T))]dt + \int_0^1\int_0^1 \lambda\, h_{xx}(\bar{x}_{\lambda\mu}^\varepsilon)(\tilde{x}^\varepsilon(T))^2 d\lambda d\mu +$$

$$+ \int_0^T\int_0^1\int_0^1 \lambda\, H_{\sigma\sigma}(\bar{w}_{\lambda\mu}^\varepsilon)(\tilde{\sigma}^\varepsilon)^2 dt d\lambda d\mu + \varepsilon\, 0(|\tilde{y}^\varepsilon|^2 + |\tilde{u}^\varepsilon|^2) +$$

$$+ \varepsilon^2\, 0(|\tilde{y}^\varepsilon| + |\tilde{u}^\varepsilon|) + \varepsilon^4\, 0(1)$$

## 4.3. The main results.

Before stating the main convergence results, let us give an asymptotic expansion for the constant term in (4.28) (i.e. the term which does not involve $\tilde{\sigma}^\varepsilon$).

We use the expression (3.4) to evaluate the Hamiltonian

$$\int_0^T H(w_o + W_o^o + W_o^T + \varepsilon w_1 + \varepsilon W_1^o + \varepsilon W_1^T)dt.$$

The term of order $\varepsilon$ is given by

$$X_1 = - q_o(T)y_o(T) - q_o(T)Z_o(0) + \int_0^T \frac{dq_o}{dt} y_o\, dt + q_o(0)\, y_o \tag{4.29}$$

$$- \int_0^\infty L_o g(\sigma_o(0) + \Sigma_o^0)d\tau - \int_0^\infty Q_o\, g(\sigma_o(T) + \Sigma_o^T)\, d\tau +$$

$$+ \int_0^\infty (H(w_o(0) + W_o^o) - H(w_o(0)))d\tau + \int_0^\infty (H(w_o(T) + W_o^T) - H(w_o(T)))d\tau$$

The term of order $\varepsilon^2$ is given by

$$X_2 = - q_o(T)(y_1(T) + Z_1(0)) + \dot{q}_o(0)\int_0^\infty Y_o d\tau + \dot{q}_o(T)\int_0^\infty Z_o d\tau \tag{4.30}$$

$$+ \int_0^T \frac{dq_o}{dt} y_1 dt + \frac{1}{2} h_{xx}(x_o(T))(x_1(T) + U_o(0))^2 - \int_0^\infty [L_1 g(\sigma_o(0) + \Sigma_o^0) +$$

$$+ L_o g_\sigma(\sigma_o(0) + \Sigma_o^0)(\sigma_1(0) + \Sigma_1^0 + \tau\,\dot{\sigma}_o(0))]d\tau - \int_0^\infty [Q_1 g(\sigma_o(T) + \Sigma_o^T) +$$

$$+ Q_0 g_\sigma(\sigma_0(T) + \Sigma_0^T)(\sigma_1(T) + \Sigma_1^T - \tau\dot{\sigma}_0(T))]d\tau + \int_0^\infty [(H_w(w_0(0) + W_0^0) -$$

$$- H_w(w_0(0)))(\tau\dot{w}_0(0) + w_1(0) + W_1^0) + (H_w(w_0(T) + W_0^T) -$$

$$- H_w(w_0(T))) (- \tau\dot{w}_0(T) + w_1(T) + W_1^T)]d\tau + \int_0^T [\frac{1}{2} H_{ww} w_1^2 -$$

$$- (p_1 f_\sigma(\sigma_0) + q_1 g_\sigma(\sigma_0))\sigma_1]dt - \int_0^\infty [M_0(f(\sigma_0(0) + \Sigma_0^0) - f(\sigma_0(0))) +$$

$$+ K_0(f(\sigma_0(T) + \Sigma_0^T) - f(\sigma_0(T)))]d\tau - p_1(0) \int_0^\infty (f(\sigma_0(0) + \Sigma_0^0) -$$

$$- f(\sigma_0(0))d\tau - q_1(0) \int_0^\infty g(\sigma_0(0) + \Sigma_0^0)\, d\tau - p_1(T) \int_0^\infty (f(\sigma_0(T) + \Sigma_0^T) -$$

$$- f(\sigma_0(T)))d\tau - q_1(T) \int_0^\infty g(\sigma_0(T) + \Sigma_0^T)\, d\tau$$

Finally the term of order $\varepsilon^3$ is the following

$$X_3 = \dot{q}_0(0) \int_0^\infty Y_1 d\tau + \dot{q}_0(T) \int_0^\infty Z_1 d\tau + \frac{1}{6} h_{xxx}(x_0(T))(x_1(T) + \tag{4.31}$$

$$+ U_0(0))^3 - \dot{y}_1(0) \int_0^\infty L_0 d\tau - \ddot{y}_0(0) \int_0^\infty \tau L_0 d\tau + \ddot{q}_0(0) \int_0^\infty \tau Y_0 d\tau -$$

$$- \int_0^\infty L_1 g_\sigma(\sigma_0(0) + \Sigma_0^0)(\Sigma_1^0 + \sigma_1(0) + \tau\dot{\sigma}_0(0))d\tau - \dot{y}_1(T) \int_0^\infty Q_0 d\tau +$$

$$+ \ddot{y}_0(T) \int_0^\infty \tau Q_0 d\tau - \ddot{q}_0(T) \int_0^\infty \tau Z_0 d\tau - \int_0^\infty Q_1 g_\sigma(\sigma_0(T) + \Sigma_0^T)(\Sigma_1^T + \sigma_1(T) -$$

$$- \tau\dot{\sigma}_0(T))d\tau - \int_0^T q_1 \frac{dy_1}{dt}\, dt + \int_0^\infty [(H_w(w_0(0) + W_0^0) -$$

$$- H_w(w_0(0)))(\frac{1}{2} \tau^2 \ddot{w}_0(0) + \tau\dot{w}_1(0)) + (H_{ww}(w_0(0) + W_0^0) -$$

$$- H_{ww}(w_0(0)))(\frac{1}{2}(\tau\dot{w}_0(0) + w_1(0))^2 + W_1^0\, \tau\dot{w}_0(0)) +$$

$$+ H_{ww}(w_0(0) + W_0^0)(W_1^0 w_1(0) + \frac{1}{2}(W_1^0)^2]d\tau \;+$$

$$+ \int_0^\infty [(H_w(w_o(T) + W_o^T) - H_w(w_o(T)))(\frac{1}{2}\tau^2\dot{w}_o(T) - \tau\dot{w}_1(T)) +$$

$$+ (H_{ww}(w_o(T) + W_o^T) - H_{ww}(w_o(T)))(\frac{1}{2}(-\tau\dot{w}_o(T) + w_1(T))^2 -$$

$$- W_1^T \tau\dot{w}_o(T)) + H_{ww}(w_o(T) + W_o^T)(W_1^T w_1(T) + \frac{1}{2}(W_1^T)^2]d\tau +$$

$$+ \frac{1}{6}\int_0^T H_{www}(w_o)w_1^3 dt - \int_0^\infty [M_o f_\sigma(\sigma_o(0))\sigma_1(0) + \tau\dot{p}_1(0)(f(\sigma_o(0) + \Sigma_o^0) -$$

$$- f(\sigma_o(0))) + \tau\dot{q}_1(0)g(\sigma_o(0) + \Sigma_o^0)) + q_1(0)g_\sigma(\sigma_o(0) + \Sigma_o^0)\Sigma_1^0 +$$

$$+ q_1(0)(g_\sigma(\sigma_o(0) + \Sigma_o^0) - g_\sigma(\sigma_o(0)))(\sigma_1(0) + \tau\dot{\sigma}_o(0)]d\tau -$$

$$- \int_0^\infty [K_o f_\sigma(\sigma_o(T))\sigma_1(T) - \tau\dot{p}_1(T)(f(\sigma_o(T) + \Sigma_o^T) - f(\sigma_o(T))) -$$

$$- \tau\dot{q}_1(T)g(\sigma_o(T) + \Sigma_o^T) + q_1(T)g_\sigma(\sigma_o(T) + \Sigma_o^T)\Sigma_1^T + q_1(T)(g_\sigma(\sigma_o(T) + \Sigma_o^T) -$$

$$- g_\sigma(\sigma_o(T)))(\sigma_1(T) - \tau\dot{\sigma}_o(T))]d\tau.$$

Using the optimization problems detailed in § 3.2, we can rewrite the quantities $X_1$, $X_2$, $X_3$ as follows

$$X_1 = q_o(0)(y_o - y_o(0)) - \int_0^T q_o \frac{dy_o}{dt} + \mathcal{G}_o(\xi_o) + \mathcal{K}_o(\eta_o) \tag{4.32}$$

$$X_2 = \dot{x}_o(0)\int_0^\infty M_o d\tau + \dot{x}_o(T)\int_0^\infty K_o d\tau - \dot{p}_o(0)\int_0^\infty X_o d\tau \tag{4.33}$$

$$- \dot{p}_o(T)\int_0^\infty U_o d\tau + \int_0^\infty [M_o(f(\sigma_o(0) + \Sigma_o^0) - f(\sigma_o(0)) +$$

$$+ K_o(f(\sigma_o(T) + \Sigma_o^T) - f(\sigma_o(T))]d\tau + \frac{1}{2} h_{xx}(x_o(T))U_o(0)^2 -$$

$$- K_o(0)U_o(0) + J_1(u_1)$$

$$X_3 = \ddot{q}_o(0)\int_0^\infty \tau\, Y_o d\tau - \ddot{q}_o(T)\int_0^\infty \tau\, Z_o d\tau - \ddot{y}_o(0)\int_0^\infty \tau\, L_o d\tau + \tag{4.34}$$

$$+ \ddot{y}_o(T) \int_0^\infty \tau\, Q_o d\tau + \dot{x}_1(0) \int_0^\infty M_o d\tau + \dot{x}_1(T) \int_0^\infty K_o d\tau +$$

$$+ \frac{1}{2} \int_0^\infty [(H_w(w_o(0) + W_o^0) - H_w(w_o(0)))\tau^2 \ddot{w}_o(0) + (H_{ww}(w_o(0) + W_o^0) -$$

$$- H_{ww}(w_o(0)))(\tau\dot{w}_o(0) + w_1(0))^2]d\tau + \int_0^\infty [(H_{xw}(w_o(0) + W_o^0) -$$

$$- H_{xw}(w_o(0)))\tau\dot{w}_o(0) + H_{xw}(w_o(0) + W_o^0)w_1(0)]X_o d\tau +$$

$$+ \int_0^\infty M_o(f_\sigma(\sigma_o(0) + \Sigma_o^0) - f_\sigma(\sigma_o(0)))(\tau\dot{\sigma}_o(0) + \sigma_1(0))d\tau +$$

$$+ \int_0^\infty [\frac{1}{2} H_{xx}(w_o(0) + W_o^0)X_o^2 + M_o f_x(\sigma_o(0) + \Sigma_o^0)X_o]d\tau -$$

$$- q_1(0)y_1(0) - \int_0^T q_1 \frac{dy_1}{dt} dt + \frac{1}{6} h_{xxx}(x_o(T))(x_1(T) + U_o(0))^3 +$$

$$+ \int_0^T \frac{1}{6} H_{www}(w_o(t))w_1^3 dt + \frac{1}{2} \int_0^\infty [(H_w(w_o(T) + W_o^T) -$$

$$- H_w(w_o(T)))\tau^2 \ddot{w}_o(T) + (H_{ww}(w_o(T) + W_o^T) -$$

$$- H_{ww}(w_o(T)))(- \tau\dot{w}_o(T) + w_1(T))^2]d\tau + \int_0^\infty [(H_{xw}(w_o(T) + W_o^T) -$$

$$- H_{xw}(w_o(T)))\tau\dot{w}_o(T) + H_{xw}(w_o(T) + W_o^T)w_1(T)]U_o d\tau +$$

$$+ \int_0^\infty K_o(f_\sigma(\sigma_o(T) + \Sigma_o^T) - f_\sigma(\sigma_o(T)))(- \tau\dot{\sigma}_o(T) + \sigma_1(T))d\tau +$$

$$+ \int_0^\infty [\frac{1}{2} H_{xx}(w_o(T) + W_o^T)U_o^2 + K_o f_x(\sigma_o(T) + \Sigma_o^T)U_o]d\tau +$$

$$+ \mathcal{Y}_1(\xi_1) + \mathcal{K}_1(\eta_1).$$

To recover in the linear quadratic case the formulas (2.75) we notice that in this case one has

$$\dot{q}_o(0) \int_0^\infty Y_o d\tau - \dot{y}_o(0) \int_0^\infty L_o d\tau = \dot{x}_o(0) \int_0^\infty M_o d\tau - \dot{p}_o(0) \int_0^\infty X_o d\tau \tag{4.35}$$

$$\dot{q}_o(T)\int_0^\infty Z_o d\tau - \dot{y}_o(T)\int_0^\infty Q_o d\tau = + \dot{x}_o(T)\int_0^\infty K_o d\tau - \dot{p}_o(T)\int_0^\infty U_o d\tau$$

and for (4.34)

$$\int_0^\infty H_{xw}(w_o(0) + W_o^0)w_1(0)X_o d\tau = \dot{p}_1(0)\int_0^\infty \tau(BY_o + G\xi_o)d\tau \tag{4.36}$$

$$\int_0^\infty H_{xw}(w_o(T) + W_o^T)w_1(T)U_o d\tau = - \dot{p}_1(T)\int_0^\infty \tau(BZ_o + G\eta_o)d\tau$$

From (4.28) we then have

$$\begin{aligned} J^\varepsilon(u^\varepsilon) &= J(u_o) + \varepsilon X_1 + \varepsilon^2 X_2 + \varepsilon^3 X_3 + \\ &+ \int_0^1\int_0^1 \lambda\, h_{xx}(\bar{x}^\varepsilon_{\lambda\mu})(\tilde{x}^\varepsilon(T))^2 d\lambda d\mu + \\ &+ \int_0^T\int_0^1\int_0^1 \lambda\, H_{\sigma\sigma}(\bar{w}^\varepsilon_{\lambda\mu})(\tilde{\sigma}^\varepsilon)^2 dt d\lambda d\mu + \\ &+ \varepsilon\, 0(|\tilde{y}^\varepsilon|^2 + |\tilde{u}^\varepsilon|^2) + \varepsilon^2\, 0(|\tilde{y}^\varepsilon| + |\tilde{u}^\varepsilon|) + \varepsilon^4\, 0(1) \end{aligned} \tag{4.37}$$

We can then assert the

Theorem 4.1. *We assume* (1.2) *and* (4.1) *and the existence of solutions to the systems* (3.5) ; (3.6) *and* (3.16) ; (3.7) *and* (3.17), *as well as* (3.62), (3.63), (3.64). *We then have*

$$|\mathrm{Inf}\, J^\varepsilon(v(.)) - J(u_o) - \varepsilon X_1 - \varepsilon^2 X_2 - \varepsilon^3 X_3| \le C\varepsilon^4. \tag{4.38}$$

*If* $u^\varepsilon$ *is a control satisfying*

$$J^\varepsilon(u^\varepsilon) \le J^\varepsilon(u_o + \xi_o + \eta_o + \varepsilon u_1 + \varepsilon\xi_1 + \varepsilon\eta_1) \tag{4.39}$$

*then defining* $\tilde{x}^\varepsilon$, $\tilde{y}^\varepsilon$, $\tilde{u}^\varepsilon$ *by the formulas* (3.2) *one has*

$$|\tilde{u}^\varepsilon|_{L^2} \le C\varepsilon^2 \quad , \quad |\tilde{y}^\varepsilon|_{L^2} \le C\varepsilon^2 \tag{4.40}$$

$$|\tilde{x}^\varepsilon|_{C(o,T)} \le C\varepsilon^2$$

and

$$|\inf J^{\varepsilon}(v(.)) - J^{\varepsilon}(u^{\varepsilon})| \leq C\varepsilon^4$$

□

We first give the analogue of Lemma (1.4), namely denoting

$$\bar{u}^{\varepsilon}(t) = u_o(t) + \xi_o(\tfrac{t}{\varepsilon}) + \eta_o(\tfrac{T-t}{\varepsilon}) + \varepsilon u_1(t) + \varepsilon\xi_1(\tfrac{t}{\varepsilon}) + \varepsilon\eta_1(\tfrac{T-t}{\varepsilon}) \tag{4.41}$$

one has

Lemma 4.2. The following estimate holds

$$|J^{\varepsilon}(\bar{u}^{\varepsilon}) - J(u_o) - \varepsilon X_1 - \varepsilon^2 X_2 - \varepsilon^3 X_3| \leq C\varepsilon^4 \tag{4.42}$$

Proof.

Let us term $\bar{x}^{\varepsilon}$, $\bar{y}^{\varepsilon}$ the trajectories corresponding to the control $\bar{u}^{\varepsilon}$ , and define

$$x_1^{\varepsilon} = \bar{x}^{\varepsilon} - x_o - \varepsilon x_1 - \varepsilon X_o - \varepsilon U_o$$

$$y_1^{\varepsilon} = \bar{y}^{\varepsilon} - y_o - Y_o - Z_o - \varepsilon y_1 - \varepsilon Y_1 - \varepsilon Z_1.$$

By analogy with Lemma 4.1, we establish

$$\frac{dx_1^{\varepsilon}}{dt} = Ax_1^{\varepsilon} + By_1^{\varepsilon} + \phi^{\varepsilon} \tag{4.43}$$

$$x_1^{\varepsilon}(0) = -\varepsilon U_o(\tfrac{T}{\varepsilon})$$

where $\phi^{\varepsilon}$ is given by (4.11)

and

$$\varepsilon\frac{dy_1^{\varepsilon}}{dt} = Cx_1^{\varepsilon} + Dy_1^{\varepsilon} + \theta^{\varepsilon} \tag{4.44}$$

$$y_1^{\varepsilon}(0) = -(Z_o + \varepsilon Z_1)(\tfrac{T}{\varepsilon})$$

where A, B, C, D are bounded matrices depending on t (and also of $\varepsilon$), and noting that

$$D = \int_0^1 g_y(\bar{\sigma}^\varepsilon - (1-\lambda)\sigma_1^\varepsilon)d\lambda$$

we have by (1.2)

$$D\eta.\eta \le -\mu|\eta|^2.$$

Moreover

$$\theta^\varepsilon(t) = g(\sigma_0 + \Sigma_0^0 + \Sigma_0^T + \varepsilon\sigma_1 + \varepsilon\Sigma_1^0 + \varepsilon\Sigma_1^T) - g(\sigma_0(0) + \Sigma_0^0) -$$

$$- g(\sigma_0(T) + \Sigma_0^T) - \varepsilon g_\sigma(\sigma_0)\sigma_1 - \varepsilon g_\sigma(\sigma_0(0) + \Sigma_0^0)\Sigma_1^0 -$$

$$- \varepsilon g_\sigma(\sigma_0(T) + \Sigma_0^T)\Sigma_1^T - \varepsilon(g_\sigma(\sigma_0(0) + \Sigma_0^0) -$$

$$- g_\sigma(\sigma_0(0)))(\sigma_1(0) + \frac{t}{\varepsilon}\dot{\sigma}_0(0)) - \varepsilon(g_\sigma(\sigma_0(T) + \Sigma_0^T) -$$

$$- g_\sigma(\sigma_0(T)))(\sigma_1(T) + \frac{t-T}{\varepsilon}\dot{\sigma}_0(T)) - \varepsilon^2\frac{dy_1}{dt}.$$

and

(4.45) $$|\theta^\varepsilon(t)| \le C\varepsilon^2.$$

Note that an estimate like (4.43) is not valid for $\phi^\varepsilon$, but introducing $\psi^\varepsilon$ as in the proof of Lemma 4.1 we have

(4.46) $$|\psi^\varepsilon(t)| \le C\varepsilon^2$$

and

(4.47) $$\frac{d}{dt}(x_1^\varepsilon - \psi^\varepsilon) = A(x_1^\varepsilon - \psi^\varepsilon) + By_1^\varepsilon.$$

Therefore, considering (4.42) and (4.45), we deduce the estimate

$$(4.48)\qquad (|x_1^\varepsilon(t) - \psi^\varepsilon(t)|^2 + \varepsilon|y_1^\varepsilon(t)|^2) + \mu \int_0^t |y_1^\varepsilon(s)|^2 ds \le$$

$$C \int_0^t |x_1^\varepsilon(s) - \psi^\varepsilon(s)|^2 ds + C\varepsilon^4$$

which implies easily

$$(4.49)\qquad \int_0^T |y_1^\varepsilon(t)|^2 dt \le C\varepsilon^4, \quad |x_1^\varepsilon(t)| \le C\varepsilon^2.$$

Now we can compute $J^\varepsilon(\bar{u}^\varepsilon)$ by a formula similar to (4.37) in which $\tilde{\sigma}^\varepsilon$ has to be replaced by $\sigma_1^\varepsilon$. Taking account of (4.49), we easily obtain (4.42).

□

Proof of Theorem 4.1.

Considering $\tilde{x}^\varepsilon$, $\tilde{y}^\varepsilon$ given by (4.2), (4.3), we start by showing like for (4.48) that

$$(4.50)\qquad |\tilde{x}^\varepsilon(t)|^2 + \varepsilon|\tilde{y}^\varepsilon(t)|^2 + \mu \int_0^t |\tilde{y}^\varepsilon(s)|^2 ds \le C\,[\int_0^t |\tilde{x}^\varepsilon(s)|^2 ds +$$

$$+ \int_0^t |\tilde{u}^\varepsilon(s)|^2 ds + \varepsilon^4].$$

Define next $Z^\varepsilon(\lambda,\mu,t)$ by the formula

$$Z^\varepsilon = \begin{pmatrix} \tilde{y}^\varepsilon \\ \tilde{u}^\varepsilon \end{pmatrix} \div \begin{pmatrix} H_{yy} & H_{yv} \\ H_{vy} & H_{vv} \end{pmatrix}^{-1} \begin{pmatrix} H_{yx} \\ H_{yv} \end{pmatrix} (\bar{w}_{\lambda\mu}^\varepsilon)\tilde{x}^\varepsilon$$

hence from (4.50)

$$|\tilde{x}^\varepsilon(t)|^2 + \varepsilon|\tilde{y}^\varepsilon(t)|^2 + \mu \int_0^t |\tilde{y}^\varepsilon(s)|^2 ds \le C\,[\int_0^t |\tilde{x}^\varepsilon(s)|^2 ds +$$

$$+ \int_0^t |Z^\varepsilon(\lambda,\mu,s)|^2 ds + \varepsilon^4].$$

therefore

$$(4.51) \qquad |\tilde{x}^{\varepsilon}(t)|^2 \le C[\int_0^T |Z^{\varepsilon}(\lambda,\mu,s)|^2 ds + \varepsilon^4]$$

$$\int_0^T |\tilde{y}^{\varepsilon}(t)|^2 dt + \int_0^T |\tilde{u}^{\varepsilon}(t)|^2 dt \le C[\int_0^T |Z^{\varepsilon}(\lambda,\mu,s)|^2 ds + \varepsilon^4] .$$

$\forall\ \lambda,\mu.$

Now from the assumptions (3.62), (3.63) we have

$$\int_0^T H_{\sigma\sigma}(\bar{w}^{\varepsilon}_{\lambda\mu})(\tilde{\sigma}^{\varepsilon})^2 dt \ge (\beta - Ce^{-k\frac{T}{\varepsilon}}) \int_0^T |Z^{\varepsilon}(\lambda,\mu,t)|^2 dt$$

$$\ge \frac{\beta}{2} \int_0^T |Z^{\varepsilon}(\lambda,\mu,t)|^2 dt , \quad \forall\ \lambda,\mu$$

and with (4.51), we easily deduce from (4.37) that one has

$$(4.52) \qquad J^{\varepsilon}(u^{\varepsilon}) \ge J(u_0) + \varepsilon X_1 + \varepsilon^2 X_2 + \varepsilon^3 X_3 + \gamma \int_0^T \int_0^1 \int_0^1 \lambda |Z^{\varepsilon}|^2 dt d\lambda d\mu - C\varepsilon^4.$$

Therefore if $u^{\varepsilon}$ satisfies (4.39), i.e

$$J^{\varepsilon}(u^{\varepsilon}) \le J^{\varepsilon}(\bar{u}^{\varepsilon})$$

taking account of Lemma 4.2, we deduce

$$\int_0^T \int_0^1 \int_0^1 \lambda |Z^{\varepsilon}|^2 dt d\lambda d\mu \le C\varepsilon^4$$

and (4.51) implies (4.40). Moreover from (4.37) it follows

$$|J^{\varepsilon}(u^{\varepsilon}) - J^{\varepsilon}(\bar{u}^{\varepsilon})| \le C\varepsilon^4.$$

Now we can find $u^{\varepsilon}$ such that

$$J^{\varepsilon}(u^{\varepsilon}) \le J^{\varepsilon}(\bar{u}^{\varepsilon})$$

and $\qquad J^{\varepsilon}(u^{\varepsilon}) \le \inf J^{\varepsilon}(v(.)) + \varepsilon^4$

hence

$$J^{\varepsilon}(\bar{u}^{\varepsilon}) - C\varepsilon^4 \le \inf J^{\varepsilon}(v(.)) + \varepsilon^4$$

$$\le J^{\varepsilon}(\bar{u}^{\varepsilon}) + \varepsilon^4$$

and thus obtain

$$|\inf J^{\varepsilon}(v(.)) - J^{\varepsilon}(\bar{u}^{\varepsilon})| \le C\varepsilon^4$$

which proves (4.38). This completes the proof of the results desired

□

## 5. DYNAMIC PROGRAMMING.

### 5.1. Setting of the problem.

Let us consider the family of control problems

$$\frac{dx^{\varepsilon}}{ds} = f(x^{\varepsilon},y^{\varepsilon},v) \quad , \quad s \in (t,T) \tag{5.1}$$

$$\varepsilon \frac{dy^{\varepsilon}}{ds} = g(x^{\varepsilon},y^{\varepsilon},v)$$

$$x^{\varepsilon}(t) = x \ , \ y^{\varepsilon}(t) = y$$

$$J^{\varepsilon}_{x,y,t}(v(.)) = \int_t^T \ell(x^{\varepsilon}(s),y^{\varepsilon}(s),v(s))ds + h(x^{\varepsilon}(T)) \tag{5.2}$$

and define

$$\Phi^{\varepsilon}(x,y,t) = \inf_{v(.)} J^{\varepsilon}_{x,y,t}(v(.)). \tag{5.3}$$

We shall assume besides (1.1), (1.2) that

$$\ell_{\sigma\sigma} \ge \gamma I \qquad (\sigma = (x,y,v)) \tag{5.4}$$

$$h_{xx} \ge 0$$

(5.5) the $2^{nd}$ derivatives of f,g are bounded by $\dfrac{C_0}{1 + |x|}$, where $C_0$ is not too large compared to $\gamma$.

Note that, thanks to the assumption (5.4) we can, without less of generality restrict the controls to satisfy

$$\int_t^T |v(s)|^2 ds \leq C_o( 1 + |x|^2 + \varepsilon|y|^2). \tag{5.6}$$

Indeed one has, by considering the controls which are better than 0

$$\gamma \int_t^T |v(s)|^2 ds - C_1 \leq J^{\varepsilon}_{x,y,t}(v(.)) \leq J^{\varepsilon}_{xy,t}(0)$$

$$\leq C(1 + |x|^2 + \varepsilon|y|^2)$$

hence (5.6).

It is standard that the function $\Phi^{\varepsilon}$ defined by (5.3) can be characterized as the maximum solution in the class of Lipschitz functions with quadratic growth, of the Hamilton-Jacobi-Bellman (HJB) equation

$$\frac{\partial \Phi^{\varepsilon}}{\partial t} + \operatorname*{Inf}_v [ D_x \Phi^{\varepsilon}.f(x,y,v) + \frac{1}{\varepsilon} D_y \Phi^{\varepsilon}.g(x,y,v) + \ell(x,y,v)] = 0 \tag{5.7}$$

$$\Phi^{\varepsilon}(x,T) = h(x).$$

Consider next the limit control problem

$$\frac{dx}{ds} = f(x,y,v) \quad x(t) = x \quad , \quad s \in (t,T) \tag{5.8}$$

$$g(x,y,v) = 0 \quad (\text{or } y = \bar{y}(x,v))$$

$$J_{x,t}(v(.)) = \int_t^T \ell(x(s),\bar{y}(x(s),v(s)),v(s))ds + h(x(T)) \tag{5.9}$$

and define

$$\Phi(x,t) = \operatorname*{Inf}_{v(.)} J_{x,t}(v(.)). \tag{5.10}$$

By virtue of (5.4), (5.5) it can be shown that $\Phi$ is the unique regular solution of

$$\frac{\partial \Phi}{\partial t} + \operatorname*{Inf}_v [ D_x \Phi.f(x,\bar{y}(x,v),v) + \ell(x,\bar{y}(x,v),v)] = 0 \tag{5.11}$$

$$\Phi(x,T) = h(x).$$

Regular means here that

$$(5.12) \qquad \Phi \text{ is } C^2 \ , \quad 0 \le D^2\Phi \le C$$

$$|D \frac{\partial\Phi}{\partial t}| \le C(1 + |x|) \ , \ |\frac{\partial^2\Phi}{\partial t^2}| \le C(1 + |x|^2).$$

Define $\bar{u}(x,t)$ to be the unique optimal feedback in (5.11), and

$$(5.13) \qquad \bar{y}(x,t) = \bar{y}(x,\bar{u}(x,t)).$$

It will be also convenient to introduce $\bar{q}(x,t)$ defined uniquely by the relation

$$(5.14) \qquad \ell_y(x,\bar{y}(x,t),\bar{u}(x,t)) + f_y^*(x,\bar{y}(x,t),\bar{u}(x,t))D\Phi + $$

$$+ g_y^*(x,\bar{y}(x,t),\bar{u}(x,t))\bar{q}(x,t) = 0$$

then one has

(5.15) $\quad \bar{y},\bar{u},\bar{q}$ are continuously differentiable and

$$|\bar{y}_x| \ , \ |\bar{u}_x| \ , \ |\bar{q}_x| \le C$$

$$|\bar{y}_t| \ , \ |\bar{u}_t| \ , \ |\bar{q}_t| \le C(1 + |x|).$$

In the sequel the regularity properties (5.12), (5.15) can be assumed, in which case (5.4), (5.5) can be omitted provided one can guarantee (5.6).

Our objective is to study the convergence of $\Phi^\varepsilon$ to $\Phi$

Remark 5.1. The regularity assumptions on the limit HJB equation play a role similar to (1.12), (1.13). Note that they imply that the optimal control for (5.8), (5.9) exists and is uniquely defined.

5.2. Expansion.

We look for an expansion of   of the form

$$\Phi(x,t) + \varepsilon\, \Phi_1(x,y,t;\frac{T-t}{\varepsilon}) + \ldots$$

we deduce (formally) by equating to 0 the first term of the expansion

(5.16) $$-\frac{\partial\Phi_1}{\partial\theta} + \frac{\partial\Phi}{\partial t} + \operatorname*{Inf}_{v}\,[\, D\Phi.f(x,y,v) + D_y\Phi_1.g(x,y,v) + \ell(x,y,v)] = 0 \;,$$

We shall see that it is possible to solve (5.16), in which the pair (x,t) is a parameter

Let us define the functions

(5.17) $$G(x,t;Y,\xi) = g(x,\bar{y}(x,t) + Y,\bar{u}(x,t) + \xi)$$

(5.18) $$\begin{aligned} F(x,t;Y,\xi) = {} & \frac{\partial\Phi}{\partial t} + D\Phi.f(x,\bar{y}(x,t) + Y,\bar{u}(x,t) + \xi) + \\ & + \bar{q}(x,t).g(x,\bar{y}(x,t) + Y,\bar{u}(x,t) + \xi) + \\ & + \ell(x,\bar{y}(x,t) + Y,\, \bar{u}(x,t) + \xi) = \\ & = H(x,\bar{y}(x,t) + Y,\bar{u}(x,t) + \xi, D\Phi,\bar{q}(x,t)) - \\ & - H(x,\bar{y}(x,t),\bar{u}(x,t),\bar{q}(x,t)) \end{aligned}$$

where the Hamiltonian has been defined in (1.11).

Then by construction, one has

(5.19) $$G(x,t;0,0) = 0 \quad , \; G_Y(x,t;Y,\xi) \leq -\,\mu\, I$$

(5.20) $$F(x,t;0,0) = 0 \quad , \; F_Y(x,t;0,0) = 0 \;;\; F_\xi(x;t;0,0) = 0$$

and by virtue of (5.4), (5.5) and (5.15) we can also assert that

$$(5.21)\qquad \begin{pmatrix} F_{YY} & F_{Y\xi} \\ F_{\xi Y} & F_{\xi\xi} \end{pmatrix} \geq \beta I.$$

We consider the following problem

$$(5.22)\qquad -\frac{\partial\chi}{\partial\theta} + \operatorname*{Inf}_{\xi}[\,F(x,t;Y,\xi) + D_Y\chi.G(x,t;Y,\xi)] = 0$$

$$\chi(x,t;Y,0) = 0.$$

This problem has a lot in common with that introduced in (3.31), which may be viewed as the limit of (5.22) as $\theta \to \infty$. We associate to (5.22) the control problem

$$(5.23)\qquad \frac{dY}{d\tau} = G(x,t;Y,\xi) \qquad\qquad Y(0) = Y$$

$$\chi(x,t;Y,\theta) = \inf_{\xi(.)} \int_0^\theta F(x,t;Y(\tau),\xi(\tau))d\tau = \inf_{\xi(.)} \mathcal{J}(x,t;Y,\theta;\xi(.))$$

We have the

Proposition 5.1. Under the assumption (1.1), (1.2), (5.4), (5.5), the function $\chi$ defined by (5.23) satisfies

$$(5.24)\qquad a|Y|^2 \leq \chi \leq C|Y|^2$$

$$|D_Y\chi| \leq C|Y| \quad , \quad 0 \leq \frac{\partial\chi}{\partial\theta} \leq C|Y|^2$$

and is the maximum solution of (5.22). The constants in (5.24) do not depend on the arguments $Y,\theta$, nor of the parameters $x,t$.

Proof.

The fact that $0 \leq \chi \leq C|Y|^2$ is clear from the definition. Noting that the admissible controls can be restricted to satisfy

$$(5.25)\qquad \int_0^\infty |\xi|^2 d\tau \leq C|Y|^2$$

one derives as for (3.34) the estimate on $D_Y\chi$.

Moreover let $\theta' > \theta$ and for any $\varepsilon$, pick $\hat{\xi}$ such that

$$\int_0^\theta F(\hat{Y},\hat{\xi})d\tau - \varepsilon \leq \inf_{\xi(.)} \int_0^\theta F(Y,\xi)d\tau$$

in which $\hat{Y}$ denotes the trajectory corresponding to $\hat{\xi}$, and suppose that (5.25) holds. We deduce

$$0 \geq \chi(\theta) - \chi(\theta') \geq \int_0^\theta F(\hat{Y},\hat{\xi})d\tau - \varepsilon - \int_0^\theta F(\hat{Y},\hat{\xi})d\tau - \int_\theta^{\theta'} F(\tilde{Y},0)d\tau$$

in which $\tilde{Y}$ denotes the extension of $\hat{Y}$ on $\theta$, $\theta'$ with a control equal to 0. From (5.25) it follows

$$|\hat{Y}(\tau)|^2 \leq C|Y|^2 \quad , \quad |\tilde{Y}(\tau)| \leq C|Y|^2$$

hence

$$0 \geq \chi(\theta) - \chi(\theta') \geq - C|Y|^2(\theta' - \theta) - \varepsilon$$

and since is arbitrary, the estimate on $\frac{\partial \chi}{\partial \theta}$ is established.

We then establish a functional equation using the optimality principle. We have for $\theta > \delta$

$$\chi(x,t;Y,\theta) = \inf_{\xi(.)} [ \int_0^\delta F(x,t;Y(\tau),\xi(\tau))d\tau + \chi(x,t;Y(\delta),\theta - \delta)].$$

from which one easily deduces (5.22) at any point of differentiability of $\chi$. The fact that $\chi$ is the maximum solution is a standard one in Dynamic Programming

□

It will be useful to derive a result on the dependance of $\chi$ with respect to the parameters x,t.

We make the additional assumptions

(5.26) the $3^{rd}$ derivatives of $\ell$ in y,v are bounded by $\frac{C}{1 + |x|}$

the $3^{rd}$ derivatives of f.g in y,v are bounded by $\frac{C}{1 + |x|^2}$

We then have

Lemma 5.1. The following estimates hold

$$(5.27) \qquad |\frac{\partial\chi}{\partial t}| \ , \ |\frac{\partial\chi}{\partial x}| \le C|Y|^2$$

Proof.

From the relations (5.20) and differentiating in t,x, we have

$$F_t(x,t;0,0) = 0 \quad , \quad F_x(x,t;0,0) = 0$$

$$F_{tY}(x,t;0,0) = 0 \quad , \quad F_{t\xi}(x,t;0,0) = 0$$

$$F_{xY}(x,t;0,0) = 0 \quad , \quad F_{x\xi}(x,t;0,0) = 0.$$

Therefore

$$F_t(x,t;Y,\xi) = \int_0^1 \int_0^1 \lambda \ [F_{tYY}(x,t;\lambda\mu Y;\lambda\mu\xi)Y^2 + 2F_{tY\xi}\xi Y + F_{t\xi\xi}\xi^2]d\lambda d\mu$$

and evaluating the derivatives $F_{tYY}$ ..., using (5.26) we deduce

$$|F_t(x,t;Y,\xi)| \le C(|Y|^2 + |\xi|^2)$$

and similarly

$$|F_x(x,t;Y,\xi)| \le C(|Y|^2 + |\xi|^2).$$

Similarly one can show that

$$|G_t(x,t;Y,\xi)| \le C(|Y| + |\xi|)$$

$$|G_x(x,t;Y,\xi)| \le C(|Y| + |\xi|).$$

Recalling the property (5.25) and using the preceding estimates, we deduce from the definition (5.23) of $\chi$ the results desired

□

Let us then define

$$(5.28) \qquad \Phi_1(x,y,t;\theta) = \bar{q}(x,t)(y - \bar{y}(x,t)) + \chi(x,t;y - \bar{y}(x,t),\theta).$$

We have

$$\frac{\partial \Phi_1}{\partial \theta} = \frac{\partial \chi}{\partial \theta}, \qquad D_y\Phi_1 = \bar{q}(x,t) + D_Y\chi$$

then clearly $\Phi_1$ satisfies (5.16).

Moreover

$$\Phi_1(x,y,t;0) = \bar{q}(x,t)(y - \bar{y}(x,t))$$

$$\frac{\partial \Phi_1}{\partial t} = \bar{q}_t(y - \bar{y}(x,t)) - \bar{q}\,\bar{y}_t + \frac{\partial \chi}{\partial t} - \frac{\partial \chi}{\partial Y}\,\bar{y}_t$$

$$\frac{\partial \Phi_1}{\partial x} = \bar{q}_x(y - \bar{y}(x,t)) - \bar{q}\,\bar{y}_x + \frac{\partial \chi}{\partial x} - \frac{\partial \chi}{\partial Y}\,\bar{y}_x$$

hence

$$(5.29) \qquad |\Phi_1(x,y,t;0)| \le C[(1 + |x|)|y| + 1 + |x|^2]$$

$$\left|\frac{\partial \Phi_1}{\partial t}(x,y,t;\theta)\right| \le C(1 + |x|^2 + |y|^2)$$

$$\left|\frac{\partial \Phi_1}{\partial x}(x,y,t;\theta)\right| \le C(1 + |x|^2 + |y|^2).$$

Let us now prove the convergence of $\Phi^\varepsilon$ to $\Phi$. We have

Theorem 5.1. We assume (1.1), (1.2), (5.4), (5.5), (5.26). Then one has

$$(5.30) \qquad |\Phi^\varepsilon(x,y,t) - \Phi(x,t)| \le C\sqrt{\varepsilon}\,[1 + |x|^3 + |y|\,|x|] + C\varepsilon|y|^2 + C\varepsilon^2|y|^3$$

Proof.

Pick any control satisfying (cf (5.6))

$$(5.31) \qquad \int_t^T |v(s)|^2 ds \le C_0(1 + |x_0|^2 + \varepsilon|y_0|^2)$$

where $x_o$, $y_o$ are fixed. Call $x^\varepsilon$, $y^\varepsilon$ the trajectories (5.1) with the initial conditions $x_o$, $y_o$ at time t. It is easy to check that

$$(5.31)\qquad |x^\varepsilon(s)|^2\ ,\ \varepsilon|y^2(s)|^2,\ \int_t^T |y^\varepsilon(s)|^2 ds \le C(1+|x_o|^2+\varepsilon|y_o|^2).$$

From (5.16) we can then assert that

$$\frac{\partial}{\partial s}(\Phi+\varepsilon\Phi_1) + D_x(\Phi+\varepsilon\Phi_1)f(x,y,v(s)) + \frac{1}{\varepsilon}D_y(\Phi+\varepsilon\Phi_1)g(x,y,v(s)) +$$

$$+\ \ell(x,y,v(s)) \ge \frac{\partial\Phi_1}{\partial s} + \varepsilon\, D_x\Phi_1.f(x,y,v(s))$$

$$(\Phi+\varepsilon\Phi_1)(x,y,T) = h(x) + \varepsilon\bar{q}(x,T)(y-\bar{y}(x,T))$$

which implies

$$(5.32)\qquad \Phi(x_o,t) + \varepsilon\Phi_1(x_o,y_o,t;\tfrac{T-t}{\varepsilon}) \le J^\varepsilon_{x_o,y_o,t}(v(.)) + \varepsilon\bar{q}(x^\varepsilon(T),T)(y^\varepsilon(T)-\bar{y}(x^\varepsilon(T)T))-$$

$$-\ \varepsilon\int_t^T \left(\frac{\partial\Phi_1}{\partial s} + D_x\Phi_1.f\right)(x^\varepsilon(s),y^\varepsilon(s),v(s))ds.$$

But from the estimates (5.29) we deduce

$$\left|\int_t^T \left(\frac{\partial\Phi_1}{\partial s} + D_x\Phi_1.f\right)(x^\varepsilon(s),y^\varepsilon(s),v(s)) \le \int_0^T C[1+|x^\varepsilon(s)|^3+|y^\varepsilon(s)|^3\right.$$

$$+\ |v(s)|(1+|x^\varepsilon(s)|^2+|y^\varepsilon(s)|^2)]ds$$

and from the estimates (5.31)

$$\le \frac{C}{\sqrt{\varepsilon}}(1+|x_o|^2+\varepsilon|y_o|^2)^{3/2}.$$

Similarly

$$|\bar{q}(x^\varepsilon(T),T)(y^\varepsilon(T)-\bar{y}(x^\varepsilon(T),T))| \le \frac{C}{\sqrt{\varepsilon}}(1+|x_o|^2+\varepsilon|y_o|^2)$$

$$\Phi_1(x_o,y_o,t;\tfrac{T-t}{\varepsilon}) \ge -\,C(1+|x_o|^2+|y_o||x_o|).$$

There estimates in (5.32) yield

$$(5.33) \qquad \Phi(x_0,t) \le \Phi^\varepsilon(x_0,y_0,t) + C[\sqrt{\varepsilon}(1+|x_0|^3) + \varepsilon|x_0||y_0| + \varepsilon^2|y_0|^3]$$

To get a reverse inequality consider the system (5.1) using as control the feedback $\bar{u}(x,t)$, which is optimal for the limit problem. The corresponding trajectory is defined by the relations

$$\frac{dx^\varepsilon}{ds} = f(x^\varepsilon,y^\varepsilon,\bar{u}(x^\varepsilon,s)) \qquad x^\varepsilon(t) = x_0$$

$$\varepsilon\frac{dy^\varepsilon}{ds} = g(x^\varepsilon,y^\varepsilon,\bar{u}(x^\varepsilon,s)) \qquad y^\varepsilon(t) = y_0 \quad .$$

Let us consider the limit problem

$$\frac{dx_0}{ds} = f(x_0,y_0,\bar{u}(x_0,s)) \qquad x_0(0) = x_0$$

$$0 = g(x_0,y_0,\bar{u}(x_0,s))$$

These relations yield as easily seen

$$|x^\varepsilon(s) - x_0(s)|^2 + \varepsilon\,|y^\varepsilon(s) - y_0(s)|^2 + \mu\int_t^s |y^\varepsilon(\lambda) - y_0(\lambda)|^2 d\lambda$$

$$\le C\,\varepsilon|y_0 - y_0(0)|^2 + C\int_t^s |x^\varepsilon(\lambda) - x_0(\lambda)|^2 d\lambda$$

$$+ C\,\varepsilon^2\int_t^T \left|\frac{dy_0}{ds}\right|^2 ds$$

hence

$$|x^\varepsilon(s) - x_0(s)|^2 \le C\,\varepsilon(|y_0|^2 + |x_0|^2 + 1)$$

$$|y^\varepsilon(s) - y_0(s)| \le C\,(|y_0|^2 + |x_0|^2 + 1)$$

$$\int_t^T |y^\varepsilon(s) - y_0(s)|^2 ds \le C\,\varepsilon(|y_0|^2 + |x_0|^2 + 1)$$

which implies

$$J^\varepsilon_{x_0,y_0,t}(\bar{u}) \le C\sqrt{\varepsilon}\,[1 + |x_0|^2 + |y_0||x_0| + \sqrt{\varepsilon}\,|y_0|^2] + \Phi(x_0,t)$$

hence

$$\Phi^{\varepsilon}(x_0,y_0,t) - \Phi(x_0,t) \leq C\sqrt{\varepsilon}[1 + |x_0|^2 + |y_0||x_0| + \sqrt{\varepsilon}|y_0|^2].$$

Combining this estimate with (5.33) yields the result (5.30)

□

5.3. Composite feedback.

Consider the function $\bar{X}$ which is the limit of X as $\theta \to \infty$ in (5.22). It is the solution of

$$\inf_{\xi} [F(x,t;Y,\xi) + D_Y\bar{X}.G(x,t;Y,\xi)] = 0. \tag{5.34}$$

The infimum is attained in a point $\bar{\xi}(x,t;Y)$ and it is easy to check that

$$|\bar{\xi}(x,t;Y)| \leq C|Y|. \tag{5.35}$$

We consider a situation where $\bar{X}$ is $C^2$ and $\bar{\xi}$ is $C^1$ in Y, with

$$0 \leq D_Y^2\bar{X} \leq C \tag{5.36}$$

$$|\bar{\xi}_Y| \leq C. \tag{5.37}$$

It is possible to show that these properties hold when we reinforce (5.5) by

(5.38) the 2nd derivatives of g are bounded by $\dfrac{C_0}{1 + |x| + |y|}$ with $C_0$ not too large.

Following a concept introduced by J. CHOW and P.V. KOKOTOVIC [1], we shall term composite feedback, the function

$$u_c(x,y,t) = \bar{u}(x,t) + \bar{\xi}(x,t;y - \bar{y}(x,t)). \tag{5.39}$$

Considering the function (cf. also (5.28))

$$\bar{\Phi}_1(x,y,t) = \bar{q}(x,t)(y - \bar{y}(x,t)) + \bar{X}(x,t;y - \bar{y}(x,t)) \tag{5.40}$$

it satisfies the equation

(5.41) $$\frac{\partial\Phi}{\partial t} + \underset{v}{\mathrm{Inf}}\, [D\Phi . f(x,y,v) + D_y\bar{\Phi}_1 . g(x,y,v) + \ell(x,y,v)] = 0$$

and the infimum is attained for the composite feedback.

Let us first consider the behaviour of the dynamic system (5.1) when one applies the composite feedback. One can expect the composite feedback to be an improvement with respect to the limit feedback.

Surprisingly the convergence of the corresponding trajectory to the limit trajectory cannot be proven as easily as in the proof of Theorem 5.1. The natural stability of g is not sufficient, since the fast state enters in the control. It is necessary to prove a priori that the fast state remains bounded. This is done via a Lyamunov type argument.

Let us denot by $x_c^\varepsilon$, $y_c^\varepsilon$ the trajectory corresponding to the composite control

Proposition 5.2. We have

(5.42) $$|x_c^\varepsilon(s)| \le C(1 + |x| + |y|)$$

(5.43) $$|y_c^\varepsilon(s)| \le C(1 + |x| + |y|)$$

Proof.

By definition $x_c^\varepsilon$, $y_c^\varepsilon$ are solutions of (we drop the index c for brevity)

(5.44) $$\frac{dx^\varepsilon}{ds} = f(x^\varepsilon, y^\varepsilon, u_c(x^\varepsilon, y^\varepsilon, s))$$

$$\varepsilon\frac{dy^\varepsilon}{ds} = g(x^\varepsilon, y^\varepsilon, u_c(x^\varepsilon, y^\varepsilon, s))$$

$$x^\varepsilon(t) = x, \quad y^\varepsilon(t) = y.$$

We then deduce

$$\frac{d}{ds}\, \bar{X}(x^\varepsilon(s), s, y^\varepsilon(s) - \bar{y}(x^\varepsilon(s), s)) = \frac{\partial\bar{X}}{\partial x} f(x^\varepsilon, y^\varepsilon, u^c) + \frac{\partial\bar{X}}{\partial s}$$

$$- D_Y\bar{X}\,\frac{d}{ds}\,\bar{y}(x^\varepsilon(s), s) - \frac{1}{\varepsilon}\, F(x^\varepsilon(s), s; y^\varepsilon - \bar{y}(x^\varepsilon, s), \bar{\xi}(x^\varepsilon, s; y^\varepsilon - \bar{y}(x^\varepsilon)))$$

Let us set

$$\xi^\varepsilon = \bar{\xi}(x^\varepsilon(s),s;y^\varepsilon - \bar{y}(x^\varepsilon)).$$

We can write

$$|f(x^\varepsilon,y^\varepsilon,u^c)| \le C[1 + |y^\varepsilon - \bar{y}(x^\varepsilon)| + |\xi^\varepsilon| + |x^\varepsilon - x_0|]$$

$$|\frac{d}{ds}\bar{y}(x^\varepsilon(s),s)| \le C[1 + |y^\varepsilon - \bar{y}(x^\varepsilon)| + |\xi^\varepsilon| + |x^\varepsilon - x_0|]$$

hence using the estimates on $\bar{X}$ as well as on F

$$\begin{aligned}(5.45)\qquad &\frac{d\bar{X}}{ds}(x^\varepsilon(s),s,y^\varepsilon(s) - \bar{y}(x^\varepsilon(s),s)) \le C|y^\varepsilon(s) - \bar{y}(x^\varepsilon)|^2[1 + |y^\varepsilon - \bar{y}(x^\varepsilon)| + \\ &+ |\xi^\varepsilon| + |x^\varepsilon - x_0|] + C|y^\varepsilon(s) - \bar{y}(x^\varepsilon)|[1 + |y^\varepsilon - \bar{y}(x^\varepsilon)| + \\ &+ |\xi^\varepsilon| + |x^\varepsilon - x_0|] - \frac{1}{\varepsilon}\beta(|y^\varepsilon(s) - \bar{y}(x^\varepsilon)|^2 + |\xi^\varepsilon|^2) \\ &\le C[1 + |y^\varepsilon(s) - \bar{y}(x^\varepsilon)|^4 + |y^\varepsilon(s) - \bar{y}(x^\varepsilon)|^2 + |\xi^\varepsilon|^2 + |x^\varepsilon(s) - x_0(s)|^2] - \\ &- \frac{\beta}{\varepsilon}(|y^\varepsilon(s) - \bar{y}(x^\varepsilon)|^2 + |\xi^\varepsilon|^2).\end{aligned}$$

On the other hand

$$\begin{aligned}(5.46)\qquad &\frac{d}{ds}e^{k(T-s)}[1 + |x^\varepsilon(s) - x_0(s)|^2] = -ke^{k(T-s)}[1 + |x^\varepsilon(s) - x_0(s)|^2] + \\ &+ 2e^{k(T-s)}c_0[|x^\varepsilon(s) - x_0(s)|^2 + |\xi^\varepsilon|^2 + |y^\varepsilon(s) - \bar{y}(x^\varepsilon)|^2].\end{aligned}$$

We first pick k such that

$$C \le k\, e^{k(T-t)}\quad,\quad C \le \frac{k}{2}e^{k(T-t)}\quad,\quad 2c_0 \le \frac{k}{2}$$

and $\varepsilon \le \varepsilon_1$ with

$$C + 2e^{kT}c_0 \le \frac{\beta}{2\varepsilon_1}$$

We deduce from (5.45), (5.46) that

$$(5.47)\qquad \frac{d}{ds}\bar{X}(x^{\varepsilon}(s),s,y^{\varepsilon}(s)-\bar{y}(x^{\varepsilon}(s),s))+\frac{d}{ds}e^{k(T-s)}[1+|x^{\varepsilon}(s)-x_0(s)|^2]$$

$$\le C|y^{\varepsilon}(s)-\bar{y}(x^{\varepsilon}(s),s)|^4-\frac{\beta}{2\varepsilon}|y^{\varepsilon}(s)-\bar{y}(x^{\varepsilon}(s),s)|^2.$$

Note that

$$a|Y|^2\le\bar{X}(Y)\le M|Y|^2$$

and let $\rho$ be chosen such that

$$(5.48)\qquad e^{k(T-t)}+M|y_0-\bar{y}(x_0,t)|^2\le a\rho^2$$

and pick $\varepsilon\le\varepsilon_2$ where

$$(5.49)\qquad C\rho^2<\frac{\beta}{2\varepsilon_2}.$$

Since from (5.48)

$$|y_0-\bar{y}(x_0,t)|<\rho$$

we can assert that for s close to t, say $s\in[t,t_1)$

$$|y^{\varepsilon}(s)-\bar{y}(x^{\varepsilon}(s),s)|<\rho$$

which implies, because of (5.49)

$$(5.50)\qquad \frac{d}{ds}[\bar{X}(x^{\varepsilon}(s),s,y^{\varepsilon}(s)-\bar{y}(x^{\varepsilon}(s),s))+e^{k(T-s)}(1+|x^{\varepsilon}(s)-x_0(s)|^2)]\le 0$$

and thus

$$\bar{X}(x^{\varepsilon}(s),s,y^{\varepsilon}(s)-\bar{y}(x^{\varepsilon}(s),s))+e^{k(T-s)}(1+|x^{\varepsilon}(s)-x_0(s)|^2)$$

$$\le\bar{X}(x_0,t,y_0-\bar{y}(x_0,t))+e^{k(T-t)},\quad\text{for } s\in[t,t_1).$$

But then also by continuity

$$\bar{\chi}(x^{\varepsilon}(t_1),t_1,y^{\varepsilon}(t_1) - \bar{y}(x^{\varepsilon}(t_1),t_1)) + e^{k(T-t_1)}(1 + |x^{\varepsilon}(t_1) - x_o(t_1)|^2)$$

$$\leq \bar{\chi}(x_o,t,y_o - \bar{y}(x_o,t)) + e^{k(T-t)}$$

which implies in particular

$$|y^{\varepsilon}(t_1) - \bar{y}(x^{\varepsilon}(t_1),t_1)| < \rho.$$

Therefore we can proceed beyond $t_1$, and prove finally that (5.50) holds on $[t,T)$. This implies the results desired

□

We can deduce from Proposition 5.2 a slight improvement of Theorem 5.1, as follows

Propostion 5.3. Under the assumptions of Theorem 5.1, one has

$$\phi^{\varepsilon}(x,y,t) \leq \phi(x,t) + C\varepsilon(1 + |x|^3 + |y|^3) \tag{5.51}$$

Proof

From (5.41) we deduce

$$\frac{\partial}{\partial s}(\phi + \varepsilon\bar{\Phi}_1) + D_x(\Phi + \varepsilon\bar{\Phi}_1)f(x,y,u_c) + \frac{1}{\varepsilon} D_y(\phi + \varepsilon\bar{\Phi}_1).g(x,y,u_c) +$$

$$+ \ell(x,y,u_c) = \varepsilon \frac{\partial\bar{\Phi}_1}{\partial s} + \varepsilon D_x\bar{\Phi}_1.f(x,y,u_c)$$

$$(\phi + \varepsilon\bar{\Phi}_1)(x,y,T) = h(x) + \varepsilon\bar{\Phi}_1(x,y,T).$$

Considering

$$\psi^{\varepsilon} = \phi^{\varepsilon} - \phi - \varepsilon\bar{\Phi}_1$$

we obtain the inequality

$$\frac{\partial\psi^{\varepsilon}}{\partial s} + D_x\psi^{\varepsilon}.f(x,y,u_c) + \frac{1}{\varepsilon} D_y\psi^{\varepsilon}.g(x,y,u_c) \geq -\varepsilon(\frac{\partial\bar{\Phi}_1}{\partial s} + D_x\bar{\Phi}_1.f(x,y,u_c)) \tag{5.52}$$

$$\psi^{\varepsilon}(x,y,T) = -\varepsilon\bar{\Phi}_1(x,y,T) \leq - \varepsilon\bar{q}(x,T)(y - \bar{y}(x,T))$$

The estimates (5.29) being valid for $\bar{\Phi}_1$ as well, we deduce from (5.52) that

$$\psi^\varepsilon(x,y,t) \le C\,\varepsilon \int_t^T [1 + |x^\varepsilon(t)|^2 + |y^\varepsilon(t)|^2](1 + |x^\varepsilon(t)| + |y^\varepsilon(t)|]dt +$$

$$+ C\varepsilon(1 + |x^\varepsilon(T)|^2 + |x^\varepsilon(T)||\,y^\varepsilon(T)|)$$

and from Proposition 5.2, it follows that

$$\psi^\varepsilon(x,y,t) \le C\varepsilon(1 + |x_0|^3 + |y_0|^3)$$

which proves the result desired

□

Remark 5.2. It does not seem evident to obtain a reverse inequality in (5.51), unless we can reduce the set of admissible controls. In particular, if we can reduce it to those for which exist a priori uniform bounds of the type (5.42), (5.43), then such a reverse inequality follows.

Remark 5.3. It would be interesting to develop in the spirit of Dynamic Programming, an expansion of the optimal cost $\phi^\varepsilon$, similar to that given in Theorem 4.1.

□

## 6. STABILITY AND INFINITE HORIZON PROBLEMS.

Synopsis.

We consider in this section optimal control problems on an infinite horizon. We can naturally adress the same questions as in the preceding sections, in particular expansions and boundary layer terms. We shall however restrict ourselves to the Dynamic Programming point of view, and consider more particularly the composite control.

### 6.1. Setting of the problem . Assumptions

Consider the system

$$(6.1) \qquad \frac{dx^\varepsilon}{dt} = f(x^\varepsilon,y^\varepsilon,v) \qquad\qquad x^\varepsilon(0) = x$$

$$\varepsilon\frac{dy^\varepsilon}{dt} = g(x^\varepsilon,y^\varepsilon,v) \qquad\qquad y^\varepsilon(0) = y$$

and the cost function

(6.2) $$J^{\varepsilon}_{x,y}(v(.)) = \int_0^{\infty} \ell(x^{\varepsilon},y^{\varepsilon},v)dt.$$

We are interested in the function

(6.3) $$\phi^{\varepsilon}(x,y) = \operatorname*{Inf}_{v(.)} J^{\varepsilon}_{x,y}(v(.))$$

which will appear as the maximum solution of

(6.4) $$\operatorname*{Inf}_{v} [D_x\phi^{\varepsilon}.f(x,y,v) + \frac{1}{\varepsilon} D_y\phi^{\varepsilon}.g(x,y,v) + \ell(x,y,v)] = 0.$$

Let us make the following assumptions

(6.5) f, g,$\ell$ are $C^2$ ; the derivatives of f,g of all orders are bounded ; the second derivatives of $\ell$ are bounded

(6.6) $$\ell(o,o,o) = 0 \qquad \ell_{\sigma}(o,o,o) = 0 \qquad (\sigma = (x,y,v)).$$

$$f(o,o,o) = g(o,o,o) = 0$$

(6.7) $$\ell_{\sigma\sigma} \geq \gamma I, \qquad \gamma > 0$$

(6.8) $$\begin{pmatrix} f_x & f_y \\ g_x & g_y \end{pmatrix} \leq -\mu I$$

(6.9) the $2^{nd}$ derivatives of f,g are bounded by $\frac{C_o}{1+|x|}$, where $C_o$ is not too large compared to $\gamma$.

For the control problem (6.2) the admissible controls satisfy

(6.10) $$\int_0^{\infty} |v(t)|^2 dt \leq C(|x|^2 + \varepsilon|y|^2)$$

and $\phi^{\varepsilon}$ defined by (6.3) is the maximum solution of (6.4) among the class of Lipschitz functions satisfying the growth condition given at the right hand side of (6.10).

Consider then the limit problem

(6.11) $\frac{dx}{dt} = f(x,y,v)$ $x(0) = x$

$g(x,y,v) = 0$

$J_x(v(.)) = \int_0^\infty \ell(x(t),y(t),v(t))dt$

and

(6.12) $\phi(x) = \operatorname{Inf}_{v(.)} J_x(v(.))$.

Denote as usual by $\bar{y}(x,v)$ the solution (in y) of the algebraic equation $g = 0$, and let us notice that

(6.13) $f_x + f_y \bar{y}_x \leq - \mu I$

therefore the limit dynamic system has a unique solution for $v \in L^2(0,\infty;R^k)$.

The function $\phi$ defined by (6.12) is smooth in the following sense

(6.14) $0 \leq \phi(x) \leq m|x|^2$ , $|D\phi(x)| \leq C|x|$, $|D^2\phi(x)| \leq C$.

In fact it will be useful to notice that

(6.15) $\phi(x) \geq \beta_0 |x|^2$ , $\beta_0 > 0$

and is the unique solution satisfying (6.14) of

(6.16) $\operatorname{Inf}_v [D\phi.f(x,\bar{y}(x,v),v) + \ell(x,\bar{y}(x,v),v)] = 0$.

Define $\bar{u}(x)$, $\bar{y}(x)$ and $\bar{q}(x)$ in a way similar to (5.13), (5.14) and note that

(6.17) $\bar{y}_x$, $\bar{u}_x$, $\bar{q}_x$ bounded

$\bar{y}(0) = \bar{u}(0) = \bar{q}(0) = 0$

6.2. Expansion

We consider the problem

(6.18) $$\operatorname*{Inf}_{v}[D\phi.f(x,y,v) + D_y\phi_1.g(x,y,v) + \ell(x,y,v)] = 0$$

which is solved as follows. Let us set

(6.19) $$G(x;Y,\xi) = g(x,\bar{y}(x) + Y,\bar{u}(x) + \xi)$$

(6.20) $$\begin{aligned} F(x;Y,\xi) &= D\phi.f(x,\bar{y}(x) + Y,\bar{u}(x) + \xi) + \\ &+ \bar{q}(x).g(x,\bar{y}(x) + Y,\bar{u}(x) + \xi) + \\ &+ \ell(x,\bar{y}(x) + Y,\bar{u}(x) + \xi) = \\ &= H(x,\bar{y}(x) + Y,\bar{u}(x) + \xi,D\phi,\bar{q}(x)) - \\ &- H(x,\bar{y}(x),\bar{u}(x),\bar{q}(x)). \end{aligned}$$

We solve the problem

(6.21) $$\operatorname*{Inf}_{\xi}[F(x;Y,\xi) + D_Y\chi.G(x,Y,\xi)] = 0$$

related to the control problem

(6.22) $$\frac{dY}{d\tau} = G(x;Y,\xi) \quad , \; Y(0) = Y$$

$$\chi(x;Y) = \inf_{\xi(.)} \int_0^\infty F(x;Y(\tau),\xi(\tau))d\tau = \inf_{\xi(.)} \mathcal{Y}(x;Y;\xi(.)).$$

We can prove as in Proposition 5.1 that

(6.23) $$a|Y|^2 \le \chi \le B|Y|^2$$

$$|D_Y\chi| \le C|Y|$$

with constants independant of x. Furthermore assuming (5.26) we also have, as in Lemma 5.1

(6.24) $\quad |\frac{\partial X}{\partial x}| \le C|Y|^2.$

The solution of (6.18) is defined by

(6.25) $\quad \phi_1(x,y) = \bar{q}(x).(y - \bar{y}(x)) + \chi(x;y - \bar{y}(x)).$

If we reinforce (6.9) by

(6.26) the $2^{nd}$ derivative of g are bounded by $\dfrac{C_0}{1 + |x| + |y|}$ with $C_0$ not too large

then we have

(6.27) $\quad 0 \le D^2_Y X \le C$

and for the optimal feedback in (6.21), $\bar{\xi}(Y)$

(6.28) $\quad |\bar{\xi}_Y| \le C \quad , \quad \bar{\xi}(0) = 0.$

The composite feedback is given by

(6.29) $\quad u_c(x,y) = \bar{u}(x) + \bar{\xi}(x;y - \bar{y}(x)).$

## 6.3. Convergence.

We want to study the convergence of $\phi^\varepsilon$ to $\phi$. Unfortunately, the method of Theorem 5.1 does not carry over to the present case (and this is the interesting novelty of the infinite horizon problem). This is due to the fact that the limit feedback $\bar{u}(x)$ used in the $\varepsilon$ system, is not sufficient to guarantee bounds on the state of the system. A stability argument has to be given, and the composite feedback will be necessary for that.

We begin by the analogue of Proposition 5.2.

Consider the system

(6.30)
$$\frac{dx^\varepsilon}{dt} = f(x^\varepsilon,y^\varepsilon,u_c(x^\varepsilon,y^\varepsilon)) \qquad x^\varepsilon(0) = x$$
$$\varepsilon\frac{dy^\varepsilon}{dt} = g(x^\varepsilon,y^\varepsilon,u_c(x^\varepsilon,y^\varepsilon)) \qquad y^\varepsilon(0) = y$$

then one has

Proposition 6.1. We have the estimates

$$(6.31) \qquad |x^\varepsilon(s)| \ , \ |y^\varepsilon(s)| \le C(|x| + |y|)$$

$$\int_0^\infty |x^\varepsilon(t)|^2 dt \ , \ \int_0^\infty |y^\varepsilon(t)|^2 dt \le C(|x|^2 + |y|^2)$$

Proof.

We set

$$\xi^\varepsilon = \bar{\xi}(x^\varepsilon(s), y^\varepsilon(s) - \bar{y}(x^\varepsilon(s)))$$

and compute

$$\begin{aligned}(6.32) \qquad \frac{d}{ds} \chi(x^\varepsilon(s), y^\varepsilon(s) - \bar{y}(x^\varepsilon(s))) &= \left(\frac{\partial \chi}{\partial x} - D_y\chi\, \bar{y}_x\right) f(x^\varepsilon, y^\varepsilon, \bar{u}(x^\varepsilon) + \xi^\varepsilon) - \\ &- \frac{1}{\varepsilon} F(x^\varepsilon(s), y^\varepsilon(s) - \bar{y}(x^\varepsilon(s))) \\ &\le \left(\frac{\partial \chi}{\partial x} - D_y\chi\bar{y}_x\right) f(x^\varepsilon, \bar{y}(x^\varepsilon), \bar{u}(x^\varepsilon)) + \\ &+ C\left|\frac{\partial \chi}{\partial x} - D_y\chi\bar{y}_x\right| \left(|y^\varepsilon - \bar{y}(x^\varepsilon)| + |\xi^\varepsilon|\right) - \\ &- \frac{\beta}{\varepsilon} \left(|y^\varepsilon - \bar{y}(x^\varepsilon)|^2 + |\xi^\varepsilon|^2\right) \\ &\le C_0\delta(|x^\varepsilon|^2 + |\bar{y}(x^\varepsilon)|^2 + |\bar{u}(x^\varepsilon)|^2) + \\ &+ \frac{C_0'}{\delta} \left(|y^\varepsilon - \bar{y}(x^\varepsilon)|^4 + |y^\varepsilon - \bar{y}(x^\varepsilon)|^2\right) + \\ &+ C_0''|\xi^\varepsilon|^2 - \frac{\beta_2}{\varepsilon} \left(|y^\varepsilon - \bar{y}(x^\varepsilon)|^2 + |\xi^\varepsilon|^2\right).\end{aligned}$$

On the other hand

$$\begin{aligned}(6.33) \qquad \frac{d}{ds} \phi(x^\varepsilon(s)) &= D\phi(x^\varepsilon).f(x^\varepsilon(s), y^\varepsilon(s), \bar{u}(x^\varepsilon) + \xi^\varepsilon) \\ &= D\phi(x^\varepsilon).f(x^\varepsilon, \bar{y}(x^\varepsilon), \bar{u}(x^\varepsilon)) +\end{aligned}$$

$$D\phi(x^\varepsilon).(f(x^\varepsilon,y^\varepsilon,\bar{u}(x^\varepsilon)+\xi^\varepsilon)-f(x^\varepsilon,\bar{y}(x^\varepsilon),\bar{u}(x^\varepsilon)))$$

$$= -\ell(x^\varepsilon,\bar{y}(x^\varepsilon),\bar{u}(x^\varepsilon)) + D\phi(x^\varepsilon).(f(x^\varepsilon,y^\varepsilon,\bar{u}(x^\varepsilon)+\xi^\varepsilon) -$$

$$- f(x^\varepsilon,\bar{y}(x^\varepsilon),\bar{u}(x^\varepsilon)))$$

$$\leq -\gamma_1(|x^\varepsilon|^2 + |\bar{y}(x^\varepsilon)|^2 + |\bar{u}(x^\varepsilon)|^2) +$$

$$+ C_1(|y^\varepsilon - \bar{y}(x^\varepsilon)|^2 + |\xi^\varepsilon|^2).$$

We first choose $\delta$ such that

$$C_o\delta \leq \frac{\gamma 1}{2}$$

and $\varepsilon \leq \varepsilon_1$, with

$$\frac{C'_o}{\delta} + C_1 \leq \frac{\beta_2}{2\varepsilon_1} \quad , \quad C''_o + C_1 \leq \frac{\beta_2}{\varepsilon_1}.$$

It follows that

$$(6.34)\qquad \frac{d}{ds}\,[\chi(x^\varepsilon(s),y^\varepsilon(s)-\bar{y}(x^\varepsilon(s))) + \phi(x^\varepsilon(s))] \leq \frac{C'_o}{\delta}\,|y^\varepsilon - \bar{y}(x^\varepsilon)|^4 - $$
$$- \frac{\beta_2}{2\varepsilon}|y^\varepsilon - \bar{y}(x^\varepsilon)|^2 - \frac{\gamma_1}{2}|x^\varepsilon|^2$$

Let then (recalling (6.14), (6.15), (6.23)) $\rho$ be such that

$$m|x|^2 + b|y - \bar{y}(x)|^2 < \mathrm{Min}(a,\beta_o)\rho^2$$

and pick $\varepsilon \leq \varepsilon_2$, with

$$\frac{C'_o}{\delta}\,\rho^2 \leq \frac{\beta_2}{4\varepsilon_2}.$$

Then for $s \in [o,s_1[$ , $|y^\varepsilon(s) - \bar{y}(x^\varepsilon(s))| \leq \rho$

$$(6.35)\qquad \frac{d}{ds}\,[\chi(x^\varepsilon(s),y^\varepsilon(s)-\bar{y}(x^\varepsilon(s))) + \phi(x^\varepsilon(s))] \leq -\frac{\beta_2}{4\varepsilon_2}|y^\varepsilon(s) - \bar{y}(x^\varepsilon(s))|^2 -$$
$$- \frac{\gamma_1}{2}|x^\varepsilon(s)|^2.$$

From the estimates on $\chi$, $\phi$, and arguing as in the proof of Proposition 5.2, we deduce that $|y^\varepsilon(s_1) - \bar{y}(x^\varepsilon(s_1))| \le \rho$ and step by step, it follows that (6.35) holds for any s. From this the results desired obtain

□

We can finally give the analogue of Theorem 5.1, namely

Theorem 6.1. We assume (6.5); (6.6), (6.7), (6.8), (6.9), (5.26), (6.26).

Then one has

$$|\phi^\varepsilon(x,y) - \phi(x)| \le C\sqrt{\varepsilon}(|x|^2 + |y|^2 + |x|^3 + |y|^3) \tag{6.36}$$

Proof.

We have the relation

$$D_x(\phi + \varepsilon\phi_1).f(x,y,u_c) + \frac{1}{\varepsilon} D_y(\phi + \varepsilon\phi_1).g(x,y,u_c) + \ell(x,y,u_c) = \tag{6.37}$$

$$= \varepsilon D_x\phi_1.f(x,y,u_c)$$

and

$$|D_x\phi_1.f(x,y,u_c)| \le C[|x|^2 + |y - \bar{y}(x)|^2 + |y - \bar{y}(x)|^3].$$

Compared to (6.4), as in Proposition 5.3, yields

$$\phi^\varepsilon(x,y) - \phi(x) - \varepsilon\phi_1(x,y) \le C\varepsilon\, E \int_0^\infty [|x^\varepsilon(s)|^2 + |y^\varepsilon(s) - \bar{y}(x^\varepsilon(s))|^2 +$$

$$+ |y^\varepsilon(s) - \bar{y}(x^\varepsilon(s))|^3]ds$$

where $x^\varepsilon, y^\varepsilon$ refer to the system when controlled by the composite feedback, i.e. (6.30). From the estimates (6.31) follows

$$\phi^\varepsilon(x,y) \le \phi(x) + C\varepsilon\,(|x|^2 + |y|^2 + |x|^3 + |y|^3). \tag{6.38}$$

On the other hand, for an admissible control of the class (6.10), which suffices, the corresponding states (also noted $x^\varepsilon, y^\varepsilon$ for brevity) satisfy

$$|x^\varepsilon(s)|^2 + \varepsilon|y^\varepsilon(s)|^2 + \int_0^\infty |x^\varepsilon(t)|^2dt + \int_0^\infty |y^\varepsilon(t)|^2dt \le C(|x|^2 + \varepsilon|y|^2). \tag{6.39}$$

Since

$$D_x(\phi + \varepsilon\phi_1)f(x,y,v(s)) + \frac{1}{\varepsilon} D_y(\phi + \varepsilon\phi_1)g(x,y,v(s)) + \ell(x,y,v(s)) +$$

$$+ \frac{\partial}{\partial s}(\phi + \varepsilon\phi_1) \geq \varepsilon D_x\phi_1 f(x,y,v(s))$$

we can assert that

$$\phi(x) + \varepsilon\phi_1(x,y) \leq J^{\varepsilon}_{x,y}(v(.)) - \varepsilon \int_0^{\infty} D_x\phi_1 f(x^{\varepsilon}(s),y^{\varepsilon}(s),v(s))ds$$

$$\leq J^{\varepsilon}_{x,y}(v(.)) + \varepsilon C \int_0^{\infty} [|x^{\varepsilon}(s)|^2 + |y^{\varepsilon}(s)|^2 +$$

$$+ |x^{\varepsilon}(s)|^3 + |y^{\varepsilon}(s)|^3]dt$$

and by virtue of (6.39)

$$\leq J^{\varepsilon}_{x,y}(v(.)) + C\sqrt{\varepsilon}\,(|x|^2 + |x|^3 + \varepsilon(|y|^2 + \sqrt{\varepsilon}|y|^3))$$

which with (6.38) completes the proof of the result desired

□

REFERENCES.


ARDEMA M. [1], Linearization of the boundary layer equations of the minimum time to climb problem, Journal Guidance and Control, vol 2, n°5, Sept-Oct 1979.

" [2], Singular Perturbations in Systems and Control (ed.) CISM Courses and Lectures n° 280, Springer Verlag 1983

BENSOUSSAN A. [1], Perturbations Methods in Optimal Control, to be published

BERTRAN P. [1], Calcul formel et perturbations, thèse à paraître

CHOW J.M. KOKOTOVIC P.V. [1], Near optimal feedback stabilization of a class of non linear singularly perturbed systems, SIAM J. Control Opt, 16, pp. 756-770, (1978)

FAURRE P., CLERGET M., GERMAIN F. [1], Operateurs rationnels positifs, Dunod, Paris, 1978

FREEDMAN M.I., GRANOFF B. [1], Formal Asymptotic solution of a singularly perturbed non linear optimal control problem, JOTA, 19, (1976), pp. 301-325

FREEDMAN M.I., KAPLAN J.L. [1], Singular perturbations of two point boundary value problem arising in optimal control, SIAM Cont.Opt , 14 (1976), pp. 189-215

HABETS P. [1], Singular Perturbations in Non linear Systems and Optimal Control, in M. ARDEMA (edition, see above), pp. 103-143

HADDAD A.M., KOKOTOVIC P.V. [1], Note on Singular perturbations of linear state regulators. IEEE Trans.Auto.Control, AC-16, 3, pp. 279-281, (1971)

HADLOCK C.A. [1], Existence and dependence on a parameter of solutions of a non linear two point boundary value problem. J. Diff Equat., 14, (1973), pp. 498-517

KOKOTOVIC P.V. [1], Applications of Singular perturbation techniques to control problems, SIAM Review, (1984)

KOKOTOVIC P.V., O'MALLEY Jr R.E., SANNUTI P., [1], Singular perturbations and order reduction in Control theory, an overview - Automatica, 12, (1976), pp. 123-132

KOKOTOVIC P.V., SAKSENA V.R., [1], Singular perturbations in Control theory, Survey 1976, 1982

KOKOTOVIC P.V., YACKEL R.A., [1], Singular perturbation of linear regulators, IEEE Trans.Auto.Control, (1972), AC-17, pp. 29-37

LIONS J.L. , [1], Quelques méthodes de résolution des problèmes aux limites non linéaires, Dunod, Paris (1969)

MINTY G.J. [1], Monotone (non linear) operators in Hilbert Spaces, Duke Math-Journal, 29 (1962), pp. 341-346

O'MALLEY Jr R.E. [1], The singularly perturbed linear state regulator problem, SIAM Cont, 10, (1972), pp. 399-413

" [2], Singular perturbations of the time invariant linear state regulator problem, J. Diff equat. (1972) 12, pp. 117-128

" [3], Boundary layer methods for certain non linear singularly perturbed optimal control problems, J. Math anal. App. 45, (1974), pp. 468-484

" [4], Introduction to Singular Perturbations, Academic Press, 1974, N.Y.

" [5], Singular perturbations in optimal control. In Mathematical Control Theory, Lecture Notes in Mathematics, 680, Springer, N.Y.

O'MALLEY Jr R.E.,KUNG CF.[1],The matrix Riccati approach to a singularly perturbed regulator problem, J. Diff equat. (1974), 17, pp. 413-427

" [2], The singularly perturbed linear state regulator problem, SIAM cont. (1974), 13, pp. 327-337

SAKSENA V.R., O'REILLY J., KOKOTOVIC P.V., Singular Perturbations and Time-scale methods in Control Theory. Survey 1976, 1982

SANNUTI P. [1], Asymptotic Solution of singularly perturbed optimal control problems, Automatica, (1974), 10, pp. 183-194

" [2], Asymptotic expansions of singularly perturbed quasi linear optimal systems, SIAM Cont. (1975), 13, 3 , pp. 572-592

SANNUTI P., KOKOTOVIC P.V., [1], Near Optimum design of linear systems by a singular perturbations method, IEEE Trans -Auto-Control (1969), AC-14, pp. 15-22

SANNUTI P., KOKOTOVIC P.V. [2], Singular perturbation method for near optimum design of high order non linear systems, Automatica, (1969), 5, pp. 773-779

WILDE R.R., KOKOTOVIC P.V. [1], Optimal open and closed loop control of singularly perturbed linear systems, IEEE Trans-Auto-Cont. (1973), Ac-18,pp. 616-625

[2], A dichotomy in linear control theory, IEEE Trans-Auto-Control , vol AC-17, pp. 382-383, June 1972

# SINGULAR PERTURBATIONS IN STOCHASTIC CONTROL

A. Bensoussan[†]

G.L. Blankenship[††]

**Abstract:** We consider a class of problems in stochastic control theory involving stochastic systems with small parameters. Using both analytical and probabilistic methods adapted to the special structures of singularly perturbed stochastic control problems, we develop a systematic methodology for their analysis.

## Introduction

In this article we address the following class of control problems. We have a system governed by

$$\begin{aligned} dx &= f\,(x\,,y\,,v\,)dt \;+\; \sqrt{2}dw \\ \epsilon dy &= g\,(x\,,y\,,v\,)dt \;+\; \sqrt{2\epsilon}\,db \\ x\,(0) &= x\,,\;\; y\,(0) = y. \end{aligned} \tag{1}$$

where $w$ and $b$ are independent Wiener processes. The state $x(t)$ represents the *slow* system, while the state $y(t)$ represents the *fast* system. The scaling is such that the variations of the fast system per unit of time, in average as well as in variance, are of order $\frac{1}{\epsilon}$. The dynamics are controlled via the parameter $v(t)$. There is full information and

---

[†]INRIA, Domaine de Voluceau, Rocquencourt, B.P. 105, 78153 LE CHESNAY CEDEX FRANCE. Also with the *Universite' de Paris - Dauphine*. The research of this author was supported in part by the U.S. Department of Energy.

[††]Electrical Engineering Department, University of Maryland, College Park, MD 20742. This research was supported in part by the U.S. Army Research Office and the Army Night Vision and Electro-Optics Laboratory under contract DAAG 29-83-C-0028 with Systems Engineering, Inc., Greenbelt, MD.

the objective is to minimize the payoff

$$J_{x,y}^{\epsilon}(v(\cdot)) = E\int_0^{\tau} e^{-\beta t}\, l(x^{\epsilon}(t), y^{\epsilon}(t))dt \tag{2}$$

where $\tau$ denotes the first exit time of the process $x$ from the boundary $\Gamma$ of a domain O (smooth, bounded). Call

$$u_{\epsilon}(x,y) = \underset{v(\cdot)}{Inf}\ \{J_{x,y}^{\epsilon}(v(\cdot))\},$$

then $u_{\epsilon}$ is the solution of the Bellman equation

$$-\Delta_x u^{\epsilon} - \frac{1}{\epsilon}\Delta_y u^{\epsilon} + \beta u^{\epsilon} = H(x, D_x u^{\epsilon}, y, \frac{1}{\epsilon} D_y u^{\epsilon}) \tag{3}$$

$$u^{\epsilon} = 0 \text{ for } x \in \Gamma$$

with

$$H(x,p,y,q) = \underset{v \in U_{ad}}{Inf}\ [\, l(x,y,v) + p \cdot f(x,y,v) + q \cdot g(x,v,v)\,] \tag{4}$$

Our objective in sections 1 through 7 is to study the behavior of the equation (3) for $\epsilon$ small, and to interpret the results as a limit control problem approximating (1), (2). Let us explain the type of results which one can expect.

Proceed formally with an asymptotic expansion

$$u^{\epsilon}(x,y) = u(x) + \epsilon\phi(x,y).$$

Equating the first order terms in $\epsilon$, we get

$$-\Delta u - \Delta_y \phi + \beta u = H(x, D_x u, y, D_y \phi) \tag{5}$$

which we try to match for any $x,y$ by a convenient choice of $u$ and $\phi$. Consider $x$ in (5) as a parameter, as well as $p = D_x u$ ; set

$$L(y,v) = l(x,y,v) + p \cdot f(x,y,v)$$

$$G(y,v) = g(x,y,v) \tag{6}$$

$$\mathbf{H}(y,q) = \underset{v \in U_{ad}}{Inf}\ [\, L(y,v) + q \cdot G(y,v)\,]$$

which also depend parametrically on $x$ and $p$.

One can then consider the Bellman equation of ergodic control relative to (6). It is defined are follows: pick a constant $\chi$ (constant with respect to $y$) and a function $\phi$ such that

$$- \Delta_y \phi + \chi = \mathbf{H}(y, D_y \phi). \tag{7}$$

Suppose that one can find a pair $\chi, \phi$ depending parametrically on $x, p$; hence,

$$\chi = \chi(x, p).$$

If we choose $u$ so that

$$- \Delta u + \beta u = \chi(x, D_x u), \tag{8}$$

then the pair $u, \phi$ will satisfy (5). One can thus expect a solution of (8), vanishing on the boundary $\Gamma$ of $\mathbf{O}$ to be the limit of $u^{\epsilon}$.

This procedure depends on the possibility of being able to solve ergodic control problems of the type (7). The control problem itself is as follows: Consider

$$dy = G(y, v) d\tau + \sqrt{2} db, \quad y(0) = 0 \tag{9}$$

$$k_y(v(\cdot)) = \lim_{T \to \infty} \frac{1}{T} E \int_0^T L(y, v) \, d\tau$$

then in general

$$\chi = \underset{v(\cdot)}{Inf} \{ k_y(v(\cdot)) \} \text{ independent of } y.$$

The interpretation of $\phi$ is more delicate (cf. sect. 2.6). Pick a feedback $v(y)$ and consider the controlled state

$$dy = G(y, v(y)) \, d\tau + \sqrt{2} \, db, \quad y(0) = y. \tag{10}$$

It seems inevitable to require ergodicity of the process $y$.

This means that as $\tau \to \infty$, $y(\tau)$ behaves like a random variable following a probability $m_x^{v(\cdot)}(y)$, depending on the choice of $v(\cdot)$ and of the parameter entering into the definition of $G$. Suppose, moreover, that $m$ is a probability density with respect to Lebesgue measure; it is possible to give another interpretation of $\chi$ as follows:

$$\chi = \underset{v(\cdot)}{Inf} \{ \int_Y L(y, v(y)) \, m_x^{v(\cdot)}(y) \, dy \}. \tag{11}$$

In fact, taking account of

$$EL(y(\tau), v(\tau)) \rightarrow \int_Y L(y, v(y)) \, m_x^{v(\cdot)}(y) \, dy \quad \text{as } \tau \rightarrow \infty$$

one understands the relations between both interpretations of $\chi$. Formula (11) permits a better interpretation of (8), which turns out to be a Bellman equation for the slow system.

Indeed

$$\chi(x, p) = \underset{v(\cdot)}{Inf} \{ \int_Y ( l(x, y, v(y)) + p \cdot f(x, y, v(y)) \, m_x^{v(\cdot)}(y) \, dy \}$$

Setting

$$\tilde{l}(x, v(\cdot)) = \int_Y l(x, y, v(y)) \, m_x^{v(\cdot)}(y) \, dy$$

$$\tilde{f}(x, v(\cdot)) = \int_Y f(x, y, v(y)) \, m_x^{v(\cdot)}(y) \, dy$$

then the limit problem is described by

$$\underset{v(\cdot)}{Inf} \; J(v) = \int_0^T e^{-\beta t} \, \tilde{l}(x, v(\cdot)) \, dt$$

$$dx = \tilde{f}(x, v(\cdot)) \, dt + \sqrt{2} \, dw \tag{12}$$

$$x(0) = 0$$

It is interesting to note that the set of controls in (12) is changed from the original definition. One must consider the set of feedbacks $v(y)$. A control defined by a feedback with respect to the slow system is thus a function $v(x, y)$. To justify these considerations, it is thus important to make assumptions in order that the ergodicity of the process (10) is guaranteed. There must be one way or another a Markov chain defined on a compact set for which Doeblin's theorem holds (see J.L. Doob [1] ). This is achieved when one assumes that $G$ is periodic in $y$ together with the feedback) or when one considers instead of (10) a reflected diffusion. In Sections 1 and 2 we consider the periodic case, while in sections 3 and 4 the case of reflected diffusions is treated.

In section 5 we consider a fast system of the form

$$dy = G(v)y \, dt + \sum_{m=1}^{\bar{m}} \sigma_m y \, db_m(t), \quad y(0) = y. \tag{13}$$

Defining the norm and the angular velocity

$$\rho(t) = |y(t)|, \quad \xi(t) = \frac{y(t)}{|y(t)|},$$

then by linearity of (13), $\xi(t)$ is itself a diffusion taking place on a sphere; hence, ergodicity will hold for $\xi(t)$. We assume that the slow system depends separately on the norm and the angular velocity of the fast system, namely

$$dx = f(x, \rho, \xi, v) \, dt + \sqrt{2} \, dw, \quad x(0) = x \tag{14}$$

and we minimize a cost function of the form

$$E \int_0^T e^{-\beta t} \, l(x, \rho, \xi, v) \, dt.$$

We treat this model in situations where $\rho \to 0$ as $t \to \infty$, and $\xi$ is ergodic. In sections 6 and 7 we consider (10) in the whole space. To get an invariant probability on the whole space, we follow the theory developed by Khas'minskii [2] in the case without control. Note that $G$ cannot be bounded; hence we consider a drift of the form $Fy + G(y, v(y))$, where $F$ is a stable matrix. This article covers most cases where a natural ergodic fast system governs the evolution of the state. There may be many other situations where different techniques of singular perturbations are used. Examples of such situations may be found in the paper of R. Jensen and P.L. Lions [3]. For other approaches to ergodic control, see [4].

## Acknowledgement

The first author has benefited from useful discussions with H. Brezis, L.C. Evans, and J.L. Lions. The second author would like to thank G.C. Papanicolaou and P.V. Kokotovic for their contributions to his work.

## 1. Ergodic Control for Periodic Diffusions

### 1.1 Assumptions and Notation

Let us consider

$$g(y,v): \mathbb{R}^d \times U \to \mathbb{R}^d \tag{1.1}$$

$$l(y,v): \mathbb{R}^d \times U \to \mathbb{R}$$

which are periodic with period 1 in each component[1] of $y$. Also, we assume that

$$U \text{ is a metric space} \tag{1.2}$$

$$g,\ l \text{ are continuous functions} \tag{1.3}$$

Next let $(\Omega, A, P)$ be a probability space, and let $F^t$ be an increasing family of $\sigma$-algebras with $F^\infty = A$. We assume that $\exists$ an $F^t$ standard Wiener process with values in $\mathbb{R}^d$, denoted by $b(t)$. We set

$$y(t) = y + \sqrt{2}\, b(t) \tag{1.4}$$

where $y$ is fixed in $\mathbb{R}^d$. An admissible control is a process $v(t)$ with values in $U_{ad}$, which is adapted to $F^t$, where

$$U_{ad} \text{ is a compact subset of } U. \tag{1.5}$$

Let us consider the process

$$b^v(t) = b(t) - \frac{1}{\sqrt{2}} \int_0^t g(y(s), v(s))\, ds \tag{1.6}$$

We make the change of probability given by the Girsanov transform

$$\frac{dP_v^y}{dP} \Big|_{F^t} = \exp\left\{ \int_0^t \frac{1}{\sqrt{2}} g(y(s), v(s)) \cdot db - \frac{1}{4} \int_0^t | g(y(s), v(s)) |^2 ds \right\} \tag{1.7}$$

where the parameter $y$ in (1.7) stands for the initial condition in (1.4). For the system

[1] The period can be different in each component.

$(\Omega, A, F^t, P_v^y)$, $b_v(t)$ becomes a Wiener process and the process $y(t)$ is the solution of

$$dy = g(y(t), v(t))\, dt + \sqrt{2}\, db_v(t), \quad y(0) = y. \tag{1.8}$$

Let us consider the cost function

$$K_y^{\alpha}(v(\cdot)) = E_v^y \int_0^{\infty} e^{-\alpha t}\, l(y(t), v(t))\, dt \tag{1.9}$$

We set

$$\phi_\alpha(y) = \underset{v(\cdot)}{Inf} \{ K_y^{\alpha}(v(\cdot)) \}. \tag{1.10}$$

We shall be interested in the behavior of $\phi_\alpha(y)$ as $\alpha$ tends to 0.

## 1.2 The Hamilton-Jacobi-Bellman Equation

Let us consider the Hamiltonian

$$H(y, q) = \underset{v \in U_{ad}}{Inf} \{ l(y, v) + q \cdot g(y, v) \} \tag{1.11}$$

$$= \underset{v \in U_{ad}}{Inf} \{ L(y, q, v) \}.$$

From the assumptions (1.3) and (1.5) there exists a Borel map

$\hat{V}$ with values in $U_{ad}$,

such that

$$H(y, q) = L(y, q, \hat{V}(y, q)). \tag{1.12}$$

The following is a classical result in stochastic control theory:

**Theorem 1.1.** *The function $\phi_\alpha$ is the unique periodic function belonging to $W^{2,p}(Y)$, $Y = [0,1]^d$, $\forall\, p$, $2 \le p < \infty$, such that*

$$-\Delta\phi_\alpha + \alpha\phi_\alpha = H(y, D\phi_\alpha). \tag{1.13}$$

*Moreover, if we set*

$$v_\alpha(y) = \hat{V}(y, D\phi_\alpha) \tag{1.14}$$

*then the process*

$$v_\alpha(t) = v_\alpha(y(t))$$

*is an optimal control for (1.10).*

### 1.3 Invariant measures

Let us consider a periodic bounded Borel function $v(y)$ with values in $U_{ad}$ and write

$$g^v = g(y, v(y)). \tag{1.15}$$

The function $v(y)$ is called a *feedback*.

An invariant measure $m^v = m$ is a solution of the following problem:

$$- \Delta m + \operatorname{div}(m g^v) = 0 \tag{1.16}$$

$$m \in H^1(Y), \text{ periodic.}$$

since $g^v$ is not differentiable, (1.16) must be interpreted in a weak sense. More precisely, we write

$$\int_Y Dm . Dq dy = \int_Y m g^v . Dq dy \tag{1.17}$$

$$q \in H^1(Y), \text{ periodic.}$$

**Theorem 1.2.** *There exists one and only one (up to a multiplicative constant) solution of (1.17).*

**Proof.** The result follows from the Fredholm Alternative. Indeed, let us consider for $\lambda$ large enough the equation

$$- \Delta z_\lambda - g . D z_\lambda + \lambda z_\lambda = \phi \tag{1.18}$$

$$\text{where } \phi \in L^2(Y) \text{ periodic, } z_\lambda \in H^1(Y) \text{ periodic.}$$

This problem is well posed, at least for $\lambda$ large enough. Hence, we have

$$z_\lambda = G_\lambda \phi$$

with $G_\lambda$, considered as a map from $L^2(Y) \to L^2(Y)$, compact. But (1.16) amounts to

$$(I - \lambda G_\lambda^*)m = 0. \tag{1.19}$$

The number of linearly independent solutions of (1.19) is the same as that of

$$(I - \lambda G_\lambda)z = 0$$

i.e.,

$$- \Delta z - g \cdot D_z = 0 \tag{1.20}$$

$$z \text{ periodic, } z \in H^1(Y).$$

But the solution of (1.20) is more regular, $z \in W_{\text{loc}}^{2,p}(\mathbb{R}^n)$, $\forall\ p, 2 \leq p \leq \infty$. The desired result follows. □

## 1.4 Fundamental solution

Let us consider the Cauchy problem

$$\frac{\partial z}{\partial t} - \Delta z - g^v \cdot Dz = 0 \tag{1.21}$$

$$z(y,0) = \phi(y)$$

where $\phi$ is Borel bounded. Then there exists the following representation formula

$$z(y,t) = \int_{\mathbb{R}^d} p^v(y,t,\eta)\phi(\eta)d\eta \tag{1.22}$$

where the function $p(y,t,\eta)$ is the fundamental solution and has the following properties

$$\forall\ \eta,\ p^v(\cdot,\cdot,\eta) \in L^2(\delta,T;H^1(\mathbb{R}^n))$$

$$\forall\ \delta > 0, \delta < T, \forall\ y,\ p^y(x;\cdot,\cdot) \in L^2(\delta,T;H^1(\mathbb{R}^n)) \tag{1.23}$$

$$\frac{C_1 \exp\left[\dfrac{-\alpha_1 \mid y-\eta \mid^2}{t}\right]}{t^{n/2}} \leq p^y(y,t;\eta) \leq \frac{C_2 \exp\left[\dfrac{-\alpha_2 \mid y-\eta \mid^2}{t}\right]}{t^{n/2}} \tag{1.24}$$

where $\alpha_1, \alpha_2 > 0$, $C_1, C_2 > 0$ and they depend only on the bound on $g^v$. In particular, they do not depend on the particular feedback $v(\cdot)$. This result is due to D. Aronson [5].

Now note that if $\phi$ is periodic, $z$ is periodic, and we can write

$$z(y,t) = \int_Y p_o^v(y,t,\eta)\phi(\eta)d\eta \tag{1.25}$$

where

$$p_o^v(y,t,\eta) = \sum_{k_1,\ldots,k_d \in Z} p^v(y,t,\eta + \sum k_i e_i) \tag{1.26}$$

is defined on $Y \times [0,\infty) \times Y$.

Clearly,

$$p_o^v(y,t,\eta) \geq p^v(y,t,\eta)$$

and from (1.24), in particular

$$\delta_1 \geq p_o^v(y,1,\eta) \geq \delta > 0, \ \forall\, y,\eta \in Y \tag{1.27}$$

where $\delta$ and $\delta_1$ do not depend on the particular control. Note that $p_o^v(y,1,\eta)d\eta$ is a probability on $Y$.

If we introduce the operator $\mathbf{P}$ on bounded Borel functions on $Y$, defined by

$$\mathbf{P}\phi(y) = \int_Y p_o^v(y,1,\eta)\phi(\eta)\,d\eta, \tag{1.28}$$

then ergodic theory can be applied to assert that

$$\mathbf{P}^n \chi_E(y) \rightarrow \Pi(E), \ \forall\, E \text{ a Borel subset of } Y, \ \forall\, y \in Y. \tag{1.29}$$

Moreover, $\Pi$ is a probability on $Y$ and one has

$$| \mathbf{P}^n \phi(y) - \int_Y \phi(\eta)\,\Pi(d\eta) | \leq K\, ||\phi||\, e^{-\rho n} \tag{1.30}$$

$$\text{where } K = \frac{2}{1-\delta}, \ \rho = \log \frac{1}{1-\delta}.$$

For details, see J.L. Doob [1] or [6]. Note that (1.30) can be interpreted as

$$| z(y) - \int_Y \phi(\eta)\,\Pi(d\eta) | \leq K\, ||\phi||\, e^{-\rho}.$$

Since for $t \geq n$ we have

$$z(y,t) = \int_Y p_o^v(y,t-n,\eta)\, z(n,\eta)\, d\eta$$

hence,

$$| z(y,t) - \int_Y \phi(\eta)\, \Pi(d\eta) | \leq K \, ||\phi|| \, e^{-\rho}.$$

Taking $n = [t]$, we deduce

$$| z(y,t) - \int_Y \phi(\eta)\, \Pi(d\eta) | \leq K \, e^{\rho} \, ||\phi|| \, e^{-\rho t}. \tag{1.31}$$

Using the invariant measure $m = m^v$ defined in (1.16), we also see easily from (1.16) and (1.21) that

$$\int_Y z(y,t)\, m(y)\, dy = \int_Y \phi(y)\, m(y)\, dy. \tag{1.32}$$

Using (1.31) in (1.32), we deduce

$$\int_Y m(y)\, dy \int_Y \phi(\eta)\, \Pi(d\eta) = \int_Y \phi(y)\, m(y)\, dy$$

which proves that $\int_Y m(y) dy \neq 0$, since $m$ is not $a.e.$ 0. Normalizing the integral to be 1, we see that

$$\Pi(dy) = m(y) dy$$

and thus (1.31) yields

$$| z^v(y,t) - \int_Y \phi(y) m^v(y) dy | \leq K \, e^{\rho} ||\phi|| e^{-\rho t}. \tag{1.33}$$

Now recalling (1.30)

$$\int_Y \phi(y) m(y) dy = \int_Y z(y,1) m(y) dy$$

and from (1.25) and (1.27)

$$\delta \int_Y \phi(y) dy \leq z(y,1) \leq \delta_1 \int_Y \phi(y) dy.$$

Therefore,

$$\delta \int_Y \phi(y) dy \leq \int_Y \phi(y) m(y) dy \leq \delta_1 \int_Y \phi(y) dy.$$

Hence,

$$0 \leq \delta \leq m^v(y) \leq \delta_1, \ a.e. \tag{1.34}$$

We have proved the following:

**Theorem 1.3** *For any feedback $v(\cdot)$ the invariant probability defined by (1.16) satisfies (1.34) and the property (1.33) holds.*

## 1.5 The HJB equation of ergodic control

We consider the following problem: Find a pair $\chi,\phi$ where

$$\chi \text{ is a scalar, } \phi \in W^{2,p}(Y), \text{ periodic} \tag{1.35}$$

$$-\Delta\phi + \chi = H(y, D\phi). \tag{1.36}$$

We then have the following

**Theorem 1.4** *Assume (1.1),(1.2),(1.3), and (1.5). Then* $\exists$ a unique pair $\chi$ and $\phi$ *($\phi$ defined up to an additive constant) which is a solution of (1.35) and (1.36).*

**Proof.** *Existence.* Let us consider $\phi_\alpha$ defined in the statement of Theorem 1.1. we first have

$$|\alpha\phi_\alpha(y)| \leq C. \tag{1.37}$$

Let us consider the invariant probability $m^{v_\alpha}$ corresponding to the optimal feedback $v_\alpha(y)$ (cf. (1.14)). We denote it by $m_\alpha(y)$ to simplify the notation. Since we have

$$-\Delta\phi_\alpha + \alpha\phi_\alpha = l(y, v_\alpha(y)) + D\phi_\alpha \cdot g(y, v_\alpha(y))$$

from the equation satisfied by $m_\alpha$ we deduce

$$\int_Y D\phi_\alpha \cdot D(\phi_\alpha m_\alpha)dy + \int_Y \alpha\phi_\alpha^2 m_\alpha dy =$$

$$\int_Y l(y, v_\alpha(y))m_\alpha\phi_\alpha dy + \int_Y D\phi_\alpha \cdot g(y, v_\alpha(y))m_\alpha\phi_\alpha dy$$

Hence, in fact,

$$\int_Y |D\phi_\alpha|^2 m_\alpha dy = \int_Y (l(y, v_\alpha(y)) - \alpha\phi_\alpha)m_\alpha\phi_\alpha dy. \tag{1.38}$$

But

$$\int_Y (l(y, v_\alpha(y)) - \alpha\phi_\alpha(y))m_\alpha(y)dy = 0.$$

Considering

$$\overline{\phi}_\alpha = \int_Y \phi_\alpha(y)dy$$

we deduce from (1.38) that

$$\int_Y m_\alpha |D\phi_\alpha|^2 dy = \int_Y (l(y, v_\alpha(y)) - \alpha\phi_\alpha(y))m_\alpha(\phi_\alpha - \overline{\phi}_\alpha)dy.$$

But from Theorem 1.3 we can assert that

$$0 < \delta \leq m_\alpha(y) \leq \delta_1$$

where $\delta$, $\delta_1$ do not depend on $\alpha$. Therefore, it follows that

$$\int_Y |D\phi_\alpha|^2 dy \leq C |\phi_\alpha - \overline{\phi}_\alpha|_{L^2(Y)}$$

$$\leq C \left( \int_Y |D\phi_\alpha|^2 dy \right)^{1/2}$$

By Poincare's inequality we thus obtain the estimate

$$|D\phi_\alpha|_{L^2} \leq C. \tag{1.39}$$

Let us consider

$$\phi_\alpha - \tilde{\phi}_\alpha = \overline{\phi}_\alpha, \quad \chi_\alpha = \alpha\overline{\phi}_\alpha.$$

Hence, $\tilde{\phi}_\alpha$ is bounded in $H^1(Y)$ and $\chi_\alpha$ is bounded. Note that $H(y, D\phi_\alpha)$ is also bounded in $L^2(Y)$.

Consider a subsequence such that

$$\chi_\alpha \to \chi, \quad \tilde{\phi}_\alpha \to \phi \text{ in } H^1(Y) \text{ weakly and a.e.} \tag{1.40}$$

$$H(y, D\phi_\alpha) \to \xi \text{ in } L^2(Y) \text{ weakly.}$$

Passing to the limit in (1.13) yields

$$-\Delta\phi + \chi = \xi, \quad \phi \text{ periodic}, \int_Y \phi dy = 0.$$

Moreover,

$$\int_Y |D\tilde{\phi}_\alpha|^2 dy \rightarrow \int_Y \xi\phi dy$$

and by a classical argument it follows that $\tilde{\phi}_\alpha \rightarrow \phi$ in $H^1(Y)$. Then $H(y, D\tilde{\phi}_\alpha) \rightarrow H(y, D\phi)$ in $L^2(Y)$. The pair $(\chi, \phi)$ is a solution, since by standard regularity arguments $\phi \in W^{2,p}(Y), \forall\, p,\ 2 \leq p < \infty$.

*Uniqueness.* Consider $\hat{v}(y) = \hat{V}(y, D\phi)$ and let $\hat{m}$ be the corresponding invariant probability. One has

$$-\Delta\phi + \chi = l(y, \hat{v}(y)) + D\phi \cdot g(y, \hat{v}(y))$$

Hence,

$$\chi = \int_Y l(y, \hat{v}(y))\hat{m}(y)dy.$$

On the other hand, for any feedback $v(\cdot)$

$$-\Delta\phi + \chi \leq l(y, v(y)) + D\phi \cdot g(y, v(y))$$

Hence,

$$\chi \leq \int_Y l(y, v(y))m^v(y)dy.$$

Therefore,

$$\chi = \underset{v(\cdot)}{Inf} \left\{ \int_Y l(y, v(y))m^v(y)dy \right\} \tag{1.41}$$

which proves that $\chi$ is uniquely determined.

Let us now assume that there are two solutions $\phi_1$ and $\phi_2$. We can write (in the obvious notation)

$$-\Delta\phi_1 + \chi = l(y, \hat{v}_1(y)) + D\phi_1 \cdot g(y, \hat{v}_1(y))$$

$$-\Delta\phi_2 + \chi = l(y, \hat{v}_2(y)) + D\phi_2 \cdot g(y, \hat{v}_2(y))$$

Hence,

$$- \Delta(\phi_1 - \phi_2) \leq D(\phi_1 - \phi_2).g(y, \hat{v}_2(y)).$$

Considering $m_2$ the invariant probability corresponding to $\hat{v}_2$, we deduce by multiplying by $m_2(\phi_1 - \phi_2)^+$,

$$\int_Y m_2 \, | D(\phi_1 - \phi_2)^+ |^2 dy = 0$$

hence, $D(\phi_1 - \phi_2)^+ = 0$. Similarly, $D(\phi_1 - \phi_2)^- = D(\phi_1 - \phi_2)^+ = 0$. Therefore, $D(\phi_1 - \phi_2) = 0$; hence, $\phi_1 - \phi_2 =$ constant. □

## 1.6 The ergodic control problem

We can now interpret the pair $(\chi, \phi)$. We already know

$$\chi = \lim_{\alpha \to 0} \alpha \, u_\alpha(y)$$

$$= \lim_{\alpha \to 0} \alpha \, Inf \, K_y^\alpha(v(\cdot)).$$

In fact, one can be slightly more precise. We have

**Theorem 1.5.** *Under the assumptions of Theorem 1.4 we have*

$$\chi = \underset{v(\cdot)}{Inf} \{ \underline{\lim}_{\alpha \to 0} \alpha \, E_v^y \int_0^\infty e^{-\alpha t} \, l(y(t), v(t)) \, dt \} \tag{1.42}$$

$$= \underset{v(\cdot)}{Inf} \{ \underline{\lim}_{T \to \infty} \frac{1}{T} \, E_v^y \int_0^T l(y(t), v(t)) \, dt \}$$

*Moreover, choosing the undetermined constant for* $\phi$ *such that* $\int_Y \phi(y) \hat{m}(y) dy = 0$, *we have*

$$\phi(y) = Inf \; \{ \underline{\lim}_{T \to \infty} E_v^y \int_0^T (l(y(t), v(t)) - \chi) \, dt \;\; | \; v(\cdot) : E_v^y \phi(y(T)) \to 0 \} \tag{1.43}$$

**Proof.** Let us simply prove (1.43). For any control $v(\cdot)$ we have

$$\phi(y) \leq E_v^y \int_0^T (l(y(t), v(t)) - \chi) \, dt + E_v^y \phi(y(T))$$

Hence, if $E_v^y \phi(y(T)) \to 0$, then

$$\phi(y) \le \lim_{T\to\infty} E_v^y \int_0^T ( l(y(t),v(t)) - \chi)\, dt.$$

Moreover, considering the feedback $\hat{v}$ and setting $\hat{v}(t) \equiv \hat{v}(y(t))$, we have

$$\phi(y) = E_{\hat{v}}^y \int_0^T (l(y(t),\hat{v}(t)) - \chi)\, dt + E_{\hat{v}}^y \phi(y(T)).$$

But from (1.33)

$$E_{\hat{v}}^y \phi(y(T)) \to \int_Y \phi(y)\hat{m}(y)dy = 0.$$

Hence,

$$\phi(y) = \lim_{T\to\infty} E_{\hat{v}}^y \int_0^T (l(y(t),\hat{v}(t)) - \chi)\, dt$$

which completes the proof of the desired result. □

## 2. Singular perturbations with periodic diffusions

### 2.1 Notation and assumptions

Let us now consider functions

$$f(x,y,v): \mathbb{R}^n \times \mathbb{R}^d \times U \to \mathbb{R}^n \tag{2.1}$$

$$g(x,y,v): \mathbb{R}^n \times \mathbb{R}^d \times U \to \mathbb{R}^d$$

$$l(x,y,v): \mathbb{R}^n \times \mathbb{R}^d \times U \to R$$

continuous and periodic in $y$ with period 1 in each component

$$U_{ad} \text{ compact subset of } U \text{ a metric space.} \tag{2.2}$$

Let $(\Omega, A, P, F^t)$ be given and $b(t)$ and $w(t)$ be two independent standard Wiener processes with values in $\mathbb{R}^d$ and $\mathbb{R}^n$, respectively. We shall define

$$x(t) = x + \sqrt{2}w(t) \tag{2.3}$$

$$y_\epsilon(t) = y + \left(\frac{2}{\epsilon}\right)^{1/2} b(t)$$

An admissible control is a process $v(t)$ with values in $U_{ad}$, which is adapted to $F^t$.

We then consider the process

$$b_v^{\epsilon}(t) = -\frac{1}{\sqrt{2\epsilon}}\int_0^t g(x(s),y_\epsilon(s),v(s))\,ds \tag{2.4}$$

$$w_v^{\epsilon}(t) = w(t) - \frac{1}{\sqrt{2}}\int_0^t f(x(s),y(s),v(s))\,ds \tag{2.5}$$

Let us consider a bounded smooth domain O of $\mathbb{R}^n$, and $\tau = \tau_x$ denotes the first exit time of the process $x(t)$ from the domain O. Since we are not going to consider the process $x$ outside O, we may without loss of generality assume that $f$, $g$, and $l$ are bounded functions.

Let us define the probability $P^{\epsilon}$ (which depends also on the control $v(\cdot)$, and $(x,y)$)

$$\frac{dP^{\epsilon}}{dP}\Big|_{F^t} = \exp\Big\{\int_0^t\Big[\frac{1}{\sqrt{2\epsilon}}\,g(x(s),y_\epsilon(s),v(s))\cdot db(s)\Big]$$

$$+\Big[\frac{1}{\sqrt{2}}\,f(x(s),y_\epsilon(s),v(s))\cdot dw(s)\Big] \tag{2.6}$$

$$-\frac{1}{4}\int_0^t\Big[\frac{1}{\epsilon}\,|g(x(s),y_\epsilon(s),v(s))|^2 + |f(x(s),y_\epsilon(s),v(s))|^2\Big]ds\Big\}$$

For the system $(\Omega, A, F^t, P^{\epsilon})$, the processes $b_\epsilon(t)$ and $w_\epsilon(t)$ become standard independent Wiener processes and the processes $x(t), y_\epsilon(t)$ appear as the solutions of

$$dx = f(x(t),y_\epsilon(t),v(t))\,dt + \sqrt{2}\,dw_\epsilon(t)$$

$$dy_\epsilon = \frac{1}{\epsilon}\,g(x(t),y_\epsilon(t),v(t))\,dt + \left(\frac{2}{\epsilon}\right)^{1/2} db_\epsilon(t) \tag{2.7}$$

$$x(0) = x, \quad y_\epsilon(0) = y$$

Our objective is to minimize the pay off function

$$J_{x,y}^{\epsilon}(v(\cdot)) = E^{\epsilon}\int_0^{\tau_x} l(x(t),y_\epsilon(t),v(t))\,e^{-\beta t}\,dt \tag{2.8}$$

where $\beta > 0$. Let us set

$$u_\epsilon(x,y) = \operatorname*{Inf}_{v(\cdot)} J_{x,y}^{\epsilon}(v(\cdot)) \tag{2.9}$$

Then $u_\epsilon$ is the unique solution of the H.J.B. equation

$$-\Delta_x u_\epsilon - \frac{1}{\epsilon}\Delta_y u_\epsilon + \beta u_\epsilon = H(x, D_x u_\epsilon, y, \frac{1}{\epsilon} D_y u_\epsilon) \tag{2.10}$$

$$u_\epsilon = 0 \text{ for } x \in \Gamma, \ \forall y$$

$$u_\epsilon \text{ periodic in } y$$

$$u_\epsilon \in W^{2,p}(O \times Y), \ 2 \leq p < \infty$$

where $\Gamma = \partial O$ is the boundary of O, and where

$$\begin{aligned} H(x,p,y,q) &= \operatorname*{Inf}_{v \in U_{ad}} [l(x,y,v) + p \cdot f(x,y,v) + q \cdot g(x,y,v)] \\ &= \operatorname*{Inf}_{v \in U_{ad}} L(x,p,y,q,v). \end{aligned} \tag{2.11}$$

Moreover, there exists a Borel map $\hat{V}(x,p,y,q)$ with values in $U_{ad}$, such that

$$H(x,p,y,q) = L(x,p,y,q,\hat{V}). \tag{2.12}$$

We can define an optimal feedback for (2.8), by setting

$$v_\epsilon(x,y) = \hat{V}(x, D_x u_\epsilon, y, D_y u_\epsilon) \tag{2.13}$$

and the process

$$v_\epsilon(t) = v_\epsilon(x(t), y_\epsilon(t)) \tag{2.14}$$

is an optimal control for (2.8). Our objective is to study the behavior of $u_\epsilon$ as $\epsilon$ tends to 0.

## 2.2 Preliminaries

Let us consider the following problem

$$-\epsilon \Delta_x m_\epsilon - \Delta_y m_\epsilon + \operatorname{div}_y (m_\epsilon g(x,y,v_\epsilon)) = 0 \tag{2.15}$$

$$\frac{\partial m_\epsilon}{\partial \nu}\Big|_\Gamma = 0, \ m_\epsilon \text{ periodic in } y.$$

$$m_\epsilon \in H^1(O \times Y).$$

This problem has a unique solution up to a multiplicative constant. The situation is similar to that of Theorem 1.2. The adjoint problem is defined by

$$-\epsilon\,\Delta_x z \;-\; \Delta_y z \;-\; g(x,y,v_\epsilon).D_y z = 0$$

$$\frac{\partial z}{\partial \nu}\Big|_\Gamma = 0, \quad z \text{ is periodic in } y$$

and the solution is necessarily a constant.

We have the following important estimate

**Lemma 2.1** *We have*

$$\delta \le m_\epsilon(x,y) \le \delta_1 \tag{2.16}$$

*where* $\delta$ and $\delta_1$ *are the same as in* (1.34).

**Proof.** Let us consider the fundamental solution of the Cauchy problem.

$$\frac{\partial z}{\partial t} - \epsilon\,\Delta_x z - \Delta_y z - g_\epsilon . Dz = 0 \tag{2.17}$$

$$\frac{\partial z}{\partial \nu}\Big|_\Gamma = 0, \quad z \text{ is periodic in } y$$

$$z(x,y,0) = \psi(x,y)$$

where we have set $g_\epsilon = g(x,y,v_\epsilon(x,y))$.

We are going to use the probabilistic interpretation of (2.17). Consider a Wiener process reflected at the boundary of **O**. Namely,

$$dx = \sqrt{2\epsilon}\; dw \;-\; \chi_\Gamma(x(t))\,\nu\, d\,\xi(t)$$

$$x(0) = x$$

where $\xi$ is an increasing process. Consider next

$$dy(t) = \sqrt{2}\; db(t), \quad y(0) = y$$

and assume $w$ and $b$ are independent. Let us define the process

$$b_\epsilon(t) = b(t) - \frac{1}{\sqrt{2}} \int_0^t g((x(s),y(s),v_\epsilon(x(s),y(s)))\; ds$$

and the change of probability

$$\frac{dP^\epsilon}{dP}\Big|_{F^t} = \exp\{\int_0^t \frac{1}{\sqrt{2}}\, g_\epsilon(x(s),y(s)) . \,db \;-\; \frac{1}{4}\int_0^t |g_\epsilon|^2\, ds\}$$

where $F^t = \sigma(w(s), b(s), s \leq t)$. For the probability $P$, we have

$$dx = \sqrt{2\epsilon}\ dw - \chi_\Gamma(x(t))\, \nu\, d\xi\,, \quad x(0) = x \tag{2.18}$$

$$dy = g_\epsilon(x(t)), y(t))\ dt + \sqrt{2}\ db_\epsilon\,, \quad y(0) = y$$

and $w$, $b_\epsilon$ are independent Wiener processes. Note that $x$ and $b_\epsilon$ are also independent for $P^\epsilon$. We then have the formula

$$z(x,y,t) = E^\epsilon\, \psi(x(t), y(t)). \tag{2.19}$$

By virtue of the independence of $x(t)$ and $b_\epsilon(t)$, we can compute (2.19) first considering $x(\cdot)$ as a deterministic function and taking the expectation with respect to the process $y$. Namely, we have

$$E^\epsilon\, [\psi(x(1), y(1) \mid \sigma(x(s),\ s \leq 1)] = \varsigma(y, 0)$$

where $\varsigma$ is the solution of

$$-\frac{\partial \varsigma}{\partial t} - \Delta_y\, \varsigma - g_\epsilon(x(t), y) . D_y\, \varsigma = 0$$

$$\varsigma \text{ periodic in } y$$

$$\varsigma(y, 1) = \psi(x(1), y).$$

There is a slight difference with the situation of section 1.4; indeed, we have now a nonstationary problem, since $x(\cdot)$ enters as a parameter. Nevertheless, the results of Aronson [5] will apply. Therefore, we can assert that

$$\delta \int_Y \psi(x(1), \eta)\, d\eta \leq \varsigma(y, 0) \leq \delta_1 \int_Y \psi(x(1), \eta)\, d\eta$$

where $\delta$, $\delta_1$ depend only on the bound of $g$. In particular, they do not depend on $x(\cdot)$. It follows that

$$\delta \int_Y E^\epsilon\, \psi(y(1), \eta)\, d\eta \leq z(x, y, 1) \leq \delta_1 \int_Y E^\epsilon \psi(x(1), \eta)\, d\eta$$

which can be expressed as

$$\delta\, \phi(x, 1) \leq z(x, y, 1) \leq \delta_1\, \phi(x, 1) \tag{2.20}$$

where

$$\frac{\partial \phi}{\partial t} - \epsilon \Delta \phi = 0, \quad \frac{\partial \phi}{\partial \nu} |_\Gamma = 0 \tag{2.21}$$

$$\phi(x,0) = \int_Y \psi(x,\eta)\, d\eta$$

But we notice that

$$\int\int m_\epsilon(x,y)\, z(x,y,1)\, dxdy = \int\int m_\epsilon(x,y)\, \psi(x,y)\, dxdy$$

hence[2]

$$\delta \int\int m_\epsilon(x,y) \phi(x,1)\, dxdy \le \int\int m_\epsilon(x,y)\, \psi(x,y)\, dxdy \tag{2.22}$$

$$\le \delta_1 \int\int m_\epsilon(x,y)\, \phi(x,1)\, dxdy$$

However, from (2.15) we have

$$-\epsilon\, \Delta_x \int_Y m_\epsilon(x,y)\, dy = 0, \quad \frac{\partial}{\partial \nu} \int_Y m_\epsilon(x,y) dy = 0 \text{ on } \Gamma$$

hence

$$\int_Y m_\epsilon(x,y)\, dy = \text{constant} = 1.$$

We then deduce from (2.22)

$$\delta \int_O \phi(x,1)\, dx \le \int\int m_\epsilon(x,y)\, \psi(x,y)\, dxdy \le \delta_1 \int_O \phi(x,1)\, dx \tag{2.23}$$

But from (2.21)

$$\int_O \phi(x,1)\, dx = \int_O \phi(x,0)\, dx = \int_O\int_Y \psi(x,y)\, dxdy$$

which, used in (2.33), implies the desired property (2.16). □

## 2.3 *A priori* estimates

We start with

**Lemma 2.2** *The following estimates hold*

$$|D_x u_\epsilon|^2_{L^2}, \quad |u_\epsilon|_{L^\infty} \le C \tag{2.24}$$

[2] Here we use the positivity of $m_\epsilon$, which follows from ergodic theory, as in section 1.4.

$$|D_y u_\epsilon|^2_{L^2} \leq C\epsilon$$

**Proof.** Using the definition of $v_\epsilon$ (c.f. (2.13)), we can write (2.10) as follows:

$$-\Delta_x u_\epsilon - \frac{1}{\epsilon}\Delta_y u_\epsilon + \beta u_\epsilon = l(x,y,v_\epsilon) + Du_\epsilon \cdot f(x,y,v_\epsilon) \tag{2.25}$$

$$+\frac{1}{\epsilon} D_y u_\epsilon \cdot g(x,y,v_\epsilon)$$

$$u_\epsilon = 0 \text{ for } x \in \Gamma, \ \forall y, \ u_\epsilon \text{ periodic in } y.$$

Considering now $m_\epsilon$ defined by (2.15), we multiply (2.25) by $m_\epsilon u_\epsilon$ and integrate. By virtue of the boundary conditions we obtain

$$\iint m_\epsilon |D_x u_\epsilon|^2 dxdy + \frac{1}{\epsilon}\iint m_\epsilon |D_y u_\epsilon|^2 dxdy + \beta \iint m_\epsilon u_\epsilon^2 dxdy \tag{2.26}$$

$$= \iint (l_\epsilon + f_\epsilon \cdot D_x u_\epsilon) m_\epsilon u_\epsilon dxdy.$$

In (2.26) we have denoted

$$l_\epsilon(x,y) = l(x,y,v_\epsilon(x,y))$$

and similarly for $f_\epsilon$. Besides we have used the relation

$$\iint D_x u_\epsilon \cdot D_x \left(\frac{u_\epsilon^2}{2}\right) dxdy + \frac{1}{\epsilon}\iint D_y m_\epsilon \cdot D_y \left(\frac{u_\epsilon^2}{2}\right) dxdy$$

$$= \frac{1}{\epsilon}\iint m_\epsilon g_\epsilon \cdot D_y \left(\frac{u_\epsilon^2}{2}\right) dxdy$$

which follows from (2.15) and multiplication by $\frac{u_\epsilon^2}{2}$.

Now using, for instance (2.9), or the Maximum Principle in (2.10) we easily deduce

$$|u_\epsilon|_{L^\infty} \leq \frac{\| l \|}{\beta}. \tag{2.27}$$

Using the result of Lemma 2.1, we deduce the desired results (2.24). □

**Lemma 2.3** *Let $\phi(x) \in H_0^1(O) \cap H^2(O)$, then the following relation holds*

$$\iint m_\epsilon |D_x(u_\epsilon - \phi)|^2 dxdy + \frac{1}{\epsilon}\iint m_\epsilon |D_y u_\epsilon|^2 dxdy$$

$$+ \iint \beta m_\epsilon (u - \phi)^2 dxdy = \iint m_\epsilon u_\epsilon (\Delta\phi - \beta\phi) dxdy \tag{2.28}$$

$$+ \int |D_x \phi|^2 \, dx + \beta \int \phi^\epsilon \, dx$$

**Proof.** We multiply (2.25) by $m_\epsilon \phi$ and integrate. We have

$$\iint D_x u_\epsilon \cdot D_x (m_\epsilon \phi) \, dxdy + \frac{1}{\epsilon} \iint D_x u_\epsilon \cdot D_y (m_\epsilon \phi) \, dxdy$$

$$+ \beta \iint m_\epsilon u_\epsilon \phi \, dxdy = \iint (l_\epsilon + Du_\epsilon \cdot f_\epsilon) m_\epsilon \phi \, dxdy \tag{2.29}$$

$$+ \frac{1}{\epsilon} \iint D_y u_\epsilon g_\epsilon m_\epsilon \phi \, dxdy.$$

Now multiply (2.15) by $u_\epsilon \phi$ and integrate. Using the fact that $\phi$ does not depend on $y$, we obtain

$$\iint D_x m_\epsilon D_x u_\epsilon \phi \, dxdy + \iint D_x m_\epsilon D_x \phi u_\epsilon \, dxdy$$

$$+ \frac{1}{\epsilon} \iint D_y m_\epsilon D_y u_\epsilon \phi \, dxdy = \frac{1}{\epsilon} \iint m_\epsilon g D_y u_\epsilon \phi \, dxdy. \tag{2.30}$$

But

$$\iint D_x m_\epsilon \cdot D_x \phi u_\epsilon \, dxdy = - \iint m_\epsilon D_x \phi \cdot D_x u_\epsilon \, dxdy \tag{2.31}$$

$$- \iint m_\epsilon \Delta\phi u_\epsilon \, dxdy.$$

Using (2.30), (2.31) in (2.29) yields

$$2 \iint D_x u_\epsilon D_x \phi m_\epsilon \, dxdy + \iint m_\epsilon u_\epsilon \Delta\phi \, dxdy \tag{2.32}$$

$$+ \beta \iint m_\epsilon u_\epsilon \phi \, dxdy = \iint (l_\epsilon + D_x u_\epsilon f_\epsilon) m_\epsilon \phi \, dxdy.$$

But now

$$\iint m_\epsilon | D_x (u_\epsilon - phi) |^2 \, dxdy + \frac{1}{\epsilon} \iint m_\epsilon | D_y u_\epsilon |^2 \, dxdy + \iint \beta m_\epsilon (u_\epsilon - \phi)^2 \, dxdy \tag{2.33}$$

$$= \iint m_\epsilon | D_x u_\epsilon |^2 \, dxdy + \frac{1}{\epsilon} \iint m_\epsilon | D_y u_\epsilon |^2 \, dxdy + \iint \beta m_\epsilon u_\epsilon^2 \, dxdy$$

$$+ \int | D_x \phi |^2 \, dx + \beta \int \phi^2 \, dx - 2 \iint m_\epsilon D_x u_\epsilon \cdot D_x \phi \, dxdy - 2\beta \iint m_\epsilon u_\epsilon \phi \, dxdy$$

Using (2.26) and (2.32) in (2.33), the result (2.28) follows. □

### 2.4 Convergence

Let us first prove the

**Lemma 2.4** *Let us consider a subsequence of* $u_\epsilon$ *such* that

$$u_\epsilon \to u \text{ in } H^1(O \times Y) \text{ weakly.} \tag{2.34}$$

*Then* $u$ *is a function of* $x$ *only, belongs to* $H_0^1(O)$*. Moreover the convergence in (2.34) is strong.*

**Proof.** From Lemma 2.2, we know that subsequences for which property (2.34) holds exist. From the third estimate (2.24) and

$$| D_y u |^2 \leq \underline{\lim}_{\epsilon \to 0} | D_y u_\epsilon |^2 = 0$$

we deduce that $u = u(x)$. Clearly $u \in H_0^1(O)$, since $u_\epsilon$ vanishes for $x$ in $\Gamma$. Let us consider a sequence $\phi_k$ of functions in $H_0^1 \cap H^2$ such that

$$||\phi_k - u||_{H_0^1} \to 0 \text{ as } k \to \infty. \qquad ?$$

We have

$$\chi_\epsilon = \iint m_\epsilon | D_x (u_\epsilon - u) |^2 dxdy + \frac{1}{\epsilon} \iint m_\epsilon | D_y u_\epsilon |^2 dxdy$$

$$+ \beta \iint m_\epsilon (u_\epsilon - u)^2 dxdy \leq 2 \iint m_\epsilon | D_x (u_\epsilon - \phi_k) |^2 dxdy$$

$$+ \frac{2}{\epsilon} \iint m_\epsilon | D_y u_\epsilon |^2 dxdy + 2\beta \iint m_\epsilon (u_\epsilon - \phi_k)^2 dxdy$$

$$+ 2 \int | D_x (\phi_k - u) |^2 dx + 2\beta \int (\phi_k - u)^2 dx$$

and from Lemma 2.3

$$\chi_\epsilon \leq 2 \iint m_\epsilon u_\epsilon (\Delta\phi_k - \beta\phi_k) dxdy + 2 \int | D_x \phi_k |^2 dx \tag{2.35}$$

$$+ 2\beta \int \phi_k^2 dx + 2 \int | D_x (\phi_k - u) |^2 dx + 2\beta \int (\phi_k - u)^2 dx$$

We can without loss of generality assume that $m_\epsilon \to m^*$ in $L^\infty(O \times Y)$ weak star. Note that

$$\int_Y m_\epsilon(x,y)\, dy = 1,$$

implies

$$\int_O \int_Y m^*(x,y)\, \psi(x)\, dx = \int_O \psi(x)\, dx \quad \forall\, \psi \in L^1.$$

Letting $\epsilon$ tend to 0 in (2.35), we deduce

$$\overline{\lim}\chi_\epsilon \leq 2 \int\int m^*\, u\, (\Delta\phi_k - \beta\phi_k)\, dxdy + 2\int |D_x\, \phi_k|^2\, dx$$

$$+ 2\beta \int \phi_k^2\, dx + 2\int |D_x\, (\phi_k - u)|^2\, dx + 2\beta\int (\phi_k - u)^2\, dx$$

$$+ 2\int u\, (\Delta\phi_k - \beta\phi_k)\, dx + 2\int |D_x\, \phi_k|^2\, dx$$

$$+ 2\beta \int \phi_k^2\, dx + 2\int |D_x\, (\phi_k - u)|^2\, dx + 2\beta\int (\phi_k - u)^2\, dx$$

$$= -2\int Du\, D\phi_k\, dx - 2\beta\int u\, \phi_k\, dx + 2\int |D_x\, \phi_k|^2\, dx$$

$$+ 2\beta \int \phi_k^2\, dx + 2\int |D_x\, (\phi_k - u)|^2\, dx + 2\beta\int (\phi_k - u)^2\, dx.$$

Letting $k$ tend to $+\infty$, we deduce

$$\overline{\lim_{\epsilon \to 0}}\ \chi_\epsilon = 0.$$

Since clearly

$$\chi_\epsilon \geq \delta \int\int |D_x\, (u_\epsilon - u)|^2\, dxdy + \beta\delta\int\int (u_\epsilon - u)^2\, dxdy$$

the strong convergence property is established. □

We are now going to identify the limit. We shall use the notation of section 1. Let us consider $x \in O$, and $p \in \mathbb{R}^n$ as parameters. We shall consider the situation of section 2, with $g(x,y,v)$ instead of $g(y,v)$ and $l(x,y,v) + p \cdot f(x,y,v)$ instead of $l(x,y,v)$.

A feedback now is a Borel function of $y$, which may be indexed by $x,p$. For any feedback $v$ consider the invariant probability $m^v(x,y)$, which is the solution of

$$-\Delta_y\, m + \mathrm{div}_y\, (m\; g(x,y,v(y))) = 0$$

$$m \in H^1(Y),\ m \text{ periodic}.$$

For $x,p$ fixed we can consider the scalar $\chi$ defined in (1.36), namely

$$\chi(x,p) = \underset{v(\cdot)}{Inf} \int_Y m^v(x,y)(l(x,y,v(y)) + p \cdot f(x,y,v(y)))\, dy \tag{2.37}$$

For any $v(y)$ given, the integral is a continuous function of $x$, and a linear function of $p$. The function $\chi$ is upper semi-continuous and uniformly Lipschitz in $p$ with linear growth. Let us then consider the problem

$$-\Delta u + \beta u = \chi(x, Du), \quad u \mid_\Gamma = 0. \tag{2.38}$$

This problem has a unique solution in $W^{2,p}(O)$, $2 \leq p \leq \infty$. Our main result in this section is the following:

**Theorem 2.1** *We assume 2.1, 2.2. Then we have*

$$u_\epsilon \rightarrow u \quad \text{in } H^1(O \times Y) \text{ strongly.} \tag{2.39}$$

Before proving the theorem, we need a few more steps. Let us consider for any feedback $v(x,y)$, the function $m_\epsilon^v$ which is the solution of

$$\begin{gathered} -\epsilon \Delta_x m_\epsilon^v - \Delta_y m_\epsilon^v + \operatorname{div}_y (m_\epsilon^v g(x,y)) = 0 \\ \frac{\partial m_\epsilon^v}{\partial \nu} \Big|_\Gamma = 0, \quad m_\epsilon^v \text{ periodic in } y \\ m_\epsilon^v \in H^1(O \times Y) \end{gathered} \tag{2.40}$$

In fact $m_\epsilon$, a solution of (2.15), coincides with $m_\epsilon^{v_\epsilon}$. Note that the estimates of Lemma 2.1 hold for $m_\epsilon^v$, uniformly with respect to $v, \epsilon$. We can consider an ergodic control problem related to (2.40). Namely let $F(x,y,v)$: $O \times Y \times U_{ad}$ be such that

$$\begin{gathered} F \text{ is periodic in } y, \quad |F| \leq \bar{F}(x) \in L^2(O). \\ F \text{ is measurable, and continuous in } y, v \end{gathered} \tag{2.41}$$

Define an Hamiltonian by

$$\mathbf{H}(x,y,q) = \underset{v \in U_{ad}}{Inf} [F(x,y,v) + q \cdot g(x,y,v)]. \tag{2.42}$$

We consider then the problem to find a pair $\phi_\epsilon(x,y)$, $\Lambda_\epsilon$ a scalar such that

$$-\epsilon \Delta_x \phi_\epsilon - \Delta_y \phi_\epsilon = \mathbf{H}(x,y,D_y \phi_\epsilon) - \Lambda_\epsilon$$

$$\frac{\partial \phi_\epsilon}{\partial \nu} \mid_\Gamma = 0, \quad \phi_\epsilon \text{ periodic in } y \tag{2.43}$$

$$\phi_\epsilon \in W^{2,p}(O \times Y)$$

This problem is studied using ergodic control theory as in section 1. In particular, we have (recalling that the integral of $m_\epsilon^v$ is $| O |$)

$$\Lambda_\epsilon = \underset{v(\cdot,\cdot)}{Inf} \int\int \frac{m_\epsilon^v(x,y)}{| O |} F(x,y,v(x,y))\, dxdy. \tag{2.44}$$

Consider now for $x$ frozen the problem

$$-\Delta_y \phi = H(x,y,D_y \phi) - \Lambda(x) \tag{2.45}$$

$$\phi \text{ periodic in } y, \ \phi \in W^{2,p}(Y), \ \forall x$$

We want to study the limit of $\Lambda_\epsilon$ as $\epsilon$ tends to 0. We shall need the

**Lemma 2.5** *We have*

$$| O | \Lambda_\epsilon \to \int_O \Lambda(x)\, dx \tag{2.46}$$

**Proof.** Note that

$$\Lambda(x) = \underset{v(\cdot)}{Inf} \int m^v(x,y) F(x,y,v(y))\, dy$$

where $m$ is defined by (2.36). Hence $\Lambda \in L^2(O)$. Since we can write

$$-\Delta_y \phi = F(x,y,\hat{v}) + D_y \phi \cdot g(x,y,\hat{v}) - \Lambda(x)$$

Considering $m^{\hat{v}}$, we deduce

$$\int_Y m^{\hat{v}} | D_y \phi |^2 dy = \int_Y m^{\hat{v}} (F(x,y,\hat{v}) - \Lambda(x))(\phi - \bar{\phi})\, dy$$

where

$$\bar{\phi}(x) = \int_Y \phi(x,y)\, dy.$$

Hence,

$$\int_Y | D_y \phi |^2 dy \leq C\bar{F}(x)^2.$$

Therefore, $D_y \phi \in L^2(O \times Y)$. From the equation (2.45) it also follows that

$$\Delta_y \phi \in L^2(O \times Y).$$

Consider then a sequence

$$\phi_k \in H^1(O \times Y),\ \Delta_y \phi_k \in L^2$$

such that

$$\phi_k \to \phi,\ D_y \phi_k \to D_y \phi,\ \Delta_y \phi_k \to \Delta_y \phi \text{ in } L^2$$

satisfying the same conditions as $\phi$. Let us set

$$\Lambda_k(x,y) = \Delta_y \phi_k + \mathbf{H}(x,y,D_y \phi_k). \tag{2.47}$$

Clearly, $\Lambda_k \to \Lambda$ in $L^2(O \times Y)$. We can write for some $\hat{v}_\epsilon$

$$-\epsilon\Delta_x \phi_\epsilon - \Delta_y \phi_\epsilon = F(x,y,\hat{v}_\epsilon) + D_y \phi_\epsilon . g(x,y,\hat{v}_\epsilon) - \Lambda_\epsilon$$

hence

$$-\epsilon\Delta_x \phi_\epsilon - \Delta_y (\phi_\epsilon - \phi_k) \geq D_y (\phi_\epsilon - \phi_k) . g(x,y,\hat{v}_\epsilon) - \Lambda_\epsilon + \Lambda_k \tag{2.48}$$

Let $m^{\hat{v}_\epsilon}$ correspond to $\hat{v}_\epsilon$. By multiplying (2.48) by $m^{\hat{v}_\epsilon}$ and integrating, we deduce

$$\epsilon \iint D_x \phi_\epsilon . D_x m^{\hat{v}_\epsilon} dxdy + \iint D_y (\phi_\epsilon - \phi_k) . D_y m^{\hat{v}_\epsilon} dxdy$$

$$\geq \iint D_y (\phi_\epsilon - \phi_k) . m^{\hat{v}_\epsilon} g(x,y,\hat{v}_\epsilon)\, dxdy - \Lambda_\epsilon |O| + \iint \Lambda_k m^{\hat{v}_\epsilon} dxdy$$

or

$$\epsilon \iint D_x \phi_k . m^{\hat{v}_\epsilon} dxdy \geq -\Lambda_\epsilon |O| + \iint \Lambda_k m^{\hat{v}_\epsilon} dxdy \tag{2.49}$$

But $\sqrt{\epsilon}\, D_x m^{\hat{v}_\epsilon}$ remains bounded in $L^2$. Therefore, in (2.49) letting $\epsilon$ tend to 0 we get

$$\underline{\lim}\, \Lambda_\epsilon |O| \geq \iint \Lambda_k m^*(x,y)\, dxdy$$

where $m^*$ is some cluster point of the sequence $m^{\hat{v}_\epsilon}$ in $L^2$ weakly. Letting $k \to \infty$, we obtain since $\int_Y m^*(x,y)\, dy = 1$

$$\lim \Lambda_\epsilon |O| \geq \int \Lambda(x)\, dx. \tag{2.50}$$

On the other hand, we have

$$-\epsilon \Delta_x \phi_\epsilon - \Delta_y (\phi_\epsilon - \phi_k) \leq D_y (\phi_\epsilon - \phi_k) . g(x,y,\hat{v}) - \Lambda_\epsilon + \Lambda_k .$$

Multiplying by $m^{\hat{v}}$, we deduce

$$\epsilon \int\int D_x \phi_k \cdot D_x m^{\hat{v}} dxdy \leq -\Lambda_\epsilon |O| \int\int \Lambda_k m^{\hat{v}} dxdy$$

hence

$$\overline{\lim} \Lambda_\epsilon |O| \leq \int\int \Lambda_k m^{\hat{v}} dxdy.$$

Letting $k \to \infty$, we obtain

$$\Lambda_\epsilon |O| \leq \int \Lambda(x) dx$$

which with (2.50) completes the proof of the desired result (2.46). □

We can now proceed with the

## Proof of Theorem 2.1

Let us consider a subsequence of $u_\epsilon$ which converges to $u$ in $H^1(O \times Y)$ weakly. Such subsequences exist, by virtue of Lemma 2.2. Moreover, from Lemma 2.4, we can assert that $u$ does not depend on $y$ and that the convergence is strong. Moreover, $u \in H_0^1(O)$. Let $\phi \in C_0^\infty(O)$, $\phi \geq 0$. Consider the relation (2.25) and multiply by $\phi m_\epsilon$. Actually, we have already done the calculation in Lemma 2.3 (c.f. (2.29), and (2.32)). We thus have (c.f. (2.32)).

$$2 \int\int D_x u_\epsilon D_x \phi m_\epsilon dxdy + \int\int m_\epsilon u_\epsilon \Delta \phi dxdy \tag{2.51}$$
$$+ \beta \int\int m_\epsilon u_\epsilon \phi dxdy = \int\int (l(x,y,v_\epsilon) + D_x u \cdot f(x,y,v_\epsilon)) m_\epsilon \phi dxdy$$
$$+ \int\int D_x (u_\epsilon - u) \cdot f(x,y,v_\epsilon) m_\epsilon \phi dxdy$$

Now consider

$$F(x,y,v) = \phi(x)(l(x,y,v) + Du \cdot f(x,y,v))$$

which satisfies (2.41). Obviously,

$$\int\int (l(x,y,v_\epsilon) + D_x u \cdot f(x,y,v_\epsilon)) m_\epsilon \phi dxdy \geq |O| \Lambda_\epsilon \tag{2.52}$$

Note that

$$\Lambda(x) = \chi(x, Du) \phi(x)$$

Hence from Lemma 2.5

$$|O| \Lambda_\epsilon \to \int \chi(x, Du) \phi(x)\, dx \tag{2.53}$$

We can also assume that $m_\epsilon \to m^*$ in $L^\infty$ weak star, and $\int_Y m^*(x,y)\, dy = 1$ a.e. From (2.51)

$$2 \int\int D_x u_\epsilon D_x \phi\, m_\epsilon\, dxdy + \int\int m_\epsilon u_\epsilon \Delta\phi\, dxdy \tag{2.54}$$

$$+ \beta \int\int m_\epsilon u_\epsilon \phi\, dxdy \geq |O| \Lambda_\epsilon + \int\int D_x (u_\epsilon - u)\cdot f(x,y,v_\epsilon)\, m_\epsilon \phi\, dxdy$$

Since $u_\epsilon \to u$ strongly in $H^1$, we can pass to the limit in equality (2.54). Recalling that $u$ does not depend on $y$, this yields

$$2 \int D_x u\, D_x \phi\, dx + \int u\, \Delta\phi\, dx + \beta \int u\, \phi\, dx$$

$$\geq \int \chi(x, Du) \phi(x)\, dx$$

hence

$$\int (-\Delta u + \beta u) \phi\, dx \geq \int \chi(x, Du) \phi(x)\, dx$$

and since $\phi$ is arbitrarily positive we have

$$-\Delta u + \beta u \geq \chi(x, Du).$$

To prove the reverse inequality is easier. Take any feedback $v(x,y)$. From (2.10) we have

$$-\Delta_x u_\epsilon - \frac{1}{\epsilon} \Delta_y u_\epsilon + \beta u_\epsilon \leq l(x,y,v(x,y)) + D_x u_\epsilon \cdot f(x,y,v)$$

$$+ \frac{1}{\epsilon} D_y u_\epsilon \cdot g(x,y,v).$$

Consider $m_\epsilon^v$ defined in (2.40) and $\phi(x)$ as before. Multiplying by $\phi\, m_\epsilon^v$, we deduce as for (2.51)

$$2 \int\int D_x u_\epsilon D_x \phi\, m_\epsilon^v\, dxdy + \int\int m_\epsilon^v u_\epsilon \Delta\phi\, dxdy \tag{2.55}$$

$$+ \beta \int\int m_\epsilon^v u_\epsilon \phi\, dxdy \leq \int\int \phi\, (l(x,y,v(x,y)) + D_x u_\epsilon \cdot f(x,y,v)) m_\epsilon^v\, dx$$

Moreover, it is easy to see that $m_\epsilon^v \to m^v$ in $L^2(O \times Y)$ weakly, where $m^v$ is defined in (2.36). This and the strong convergence of $u_\epsilon$, allow us to pass the limit in (2.55). We deduce

$$2 \int D_x u \, D_x \phi \, dx + \int u \, \Delta\phi \, dx + \beta \int u \, \phi \, dx \tag{2.56}$$
$$\leq \int\int (l(x,y,v(x,y)) + D_x u \cdot f(x,y,v)) \, m^v \, \phi \, dxdy.$$

Therefore,

$$- \Delta u + \beta u \leq \int_Y (l(x,y,v(x,y)) + D_x u \cdot f(x,y,v)) \, m^v \, dy$$

and since $v$ is arbitrary, we deduce

$$- \Delta u + \beta u \leq \chi(x, Du).$$

Therefore, $u$ coincides with the solution of (2.38). By uniqueness the desired result obtains. □

## 2.5 Asymptotic expansion

The result of Theorem 2.1 does not give any estimate of the rate of convergence. But its advantage is that it requires no particular regularity assumptions. It is possible to give a rate of convergence and to proceed differently relying on maximum principle type of arguments, when additional smoothness is available. The situation is somewhat similar to what arises in homogenization (c.f. Bensoussan Lions Papanicolaou [6] ). We shall outline the argument, and in particular we shall not try to present assumptions under which the regularity requirements are satisfied (see for that aspect [7] ). Consider $\phi(x,y)$ such that

$$- \Delta_y \phi + \chi(x, Du) = H(x, D_x u, y, D_y \phi) \tag{2.57}$$
$$\phi \text{ periodic in } y.$$

We define $\tilde{u}_\epsilon$ by

$$u_\epsilon = u + \epsilon \phi + \tilde{u}_\epsilon \tag{2.58}$$

Then $\tilde{u}_\epsilon$ satisfies

$$- \epsilon \Delta_x \tilde{u}_\epsilon - \frac{1}{\epsilon} \Delta_y \tilde{u}_\epsilon + \beta \tilde{u}_\epsilon = \epsilon \Delta_x \phi - \epsilon \beta \phi \tag{2.59}$$
$$+ H(x, D_x u + \epsilon D_x \phi + D_x \tilde{u}_\epsilon, y, D_y \phi + \frac{1}{\epsilon} D_y \tilde{u}_\epsilon)$$

$$- H(x, D_x u, y, D_y \phi)$$

$$\tilde{u}_\epsilon |_\Gamma = - \epsilon \phi |_\Gamma$$

assuming, of course, that quantities like $\Delta_x \phi$ are defined. It follows that for some feedback $v_\epsilon(x, y)$

$$- \epsilon \Delta_x \tilde{u}_\epsilon - \frac{1}{\epsilon} \Delta_y \tilde{u}_\epsilon + \beta \tilde{u}_\epsilon \leq \epsilon \Delta_x \phi - \epsilon \beta \phi$$

$$+ f(x, y, v_\epsilon) \cdot (D_x \tilde{u}_\epsilon + \epsilon D_x \phi) + \frac{1}{\epsilon} g(x, y, v) \cdot D_y \tilde{u}_\epsilon$$

Hence, if $D_x \phi$ and $\Delta_x \phi$ are bounded

$$\tilde{u}_\epsilon \leq C \epsilon$$

A similar argument goes the other way round, hence

$$| u_\epsilon - u - \epsilon \phi |_{L^\infty} \leq C \epsilon$$

In particular,

$$| u_\epsilon - u |_{L^\infty} \leq C \epsilon. \tag{2.60}$$

## 2.6 Interpretation of the limit problem

The limit problem is written as

$$- \Delta u + \beta u = \underset{v(\cdot)}{Inf} \{ \tilde{l}(x, v(\cdot)) + Du \cdot \tilde{f}(x, v(\cdot)) \} \tag{2.61}$$

$$u |_\Gamma = 0$$

where we have set

$$\tilde{l}(x, v(\cdot)) = \int_Y m^v(x, y) l(x, y, v(y)) \, dy \tag{2.62}$$

$$\tilde{f}(x, v(\cdot)) = \int_Y m^v(x, y) f(x, y, v(y)) \, dy.$$

It is clear that (2.61) is an H.J.B. equation for a slow system whose drift if $\tilde{f}$, and integral cost is $\tilde{l}$. For this problem the set of controls is the set of periodic Borel functions $v(y)$ with values in $U_{ad}$. A feedback on the slow system is thus still a function $v(x, y)$. There exists an optimal feedback for the limit problem, namely $\hat{v}(x, y)$ obtained in (2.57). Indeed consider the function $\hat{V}$ defined in (2.12), then

$$\hat{v}(x,y) = \hat{V}(x,Du,y,D_y\,\phi) \tag{2.63}$$

is an optimal feedback for the limit problem. In fact, this is the feedback to be applied on the real system as a surrogate for $v_\epsilon(x,y)$ defined in (2.13). One can show by techniques similar to those used in previous paragraphs to obtain Theorem 2.1, that the corresponding cost function will converge as $\epsilon$ tends 0 to $u$ in $H^1(O \times Y)$. Note that unlike the deterministic situation the optimal feedback for the limit problem is not a function of $x$ only. In fact (2.63) corresponds to the composite feedback of Chow-Kokotovic [8] (c.f. also [7] in the deterministic case).

## 3. Ergodic Control for Reflected Diffusions

### 3.1 Assumptions and notation

Our objective in this section is to describe another class of ergodic control problems and to consider stochastic control problems with singular perturbations which can be associated to them, as in section 2. We shall consider diffusions with reflection.

Let $B$ be a smooth bounded domain in $\mathbb{R}^d$, whose boundary is denoted by $\partial B$. Let $g$ and $l$ be continuous functions

$$\begin{aligned} g(y,v)&: \overline{B} \times U \to \mathbb{R}^d \\ l(y,v)&: \overline{B} \times U \to \mathbb{R}^d \end{aligned} \tag{3.1}$$

where $U$ is as in (1.2) and $U_{ad}$ is as in (1.5).

Consider $(\Omega, A, P, F^t, b(t))$ as in section 1.1, and let $y(t)$ represent the diffusion process reflected at the boundary of $B$

$$\begin{aligned} dy &= \sqrt{2}\, db - \chi_{\partial B}(y(t))\mu\, d\eta \\ y(0) &= y. \end{aligned} \tag{3.2}$$

where $\mu$ is the outward unit normal at the boundary of $B$, and $\eta(t)$ is an increasing process.

Admissible controls are defined as in section 1.1. Let us consider next the process $b_v(t)$ defined in (1.6), and the change of probability defined in (1.7). For the system

$(\Omega, A, F^t, P_v^y)$, $b_v(t)$ becomes a standard Wiener process and the process $y(t)$ is the solution of

$$dy = g(y(t), v(t))dy + \sqrt{2}\, db_v(t) - \chi_{\partial B}(y(t))\, \mu\, d\eta \tag{3.3}$$

One then considers the functional $K_y^\alpha(v\,\cdot))$ given by (1.9) and $\phi_\alpha(y)$ defined by (1.13). We then have the analogue of Theorem 1.1, namely

$$\begin{gathered} -\Delta\phi_\alpha + \alpha\phi_\alpha = H(y, D\,\phi_\alpha) \\ \frac{\partial\phi_\alpha}{\partial\mu}\,|_{\partial B} = 0 \\ \phi_\alpha \in W^{2,p}(B),\ 2 \leq p < \infty. \end{gathered} \tag{3.4}$$

The feedback $v_\alpha(y)$ and the process $v_a(t)$ defined in Theorem 1.1 solve the problem of minimizing $K_y^\alpha(v(\cdot))$.

### 3.2 Invariant measures

Consider a feedback $g^v$ as in (1.15). An invariant measure $m^v$ will be defined as the solution of the variational problem

$$\int_B Dm \cdot Dq dy = \int_B m g^v \cdot Dq dy, \ \forall\ q \in H^1(B),\ m^v \in H^1(B). \tag{3.5}$$

As is well known, this is the variational formulation of a Neumann problem. The adjoint of (3.5) is the following

$$\int_B Dz \cdot Dq dy = \int_B g^v \cdot Dz q dy, \ \forall\ q \in H^1(B),\ z \in H^1(B) \tag{3.6}$$

and (3.6) has only constants as solutions. The Fredholm Alternative applies to imply the existence and uniqueness of the solution of (3.5), up to a multiplicative constant.

The next important step is to obtain the analogue of (1.34), the uniform bounds from above and below on $m^v$. A preliminary step is to prove that $m^v \geq 0$, $\int_B m^v(y)dy = 1$ can be achieved. This will follow from ergodic theory as in section 1.4. We shall not, however, proceed in this way, since we do not want to use the explicit properties of the Green operators corresponding to the Neumann problem, when the coefficients are not smooth. These properties have not been clearly stated in the literature. We shall proceed differently using some ideas of Y. Kogan [9].

Consider the parabolic problem

$$\frac{\partial z}{\partial t} - \Delta z - g^{v}.Dz = 0 \tag{3.7}$$

$$\frac{\partial z}{\partial \mu} \mid_{\partial B} = 0, \quad z(y,0) = \phi(y), \ \phi \text{ Borel bounded}$$

and we shall define the operator, as in (1.28),

$$\mathbf{P}\phi(y) = z(y,1). \tag{3.8}$$

Let us write for $\Gamma$ a Borel subset of $\overline{B}$

$$\lambda_{xy}^{v}(\Gamma) = \mathbf{P}\,\chi_\Gamma(x) - \mathbf{P}\,\chi_\Gamma(y), \quad y,z \in \overline{B} \tag{3.9}$$

We have

**Lemma 3.1**

$$sup\ \{ \lambda_{xy}^{v}(\Gamma) \mid v,x,y,\Gamma \} < 1. \tag{3.10}$$

**Proof.** Suppose that (3.10) does not hold. Then there exists a sequence $\{ v_k, x_k, y_k, \Gamma_k \}$ such that

$$\mathbf{P}^{v_k}\,\chi_{\Gamma_k}(x_k) - \mathbf{P}^{v_k}\,\chi_{\Gamma_k}(y_k) \to 1.$$

Hence,

$$\mathbf{P}^{v_k}\,\chi_{\Gamma_k}(x_k) \to 1, \quad \mathbf{P}^{v_k}\,\chi_{\Gamma_k}(y_k) \to 0.$$

Define $z_k(y,t)$ to be the solution of (3.7) corresponding to the control $v_k(\cdot)$, and to the initial data $\chi_{\Gamma_k}(y) = \phi_k(y)$. We then have

$$z_k(x_k,1) \to 1, \quad z_k(y_k,1) \to 0. \tag{3.11}$$

But the classical estimates on parabolic equations yield

$$z_k \text{ bounded in } W^{2,p,1}(B \times (\delta,T)), \ \forall\ \delta > 0 \tag{3.12}$$[3]

$$z_k \text{ bounded in } L^{\infty}(B \times (\delta,T)), \ \forall \ 2 \leq p < \infty.$$

Noting that $\phi_k$ is bounded in $L^{\infty}$ and $g_k = g(y, v_k)$ is bounded in $L^{\infty}$, we have for subsequences

$$\phi_k \to \phi^* \text{ in } L^{\infty} \text{ weak star}$$

$$g_k \to g^* \text{ in } L^{\infty} \text{ weak star} \tag{3.13}$$

$$z_k \to z^* \text{ in } L^{\infty} \text{ weak star and } W^{2,p,1}(B \times (\delta,T)) \text{ weakly}$$

$$\forall \ \delta, \ \forall \ p, \ 2 \leq p < \infty.$$

Note that $z_k$ is also bounded in $L^2(0,T;H^1(B))$, $\frac{\partial z_k}{\partial t}$ is bounded in $L^2(0,T;H^{-1}(B))$. These estimates are sufficient to pass to the limit in $k$, and obtain the result

$$\frac{\partial z^*}{\partial t} - \Delta z^* - g^* . Dz^* = 0 \tag{3.14}$$

$$\frac{\partial z^*}{\partial \mu} \Big|_{\partial B} = 0, \quad z^*(x,0) = \phi^*(x).$$

From the estimate (3.12) we can assert that

$$z_k(x,1) \to z^*(x,1) \text{ in } C^0(\bar{B}). \tag{3.15}$$

But we can also pick a subsequence of $x_k$, $y_k$ which converges in $\bar{B}$ to $x^*$, $y^*$. From (3.15) it follows that

$$z_k(x_k,1) \to z^*(x^*,1)$$

$$z_k(y_k,1) = z^*(y^*,1)$$

and thus

$$z^*(x^*,1) = 1, \ z^*(y^*,1) = 0.$$

But since $0 \leq z^* \leq 1$, $(x^*,1)$ and $(y^*,1)$ will be a maximum and a minimum of $z^*$. That will contradict the maximum principle for (3.14), since $z^*$ cannot be a constant. □

The importance of Lemma 3.1 is that it suffices to obtain the properties (1.29), (1.30), and (1.31) with constants which are uniform with respect to the control.

---

[3]Here $W^{2,p,1}(B \times (\delta,T)) = \{z \in L^p \mid \frac{\partial z}{\partial t}, \frac{\partial z}{\partial y_i}, \frac{\partial^2 z}{\partial y_i \partial y_j} \in L^p\}$. (See [10] ).

Therefore, (1.33) also holds. This estimate implies in particular that $m^v \geq 0$ and can be taken to be a probability.

To prove the estimate (1.34), we rely on the following Lemma.

**Lemma 3.2.** *The solution of (3.5) (normalized as a probability) satisfies*

$$m^v \in L^\lambda(B), \ \forall \ 1 \leq \lambda < \infty, \tag{3.17}$$

*with a norm bounded in* $v(\cdot)$.

**Proof.** Let us consider the problem

$$-\Delta\phi - g^v . D\phi + \phi = \psi, \quad \frac{\partial \phi}{\partial \mu} |_{\partial B} = 0$$

If $\psi \in L^s(B)$ then $\phi \in W^{2,s}(B)$, $1 \leq s < \infty$. Moreover, we have

$$\int_B m\,\phi dy = \int_B m\,\psi dy. \tag{3.18}$$

We deduce, since $m$ is a probability,

$$\begin{aligned} |\int_B m\,\psi dy\,| &\leq |\phi|_{C^0(\bar{B})} \\ &\leq C\,||\phi||_{W^{2,s'_0}} \quad \text{with } s'_0 > \frac{d}{2} \\ &\leq C\,|\psi|_{L^{s'_0}}. \end{aligned}$$

Therefore, $m \in L^{s_0}$, with $\frac{1}{s_0} = 1 - \frac{1}{s'_0}$, $1 > \frac{1}{s_0} > 1 - \frac{2}{d}$. If $d = 1,2$, the result (3.17) is proved. Suppose $d = 3$, then we have $m \in L^{s_0}$, $1 \leq s_0 < 3$. We proceed using (3.18) together with a boot strapping argument. We have

$$|\int_B m\,\psi dy\,| \leq C\,|\phi|_{L^{s'_0}} \leq C\,||\phi||_{W^{2,s'_1}} \leq C\,|\psi|_{L^{s'_1}}$$

with $\frac{1}{s'_1} < \frac{1}{s'_0} + \frac{2}{d}$; hence, $\frac{1}{s_0} > \frac{1}{s_1} > \frac{1}{s_0} - \frac{2}{d}$. Since $d = 3$, $s_1$ is arbitrary and (3.17) is proved. If $d = 4, \cdots$, we proceed in the same way. For any value of $d$, a finite number of steps suffice to imply (3.17).

The fact that the norm is bounded with respect to $v(\cdot)$, follows from the above argument. □

We then prove that the estimate (1.34) holds.

**Theorem 3.1.** *For any feedback* $v(\cdot)$ *the invariant probability defined by (3.5) satisfies the estimates (1.34).*

**Proof.** Let us prove the second estimate first. We rely on ideas from [11], p. 62. For $k > 0$ define

$$\eta = (m - k)^+, \quad A(k) = \{y \mid m(y) \geq k\}.$$

We deduce from (3.5)

$$\int_B |D\eta|^2 dy = \int_B mg^v . D\eta dy = \int_{A(k)} mg^v . D\eta dy$$

Hence

$$|D\eta|_{L^2} \leq C \left[\int_{A(k)} m^2 dy\right]^{\frac{1}{2}} \leq C |m|_{L^s} (Meas\ A(k))^{\frac{1}{2} - \frac{1}{s}}, \quad \forall\ 2 \leq s < \infty \tag{3.19}$$

Moreover,

$$\int_B \eta^2 dy = \int_B (m - k)\eta\, dy \leq \int_B m\,\eta\, dy = \int_{A(k)} m\,\eta\, dy$$

which implies as in (3.19)

$$|\eta|_{L^2} \leq |m|_{L^s} (Meas\ A(k))^{\frac{1}{2} - \frac{1}{s}}.$$

Therefore, taking account of (3.19) we can write

$$||\eta||_{H^1} \leq C |m|_{L^s} (Meas\ A(k))^{\frac{1}{2} - \frac{1}{s}}.$$

From Sobolev's inequality it follows

$$\left[\int_{A(k)} (m - k)^{s^*} dy\right]^{\frac{1}{s^*}} \leq C |m|_{L^s} (Meas\ A(k))^{\frac{1}{2} - \frac{1}{s}}. \tag{3.20}$$

where $\frac{1}{s^*} = \frac{1}{2} - \frac{2}{d}$.

If $h > k > 0$, then $A(h) \subset A(k)$, and we have

$$(h - k)(Meas\ A(h))^{\frac{1}{s^*}} \leq [\int_A (h)(m - k)^{s^*} dy]^{s^*}$$

$$\leq [\int_A (k)(m - k)^{s^*} dy]^{s^*}$$

and from (3.20) it follows that

$$(h - k)(Meas\ A(h))^{\frac{1}{s^*}} \leq C \mid m \mid_{L^s} (Meas\ A(k))^{\frac{1}{2} - \frac{1}{s}}$$

or

$$Meas\ A(h) \leq \frac{C \mid m \mid_{L^s}^{s^*}}{(h - k)^{s^*}} (Meas\ A(k))^{(\frac{1}{2} - \frac{1}{s})s^*}.$$

Pick $s > d$, then $(\frac{1}{2} - \frac{1}{s})s^* > 1$. We use the following result from [11] (p. 63): Let $\psi(t)$, $k_o \leq t < \infty$ be nonnegative and nonincreasing, such that

$$\psi(h) \leq \frac{C}{(h - k)^\alpha} \psi(k)^\beta, \ h > k > k_o \tag{3.21}$$

where $C$, $\alpha$, and $\beta$ are positive constants with $\beta > 1$. Then

$$\psi(k_o + \bar{k}) = 0 \tag{3.22}$$

where

$$\bar{k}^\alpha = C \mid \psi(k_o) \mid^{\beta - 1} 2^{\alpha \frac{\beta}{\beta - 1}}. \tag{3.23}$$

It is clear that this result applies, and thus $Meas\ A(\bar{k}) = 0$ where

$$\bar{k} = C \mid m \mid_{L^s} (Meas\ B)^{\frac{1}{d} - \frac{1}{s}}. \tag{3.24}$$

The second estimate is thus proved.

The proof of the second estimate is more involved. We refer to [7].

**Remark 3.1.** The function $m^v \in W^{1,p}(B)$, $\forall\ p \in (1,\infty)$. This follows from a general result in Lions - Magenes [12] Teo. 6.1, p. 33. Indeed, we write (3.5) as follows

$$\int_B mq\ dy + \int_B Dm \cdot Dq\ dy = L(q) \tag{3.25}$$

where

$$L(q) = \int_B m(q + g^v \cdot Dq)\ dy\ ,$$

in which $m$ is given ( in $L^\infty$). The functional $q \rightarrow L(q)$ is clearly continuous on $W^{1,\gamma'}$ , $\forall$ $1 < \gamma' < \infty$. From a representation theorem in [12] it can be written in a unique way as (3.22) with $m \in W^{1,\gamma}$, hence the result. □

### 3.3 The Hamilton - Jacobi - Bellman equation of ergodic control

We have the same result as Theorem 1.4, namely,

**Theorem 3.2.** *Assume (3.17) and (1.2)(1.5) then there exists a unique pair* $\chi, \phi$ *where* $\chi$ *is a scalar and* $\phi \in W^{2,p}(B)$ *such that*

$$-\Delta\phi + \chi = H(y, D\phi), \quad \frac{\partial\phi}{\partial\mu}\Big|_{\partial B} = 0, \quad \int_B \phi\ dy = 0. \tag{3.26}$$

The proof is similar to that of Theorem 1.4. This result has been given by Y. Kogan [9] without relying on the convergence of the sequence $\alpha\phi_\alpha$. This approach is more akin to the usual one in stochastic control, based on the method of successive approximations (cf. Fleming and Rishel [13] ); see also [4] In this situation one has three possible interpretations of the constant $\chi$, namely (1.41) or (1.42).

### 3.4 Additional results on the Cauchy problem

Unlike the method of sections 1.3 and 1.4, where we have proven the estimates (1.27) before proving (1.34), our approach for the Neumann case has been to prove the estimate (1.34) on the invariant measure directly. It remains to prove the estimates (1.27) for the Cauchy problem. They will be useful in treating the singular perturbation problem.

Consider the Cauchy problem

$$\frac{\partial z}{\partial t} - \Delta z - g^v . Dz = 0 \tag{3.27}$$

$$\frac{\partial z}{\partial \mu} \mid_{\partial B} = 0, \quad z(y,0) = \phi(y)$$

with $\phi \in L^1(B)$, $\phi \geq 0$. Let us assume that

$$z(y,t) \leq c_\delta |\phi|_{L^1}, \quad \forall \ t \in [\delta, T] \tag{3.28}$$

where $c_\delta$ does not depend on $\phi$, nor on $v(\cdot)$, $y$. We then have

**Proposition 3.1.** *The following estimate holds*

$$z(y,1) \geq c \ |\phi|_{L^1} \tag{3.29}$$

*where c is independent of* $\phi$, $v(\cdot)$, and $y$.

**Proof.** We shall prove that

$$Inf \ \{ z(y,1) \mid y, v(\cdot), \phi \geq 0, \ |\phi|_{L^1} = 1 \} \tag{3.30}$$

if this is false, there exists a sequence $\phi_k \geq 0$, $|\phi_k|_{L^1} = 1$, $y_k$, $v_k(\cdot)$ such that, denoting $z_k$ the solution of (3.27) corresponding to $\phi_k$, $v_k(\cdot)$, then one has

$$z_k(y_k, 1) \to 0. \tag{3.31}$$

Writing

$$\frac{\partial z_k}{\partial t} - \Delta z_k - g_k . Dz_k = 0, \quad \frac{\partial z_k}{\partial \mu} \mid_{\partial B} = 0, \quad z_k(y,0) = \phi_k(y)$$

and making use of (3.28), we can assert that $z_k(y,t)$ is bounded in $L^\infty(B \times (\delta, T))$. But then $z_k$ remains bounded in $W^{2,1,p}(B \times (\delta, T))$, $\forall \ p$, $2 \leq p < \infty$.

Reasoning as in Lemma 3.2, we identify a limit function $z^*$ such that

$$\frac{\partial z^*}{\partial t} - \Delta z^* - g^* . Dz^* = 0, \quad t \in [\delta, T] \tag{3.32}$$

$$\frac{\partial z^*}{\partial \mu} \mid_{\partial B} = 0$$

and as a consequence of (3.31) we have

$$z^*(y^*, 1) = 0.$$

By the maximum principle, $z^*$ is necessarily 0. Consider the invariant measure $m_k$ corresponding to $v_k$, we have

$$\int_B m_k(y)\, z_k(y,1)\, dy \to \int_B m^*(y)\, z^*(y,1)\, dy = 0.$$

But

$$\int_B m_k(y)\, z_k(y,1)\, dy = \int_B m_k(y)\, \phi_k(y)\, dy \geq c \mid \phi_k \mid = c > 0$$

which is a contradiction. □

To prove (3.28), it is sufficient to prove an estimate for a fixed positive time. We shall prove that, for instance

$$z(y,1) \leq c \mid \phi \mid_{L^1}. \tag{3.33}$$

Consider the dual problem

$$\begin{aligned} -\frac{\partial q}{\partial t} - \Delta q + \operatorname{div}(q g^v) &= 0 \\ \frac{\partial q}{\partial \mu} - q g \cdot \mu \mid_{\partial B} &= 0 \\ q(y, \tfrac{1}{2}) &= \psi. \end{aligned} \tag{3.34}$$

We shall prove the following

**Proposition 3.2** *Let $\psi \in L^\lambda(B)$, $\lambda > \frac{n}{2}$, then the solution of (3.34) belongs to $L^\infty(B \times [0, \frac{1}{2} - \delta])$, $\forall\, \delta > 0$.*

If the result of Proposition 3.2 holds, then we can write

$$\int_B z(y, \tfrac{1}{2})\, \psi(y)\, dy = \int_B q(y,0)\, \phi(y)\, dy$$

$$\leq c \mid \psi \mid_{L^\lambda} \mid \phi \mid_{L^1}.$$

Therefore, $z(y, \frac{1}{2}) \in L^{\lambda'}(B)$. Since $\lambda' > 1$, the usual regularity theory for partial differential equations implies the desired result (3.33).

To prove (3.34), we consider

$$r(y,t) = (\frac{1}{2} - t)\, q(y,t)$$

which is the solution of

$$
\begin{aligned}
-\frac{\partial r}{\partial t} - \Delta r + \operatorname{div}(rg) &= q \\
\frac{\partial r}{\partial \mu} - r\, g \cdot \mu \mid_{\partial B} &= 0 \qquad (3.35)\\
r(y, \frac{1}{2}) &= 0
\end{aligned}
$$

and we know *a priori* that

$$q \in L^2(0, \frac{1}{2}; H^1(B)) \cap L^\infty(0, \frac{1}{2}; L^\lambda(B)). \qquad (3.36)$$

The result of Proposition 3.2 follows from the following result of Ladyzenskaya, Solonnikov, and Ural'tseva [10].

**Lemma 3.3.** *The solution $r$ of (3.35) belongs to $L^\infty(B \times (0, \frac{1}{2}))$.*

**Proof.** The argument is in the spirit of Theorem 3.1 of [11]. For notational convenience we take $T$ instead of $\frac{1}{2}$, and assume $q \geq 0$. Note that $r \in L^\infty(0,T; L^\lambda(B))$, by the definition of $r$. Let $k$ be a constant and

$$\eta = (r - k)^+ .$$

We easily deduce from (3.5) that

$$
\begin{aligned}
\frac{1}{2} \mid \eta(t) \mid^2 + \int_0^t \mid D\eta \mid^2 ds &= \int_0^t ds \int_B (rg + q)\, \eta\, dx \qquad (3.37)\\
&= \int_0^t ds \int_B \chi\, \eta\, dx
\end{aligned}
$$

where $\chi \in L^\infty(0,T; L^\lambda(B))$.

Let us introduce the following norms

$$|||\eta|||^2 = \sup_{t \in [0,T]} \{ |\eta(t)|^2 + \int_0^T |D\eta|^2 dt \}$$

and define as in Theorem 3.1

$$A_k(t) = \{ x \in B \mid \eta(x,t) \geq k \}.$$

We deduce from (3.37), since $\eta = 0$ outside $A_k(t)$, *a.e.* $t$

$$|||\eta|||^2 \leq C \int_0^T dy \left( \int_{A_k(t)} \eta^{\frac{\lambda}{\lambda - 1}} dx \right)^{\frac{\lambda - 1}{\lambda}}$$

$$\leq C \int_0^t dt \left( \int_B \eta^{\frac{s\lambda}{\lambda - 1}} dx \right)^{\frac{\lambda - 1}{s\lambda}} (Meas\ A_k(t))^{\frac{\lambda - 1}{s'\lambda}}$$

$$\leq C \left[ \int_0^T dt \left( \int_B \eta^{\frac{s\lambda}{\lambda - 1}} dx \right)^{\frac{\rho(\lambda - 1)}{s\lambda}} \right]^{\frac{1}{\rho}} \left[ \int_0^T dt\, (Meas\ A_k(t))^{\frac{(\lambda - 1)}{s'\lambda}\rho'} \right]^{\frac{1}{\rho'}}$$

Introducing the notation

$$|\eta|_{q,r} = \left[ \int_0^T \left( \int_B \eta^q\, dx \right)^{\frac{r}{q}} dt \right]^{\frac{1}{r}}$$

we deduce the estimate

$$|||\eta|||^2 \leq C\ |\eta|_{\frac{s\lambda}{\lambda - 1},\rho} \left[ \int_0^T dt\ \ (Meas\ A_k(t))^{\frac{\lambda-1}{s'\lambda}\rho'} \right]^{\frac{1}{\rho'}} \tag{3.38}$$

We then use the embedding theorem (cf. Ladyzenskaya, Solonnikov, Ural'tseva [10] , p. 75) to conclude

$$|\eta|_{q,r} \leq C|||\eta|||$$

if $q,r$ satisfy $\frac{1}{r} + \frac{d}{2q} = \frac{d}{4}$, $(q \geq 2)$. Therefore, if we pick $s,\rho$ so that

$$\frac{1}{\rho} + \frac{d}{2}\frac{\lambda - 1}{s\lambda} = \frac{d}{4} \tag{3.39}$$

we can assert from (3.38) that

$$|||\eta||| \leq C \left[ \int_0^T dt \, (Meas \ A_k(t))^{\frac{\lambda - 1}{s' \lambda}\rho'} \right]^{\frac{1}{\rho'}} \tag{3.40}$$

For $k \geq 1$ we can also write

$$|||\eta||| \leq Ck \left[ \int_0^T dt \, (Meas \ A_k(t))^{\frac{\lambda - 1}{s' \lambda}\rho'} \right]^{\frac{1}{\rho'}} .$$

It then follows from the theorem in [10] *loc. cit.*, p. 102, (Theorem 6.1) that if we can write

$$\frac{\lambda - 1}{s' \lambda}\rho' = \frac{r}{q} \, , \quad \frac{1}{r} + \frac{d}{2q} = \frac{d}{4}$$

with $\frac{1}{\rho'} > \frac{1}{r}$, then $\eta$ is bounded. Expressing $\frac{1}{q} = \frac{\lambda - 1}{\lambda} \frac{\rho'}{s' r}$, then

$$1 + \frac{d}{2} \frac{\lambda - 1}{\lambda} \frac{\rho'}{s'} = \frac{d}{4} r. \tag{3.41}$$

Now from (3.37) it follows easily that

$$1 + \frac{d}{2} \frac{\lambda - 1}{\lambda} \frac{\rho'}{s'} = \rho' \left[ 1 + \frac{d}{2} \left( \frac{1}{2} - \frac{1}{\lambda} \right) \right]$$

hence

$$\frac{r}{\rho'} = \frac{4}{d} \left[ 1 + \frac{d}{4} - \frac{d}{2\lambda} \right] > 1$$

since $\lambda > \frac{d}{2}$. □

## 4. Singular Perturbations with Reflected Diffusion

### 4.1 Assumptions and notation

One can apply the ergodic theory of Theorem 3.2 to solve some problems of singular perturbations in a similar way as in section 2. Let us consider $f$, $g$, and $h$ continuous functions

$$f(x, y, v): \mathbb{R}^n \times \mathbb{R}^d \times U \to \mathbb{R}^n$$

$$g(x,y,v): \mathbb{R}^n \times \mathbb{R}^d \times U \to \mathbb{R}^d \tag{4.1}$$

$$l(x,y,v): \mathbb{R}^n \times \mathbb{R}^d \times U \to R$$

$$U_{ad} \text{ compact subset of } U \text{ (metric space)} \tag{4.2}$$

We consider on a system $(\Omega, A, P, F^t)$ where $b(t)$ and $w(t)$ are two independent, standard Wiener processes with values in $\mathbb{R}^d$ and $\mathbb{R}^n$, respectively,

$$x(t) = x + \sqrt{2}w(t) \tag{4.3}$$

$$dy_\epsilon(t) = \left(\frac{2}{\epsilon}\right)^{1/2} db - \chi_{\partial B}(y(t))\,\mu\, d\eta, \quad y_\epsilon(0) = y$$

An admissible control $v(t)$ is a process with values in $U_{ad}$ which is adapted to $F^t$. We consider the processes $b_v^\epsilon(t)$ and $w_v^\epsilon(t)$ as in (2.4) (2.5), and the change of probability $P^\epsilon$ given by (2.6). For the system $(\Omega, A, P, F^t)$ we thus have

$$dx = f(x(t), y_\epsilon(t), v(t))\, dt + \sqrt{2}\, dw_\epsilon(t)$$

$$dy_\epsilon = \frac{1}{\epsilon} g(x(t), y_\epsilon(t), v(t))\, dt + \left(\frac{2}{\epsilon}\right)^{1/2} db_\epsilon(t) \tag{4.4}$$

$$x(0) = x, \quad y_\epsilon(0) = y$$

Considering $\tau_x$ the first exit time of $x(t)$ from O, we compute

$$J^\epsilon_{x,y}(v(\cdot)) = E^\epsilon \int_0^{\tau_x} l(x(t), y_\epsilon(t), v(t))\, e^{-\beta t}\, dt \tag{4.5}$$

and define

$$u_\epsilon(x,y) = \operatorname*{Inf}_{v(\cdot)} \{ J^\epsilon_{x,y}(v(\cdot)) \} \tag{4.6}$$

Then $u_\epsilon$ is the unique solution of the H.J.B. problem

$$-\epsilon\Delta_x u_\epsilon - \frac{1}{\epsilon}\Delta_y u_\epsilon + \beta u_\epsilon = H(x, D_x u_\epsilon, y, \frac{1}{\epsilon} D_y u_\epsilon)$$

$$u_\epsilon = 0 \text{ for } x \in \Gamma, \ \forall\ y \tag{4.7}$$

$$\frac{\partial u_\epsilon}{\partial \mu}\Big|_{\partial B} = 0 \quad \forall\ x$$

$$u_\epsilon \in H^1(O \times B), \quad u_\epsilon \in C^0(\overline{O \times B}).$$

One can also choose a Borel function $v_\epsilon(x,y)$ such that

$$L(x, D_x u_\epsilon, y, \frac{1}{\epsilon} D_y u_\epsilon, v_\epsilon(x,y)) = H(x, D_x u_\epsilon, y, \frac{1}{\epsilon} D_y u_\epsilon), \text{ a.e.} \tag{4.8}$$

### 4.2 Convergence

Now the same theory as the one developed from section 2.2 to 2.4 can be carried over to study the limit of (4.7). We just state the result.

For $x, p$ parameters we solve the ergodic problem of the type (3.26)

$$-\Delta_y \phi + \chi(x,p) = H(x,p,y,D_y \phi), \quad \frac{\partial \phi}{\partial \mu} \mid_{\partial B} = 0 \tag{4.9}$$

and the limit problem is given by

$$-\Delta u + \beta u = \chi(x, Du), \quad u_\Gamma = 0, \quad u \in W^{2,p}(O) \tag{4.10}$$

**Theorem 4.1.** *Assume (4.1)(4.2). Then one has*

$$u_\epsilon \to u \quad \text{in } H^1(O \times B) \text{ strongly} \tag{4.11}$$

The same considerations as in sections 2.5 and 2.6 carry over to this case.

## 5. Singular perturbations in the case of a linear fast system

### 5.1 Study of a linear system

Let us consider the following linear system

$$dy = G(v)\, y\, dt + \sum_{r=1}^{\bar{r}} \sigma_r\, y\, db_r(t), \quad y(0) = y. \tag{5.1}$$

where

$$G(v): U_{ad} \to L(\mathbb{R}^d; \mathbb{R}^d); \text{ is a continuous bounded function} \tag{5.2}$$

$$\sigma_r \in L(\mathbb{R}^d : \mathbb{R}^d)$$

$$U_{ad} \text{ compact subset of a metric space } U. \tag{5.3}$$

$b_r(t)$ independent standard scalar Wiener processes on $(\Omega, A, P, F^t)$

An admissible control is a process $v(t)$ which is adapted to $F^t$ and takes values in $U_{ad}$.

Note that for any admissible control the solution of (5.1) is defined in a strong sense.

It is convenient to associate to the process $y(t)$, its norm and its angular velocity. They are defined by

$$\rho(t) = |y(t)|, \quad \xi(t) = \frac{y(t)}{|y(t)|} \tag{5.4}$$

Because of the linearity of equation (5.1), it turns out that $\xi(t)$ is a diffusion in itself. Namely one has

$$d\,\xi(t) = [\, G(v(t)) - \xi.G\,\xi - \frac{1}{2}\sum_r |\sigma_r\,\xi|^2 + \frac{3}{2}\sum_r |\xi.\sigma_r\,\xi|^2$$

$$- \sum_r \xi.\sigma_r\,\xi\,]\,\xi\,dt + \sum_r (\sigma_r - \xi.\sigma_r\,\xi)\,\xi\,db_r \tag{5.5}$$

$$\xi(0) = \frac{y}{|y|}$$

Note that $|\xi(t)| = 1$, as can be easily be checked from (5.5). One can obtain (5.5) by applying Itô's formula. Moreover, $\rho(t)$ satisfies

$$d\,\rho(t) = \rho\left[\xi.G\,\xi + \frac{1}{2}\sum_r |\sigma_r\,\xi|^2 - \frac{1}{2}\sum_r |\xi.\sigma_r\,\xi|^2\right]dt \tag{5.6}$$

$$+ \sum_r \xi.\sigma_r\,\xi\,db_r$$

Since $\xi(t)$ is a diffusion on the $d$ dimensional sphere $S_d$, it is useful to use local charts in order to write the equation as a diffusion in $\mathbb{R}^{d-1}$. For $d > 2$, one chart is not sufficient. Therefore, to simplify we shall assume that $d = 2$, and give some indications for the general case.

If $d = 2$, we can write

$$\xi_1(t) = \cos\theta(t), \quad \xi_2(t) = \sin\theta(t) \tag{5.7}$$

and we can derive an equation for $\theta(t)$.

We have

$$d\,\xi_1(t) = -\sin\theta\; d\,\theta(t) - \frac{1}{2}\cos\theta(t)\,(d\,\theta)^2$$

$$d\,\xi_2(t) = \cos\theta\; d\,\theta(t) - \frac{1}{2}\sin\theta(t)\,(d\,\theta)^2$$

Hence,

$$d\,\theta = -\,d\,\xi_1 \sin\theta + d\,\xi_2 \cos\theta. \tag{5.8}$$

It is convenient to introduce the vector

$$\tilde{\xi}(t) = \begin{pmatrix} -\sin\theta \\ \cos\theta \end{pmatrix} \tag{5.9}$$

which is orthogonal to $\xi(t)$. We deduce

$$d\,\theta(t) = \Big[\tilde{\xi}\cdot g\,\xi - \sum_r (\tilde{\xi}\cdot\sigma_r\,\xi)\,(\tilde{\xi}\cdot\sigma_r\,\xi)\Big]\,dt \tag{5.10}$$

$$+ \sum_r \tilde{\xi}\cdot\sigma_r\,\xi\,db_r\,.$$

We shall assume that

$$\sum_r (\tilde{\xi}\cdot\sigma_r\,\xi)^2 \geq \alpha > 0,\ \forall\ \theta. \tag{5.11}$$

Therefore, $\theta(t)$ is a nondegenerate diffusion, which is periodic with period $2\pi$.

**Remark 5.1** When $d > 2$, we shall have a local representation

$$\xi = \Gamma(\theta),\ \Gamma \in \mathbb{R}^{d-1} \rightarrow \mathbb{R}^d$$

Consider

$$D\Gamma \in L(\mathbb{R}^{d-1}, \mathbb{R}^d),\ D^2\Gamma \in L(\mathbb{R}^{d-1}; L(\mathbb{R}^{d-1}; \mathbb{R}^d)),$$

then

$$d\,\xi(t) = D\Gamma\,d\,\theta + \frac{1}{2}\,D^2\Gamma\,d\,\theta\,d\,\theta$$

hence

$$d\,\theta = (D\Gamma^*\,D\Gamma)^{-1}\,D\Gamma^*\,d\,\xi - \frac{1}{2}\,(D\Gamma^*\,D\Gamma)^{-1}\,D\Gamma^*\,D^2\Gamma\,d\,\theta\,d\,\theta.$$

Note that the relation $|\Gamma(\theta)|^2 = 1$, implies

$$D\Gamma^*\,\Gamma = 0,\ \text{ i.e., } D\Gamma^*\,\xi = 0.$$

We finally obtain

$$d\,\theta = (D\Gamma^*\,D\Gamma)^{-1}\,D\Gamma^*\,[\,(G\,\xi - \sum_r \xi\sigma_r\,\xi\,\sigma_r\,\xi \tag{5.12}$$

$$-\frac{1}{2}\,D^2\Gamma\,(D\Gamma^*\,D\Gamma)^{-1}\,D\Gamma^*\,\sigma_r\,(D\Gamma^*\,D\Gamma)^{-1}\,D\Gamma^*\,\sigma_r\,)\,dt + \sigma_r\,\xi\,db_r\,].$$

In the case (5.10) we have $\tilde{\xi} = D\Gamma$ and $D\Gamma^*\,D\Gamma = 1$, which implies $D\Gamma^*\,D^2\Gamma = 0$. The formulas simplify considerably.

To get a nondegenerate diffusion, we must assume that

$$\sum_r (D\Gamma (D\Gamma^* D\Gamma)^{-1} \phi.\sigma_r \Gamma)^2 \geq c \; |\phi|^2, \; \forall \phi \in \mathbb{R}^{d-1}. \tag{5.13}$$

In particular, assuming Khas'minskii's condition

$$\sum_r (\sigma_r \, y . \psi)^2 \geq m \; |\psi|^2 \; |y|^2, \; \forall y, \psi \in \mathbb{R}^d \tag{5.14}$$

the left side of (5.13) is larger or equal to $m((D\Gamma^* D\Gamma)^{-1} \phi . \phi)^2 \geq c \; |\phi|^2$ by virtue of the properties of the local chart.

## 5.2 A singular perturbation result

Let us consider the following model

$$f(x, \rho, \theta, v): \text{continuous bounded on } \overline{O} \times [0,\infty) \times [0,2\pi] \times U_{ad} \tag{5.15}$$

$$f_x, \; f_\rho, \; f_\theta \text{ bounded}$$

$$G(x, v) \text{ continuous bounded, } \; G_x \text{ bounded} \tag{5.16}$$

$$l(x, \rho, \theta, v): \text{continuous bounded on } \overline{O} \times [0,\infty) \times [0,2\pi] \times U_{ad}; \tag{5.17}$$

$$l_x, \; l_\rho \text{ bounded}$$

Let $(\Omega, A, P, F^t)$ be given and $w(t)$ be a standard $F^t$ $n$-dimensional Wiener process. Assume also that $b_1, \ldots, b_r$ are scalar Wiener processes which are independent of $w(\cdot)$. An admissible control is a process which is adapted to $F^t$, with values in $U_{ad}$. Let $v(t)$ be an admissible control, we solve the equation

$$dx_\epsilon = f(x_\epsilon(t), \rho_\epsilon(t), \theta_\epsilon(t), v(t)) \, dt + \sqrt{2} \, dw, \; x(0) = x \tag{5.18}$$

$$\epsilon \, dy_\epsilon = G(x_\epsilon(t), v(t)) \, y_\epsilon(t) \, dt + \sqrt{\epsilon} \sum_{r=1}^{\bar{r}} \sigma_r \, y_\epsilon(t) \, db_r(t), \; y_\epsilon(0) = y.$$

and $\rho_\epsilon(t)$, $\theta_\epsilon(t)$ are polar coordinates corresponding to $y_\epsilon(t)$. Considering $\xi_\epsilon(t) = \dfrac{y_\epsilon(t)}{\rho_\epsilon(t)}$, we have as in (5.6) (5.10)

$$\epsilon \, d\rho_\epsilon = \rho_\epsilon [\, \xi_\epsilon \cdot G(x_\epsilon, v) \, \xi_\epsilon + \frac{1}{2} \sum_r |\sigma_r \, \xi_\epsilon|^2$$

$$- \frac{1}{2} \sum_r | \xi_\epsilon \sigma_r \xi_\epsilon |^2 ] dt + \rho_\epsilon \sqrt{\epsilon} \sum_r \xi_\epsilon \sigma_r \xi_\epsilon db_r$$

$$\epsilon d \theta_\epsilon = [ \xi_\epsilon \cdot \tilde{G} \xi_\epsilon - \sum_r (\xi_\epsilon \sigma_r \xi_\epsilon) (\xi_\epsilon \tilde{\sigma}_r \xi_\epsilon) ] dt \tag{5.19}$$

$$+ \sqrt{\epsilon} \sum_r \xi_\epsilon \tilde{\sigma}_r \xi_\epsilon db_r$$

Let $\tau_\epsilon$ be the first exit time of $x_\epsilon(t)$ from O ; we consider the cost functional

$$J^\epsilon_{x,\rho,\theta} (v(\cdot)) = E \int_0^{\tau_\epsilon} e^{-\beta t} l (x_\epsilon(t), \rho_\epsilon(t), \theta_\epsilon(t), v(t)) dt \tag{5.20}$$

and we set

$$u_\epsilon(x,\rho,\theta) = J^\epsilon_{x,\rho,\theta} (v(\cdot)) \tag{5.21}$$

**Remark 5.2** The assumption on the derivatives of $f$, $G$, and $l$ allows us to consider a strong formulation of the state equations.

Consider now an ergodic control problem related to $\theta$, according to the theory developed in section 1. Let us write

$$F(x,p,\theta,v) = l(x,0,\theta,v) + p \cdot f(x,0,\theta,v) \tag{5.22}$$

where $x,p$ are parameters. Let us also define

$$g(x,\theta,v) = \xi \cdot \tilde{G}(x,v) \xi - \sum_r (\xi \sigma_r \xi) (\xi \cdot \tilde{\sigma}_r \xi) \tag{5.23}$$

$$a(\theta) = \frac{1}{2} \sum_r (\xi \tilde{\sigma}_r \xi)^2$$

$$H(x,p,\theta,q) = \inf_{v \in U_{ad}} \left[ F(x,p,\theta,v) + q \cdot g(x,\theta,v) \right]$$

(here $q$ is a scalar).

A feedback is a Borel function $v(\theta)$ with values in $U_{ad}$. To any $v(\cdot)$ we associate the invariant probability $m^v \equiv m^v(x,\theta)$, which is the solution of

$$- \frac{\partial^2}{\partial \theta^2} (a(\theta) m^v) + \frac{\partial}{\partial \theta} (m^v g) = 0 \tag{5.26}$$

$m$ periodic.

Since we are considering a one dimensional problem, we have in fact an explicit formula of the form

$$a\, m^{v} = K_1 \exp \int_0^{\theta} \frac{g}{a}\, d\beta + K \int_0^{\theta} (\exp \int_{\tau}^{\theta} \frac{g}{a}\, d\sigma)\, d\tau \tag{5.27}$$

in which $K_1$, and $K$ are fixed in order that $m$ be periodic in $\theta$ and its integral over the interval $(0,2\pi)$ be equal to one.

We then consider

$$\chi(x,p) = \underset{v(\cdot)}{Inf} \int_0^{2\pi} m^{v}(x,\theta)\, F(x,p,\theta,v(\theta))\, d\theta \tag{5.28}$$

and consider the Dirichlet problem

$$-\Delta u + \beta u = \chi(x, Du),\quad u\,|_{\Gamma} = 0,\quad u \in W^{2,p}(O). \tag{5.29}$$

To get the convergence of $u_\epsilon$ to $u$, we shall need an additional assumption. Let us consider

$$k(x,\theta,v) = \xi(\theta)\cdot G(x,v)\, \xi(\theta) + \frac{1}{2}\sum_r |\sigma_r \xi|^2 - \frac{1}{2}\sum_r |\xi\sigma_r \xi|^2 \tag{5.30}$$

and assume that

$$k(x,\theta,v) \leq -\gamma,\ \gamma > 0,\ \forall\, x,\theta,v. \tag{5.31}$$

We shall also assume Khas'minskii's condition (5.14).

Let us set, besides the notation (5.24),

$$b(\theta) = \frac{1}{2}\sum_r (\xi\sigma_r \xi)^2 \tag{5.32}$$

$$c(\theta) = \frac{1}{2}\sum_r (\xi\sigma_r \xi)\,(\tilde{\xi}\sigma_r \xi)$$

We have

$$\begin{pmatrix} a(\theta) & c(\theta) \\ c(\theta) & b(\theta) \end{pmatrix} \begin{pmatrix} \lambda \\ \mu \end{pmatrix} \begin{pmatrix} \lambda \\ \mu \end{pmatrix} = \sum_r ((\lambda \tilde{\xi} + \mu \xi)\sigma_r\, \xi)^2 \tag{5.33}$$

$$\geq m\, |\lambda \tilde{\xi} + \mu \xi|^2 = m(\lambda^2 + \mu^2)$$

and hence, it is not degenerate. The function $u_\epsilon$ defined by (5.21) is thus the solution of the Bellman equation

$$-\Delta u_\epsilon - \frac{1}{\epsilon}\rho^2 b(\theta) \frac{\partial^2 u_\epsilon}{\partial \rho^2} - \frac{1}{\epsilon} a(\theta) \frac{\partial^2 u_\epsilon}{\partial \rho^2} - \frac{2}{\epsilon}\rho\, c(\theta) \frac{\partial^2 u_\epsilon}{\partial \rho \partial \theta} + \beta u_\epsilon$$

$$= H(x, Du_\epsilon, \theta, \frac{1}{\epsilon}\frac{\partial u_\epsilon}{\partial \theta}, \rho, \frac{\rho}{\epsilon}\frac{\partial u_\epsilon}{\partial \rho}) \tag{5.34}$$

$$u_\epsilon |_\Gamma = 0, \quad u_\epsilon \text{ periodic in } \mathbf{O}$$

where we have set

$$H(x, p, \theta, q, \rho, \lambda) = \operatorname*{Inf}_{v \in U_{ad}} [\, l(x, \rho, \theta, v) + p \cdot f(x, \rho, \theta, v) \tag{5.35}$$

$$+ q\, g(x, \theta, v) + \lambda\, k(x, \theta, v)\,]$$

and clearly the function defined in (5.35) coincides with $H(x, p, \theta, q, 0, 0)$.

Let us set

$$\tilde{u}_\epsilon(x, \theta) = u_\epsilon(x, 0, \theta)$$

which is the solution of

$$-\Delta \tilde{u}_\epsilon - \frac{1}{\epsilon}\, a(\theta) \frac{\partial^2 \tilde{u}_\epsilon}{\partial \theta^2} + \beta \tilde{u}_\epsilon = H(x, D\tilde{u}_\epsilon, \theta, \frac{1}{\epsilon}\frac{\partial \tilde{u}_\epsilon}{\partial \theta}, 0, 0) \tag{5.36}$$

$$\tilde{u}_\epsilon |_\Gamma = 0, \quad \tilde{u}_\epsilon \text{ periodic in } \mathbf{O}.$$

Let us also define

$$\varsigma_\epsilon(x, \rho, \theta) = \frac{u_\epsilon(x, \rho, \theta) - \tilde{u}_\epsilon(x, \theta)}{\rho} \tag{5.39}$$

Then we have the following

**Lemma 5.1** *Assume $\gamma$ sufficiently large, then one has*

$$|\, \varsigma_\epsilon\, e^{-\mu\rho}\, |_{L^2} \le c\, \sqrt{\epsilon}, \quad |\, \rho \frac{\partial \varsigma_\epsilon}{\partial \rho}\, e^{-\mu\rho}\, |_{L^2} \le c\, \sqrt{\epsilon} \tag{5.38}$$

$$|\, \frac{\partial \varsigma_\epsilon}{\partial \theta}\, e^{-\mu\rho}\, |_{L^2} \le c\, \sqrt{\epsilon}, \quad |\, D\varsigma_\epsilon\, e^{-\mu\rho}\, |_{L^2} \le c.$$

**Proof.** The function $\varsigma_\epsilon$ satisfies

$$- \Delta \varsigma_\epsilon - \frac{1}{\epsilon}\rho^2 b \frac{\partial^2 \varsigma_\epsilon}{\partial \rho^2} - \frac{1}{\epsilon} a \frac{\partial^2 \varsigma_\epsilon}{\partial \theta^2} - \frac{c\,\rho}{\epsilon} \frac{\partial^2 \varsigma_\epsilon}{\partial \theta\, \partial \rho}$$

$$- \frac{2}{\epsilon}\rho\, b \frac{\partial \varsigma_\epsilon}{\partial \rho} - \frac{c}{\epsilon}\frac{\partial \varsigma_\epsilon}{\partial \theta} + \beta \varsigma_\epsilon \tag{5.39}$$

$$= \frac{1}{\rho}\, [\, H(x, D\tilde{u}_\epsilon + \rho D\varsigma_\epsilon, \theta, \frac{1}{\epsilon}\frac{\partial \tilde{u}_\epsilon}{\partial \theta} + \frac{1}{\epsilon}\rho\frac{\partial \varsigma_\epsilon}{\partial \theta}, \rho, \frac{\rho}{\epsilon}(\varsigma_\epsilon + \rho \frac{\partial \varsigma_\epsilon}{\partial \rho}))$$

$$- H(x, D\tilde{u}_\epsilon, \theta, \frac{1}{\epsilon}\frac{\partial \tilde{u}_\epsilon}{\partial \theta}, 0, 0)\, ].$$

Therefore, for a convenient feedback $v_\epsilon(x, \rho, \theta)$

$$- \Delta \varsigma_\epsilon - \frac{1}{\epsilon} \rho^2 b \frac{\partial^2 \varsigma_\epsilon}{\partial \rho^2} - \frac{1}{\epsilon} a \frac{\partial^2 \varsigma_\epsilon}{\partial \theta^2} - \frac{c\,\rho}{\epsilon} \frac{\partial^2 \varsigma_\epsilon}{\partial \rho\, \partial \theta} \tag{5.40}$$

$$- \frac{2}{\epsilon} \rho\, b \frac{\partial \varsigma_\epsilon}{\partial \rho} - \frac{c}{\epsilon} \frac{\partial \varsigma_\epsilon}{\partial \theta} + \beta \varsigma_\epsilon$$

$$\leq \left[ \frac{l(x,\rho,\theta,v_\epsilon) - l(x,0,\theta,v_\epsilon)}{\rho} \right] + D\bar{u}_\epsilon \cdot \left[ \frac{f(x,\rho,\theta,v_\epsilon) - f(x,0,\theta,v_\epsilon)}{\rho} \right]$$

$$+ D \varsigma_\epsilon \cdot f(x,\rho,\theta,v_\epsilon) + k(x,\theta,v_\epsilon) \left( \frac{\varsigma_\epsilon}{\epsilon} + \frac{\rho}{\epsilon} \frac{\partial \varsigma_\epsilon}{\partial \rho} \right) + \frac{1}{\epsilon} \frac{\partial \bar{\varsigma}_\epsilon}{\partial \theta} g(x,\theta,v_\epsilon).$$

We multiply by $\varsigma_\epsilon^+ e^{-2\mu\rho}$ and integrate. We get using (5.31) and canceling terms

$$\int |D \varsigma_\epsilon^+|^2 e^{-2\mu\rho}\, dn d\theta d\rho + \frac{1}{\epsilon} \int b \left( \frac{\partial \varsigma_\epsilon^+}{\partial \rho} \right)^2 \rho^2 e^{-2\mu\rho}$$

$$- \frac{2\mu}{\epsilon} \int b \frac{\partial \varsigma_\epsilon^+}{\partial \rho} \varsigma_\epsilon^+ \rho^2 e^{-2\mu\rho} + \frac{1}{\epsilon} \int a \left( \frac{\partial \varsigma_\epsilon^+}{\partial \theta} \right)^2 e^{-2\mu\rho}$$

$$+ \frac{1}{\epsilon} \int a' \frac{\partial \varsigma_\epsilon^+}{\partial \theta} \varsigma_\epsilon^+ e^{-2\mu\rho} + \frac{1}{\epsilon} \int c \frac{\partial \varsigma_\epsilon^+}{\partial \theta} \frac{\partial \varsigma_\epsilon^+}{\partial \rho} \rho\, e^{-2\mu\rho} - \frac{2\mu}{\epsilon} \int c \frac{\partial \varsigma_\epsilon^+}{\partial \theta} \varsigma_\epsilon^+ \rho\, e^{-2\mu\rho}$$

$$+ \beta \int (\varsigma_\epsilon^+)^2 e^{-2\mu\rho} \leq K \left[ \int e^{-2\mu\rho} \varsigma_\epsilon^+ + \int |D\bar{u}_\epsilon| \varsigma_\epsilon^+ e^{-2\mu\rho} \right.$$

$$+ \int |f|\, |D \varsigma_\epsilon^+|\, |\varsigma_\epsilon^+|\, e^{-2\mu\rho} + \int \frac{k}{\epsilon} (\varsigma_\epsilon^+)^2 \left(\frac{1}{2} + \mu\rho\right) e^{-2\mu\rho}$$

$$+ \frac{1}{\epsilon} \int g \frac{\partial \bar{\varsigma}_\epsilon^+}{\partial \theta} z_\epsilon^+ e^{-2\mu\rho}$$

hence also

$$\int |D \varsigma_\epsilon^+|^2 e^{-2\mu\rho} + \frac{1}{\epsilon} \int b \left( \frac{\partial \varsigma_\epsilon^+}{\partial \rho} \right)^2 \rho^2 e^{-2\mu\rho} + \frac{1}{\epsilon} \int a \left( \frac{\partial \varsigma_\epsilon^+}{\partial \theta} \right)^2 e^{-2\mu\rho}$$

$$+ \frac{1}{\epsilon} \int c \frac{\partial \varsigma_\epsilon^+}{\partial \theta} \frac{\partial \varsigma_\epsilon^+}{\partial \rho} \rho\, e^{-2\mu\rho} \leq \frac{1}{\epsilon} \int (g - a') \frac{\partial \varsigma_\epsilon^+}{\partial \theta} \varsigma_\epsilon^+ e^{-2\mu\rho}$$

$$+ \frac{2\mu}{\epsilon} \int b \frac{\partial \varsigma_\epsilon^+}{\partial \rho} \varsigma_\epsilon^+ \rho^2 e^{-2\mu\rho} + \int \left( \frac{k}{2\epsilon} - \beta \right) (\varsigma_\epsilon^+)^2 e^{-2\mu\rho}$$

$$+ \frac{1}{\epsilon} (k - c')\, \mu\rho\, (\varsigma_\epsilon^+)^2 e^{-2\mu\rho} + \int |f|\, |D \varsigma_\epsilon^+|\, |\varsigma_\epsilon^+|\, e^{-2\mu\rho}$$

$$+ K \left[ \int e^{-2\mu\rho} \varsigma_\epsilon^+ + \int |D\bar{u}_\epsilon| \varsigma_\epsilon^+ e^{-2\mu\rho} \right].$$

Note that

$$g - a' = \bar{\xi} \cdot G \xi - \sum_r \xi \sigma_r \bar{\xi} \sigma_r \bar{\xi}$$

$$k - c' = \xi . G \xi + \frac{1}{2} \sum_r |\sigma_r \xi|^2$$

$$- \frac{1}{2} \sum_r [(\tilde{\xi} \sigma_r \xi)^2 + (\xi \sigma_r \tilde{\xi})(\tilde{\xi} \sigma_r \xi) + (\tilde{\xi} \sigma_r \xi)(\xi \sigma_r \tilde{\xi})].$$

For $\gamma$ sufficiently large we can then deduce that

$$|\varsigma_\epsilon^+|_{L^2} \leq C \sqrt{\epsilon}, \quad |\rho \frac{\partial \varsigma_\epsilon^+}{\partial \rho}|_{L^2} \leq C \sqrt{\epsilon} \tag{5.41}$$

$$|\frac{\partial \varsigma_\epsilon^+}{\partial \theta}| \leq C \sqrt{\epsilon}, \quad |D \varsigma_\epsilon^+| \leq C .$$

Similarly, we can pick a feedback for which the reverse of (5.40) holds. Multiplying by $\varsigma_\epsilon^- e^{-2\mu\rho}$ and making similar calculations, we obtain the same estimates as (5.41) for $\varsigma_\epsilon^-$. The desired result follows. □

We can then state the following

**Theorem 5.1.** *Assume (5.14), (5.15), (5.16), (5.17) and (5.31) with $\gamma$ sufficiently large. Then we have*

$$(u_\epsilon(x,\rho,\theta) - u(x))\, e^{-\mu\rho} \to 0 \text{ in } L^2 \tag{5.42}$$

*where $\mu > 0$ is arbitrary.*

**Proof.** From Theorem 2.1, we have $\tilde{u}_\epsilon(x,\theta) - u(x) \to 0$ in $H^1$. This result combined with the estimates of Lemma 5.1 implies the desired result (5.42). Note that $\mu$ is arbitrary $> 0$. □

## 5.3 The smooth case

We shall make some formal calculations in the smooth case and derive estimates on the rate of convergence.

We define $\phi_0(x,\theta)$, $\phi_1(x,\theta)$ as follows:

$$- \Delta u + \beta u - a \frac{\partial^2 \phi_0}{\partial \theta^2} = H(x, Du, \theta, \frac{\partial \phi_0}{\partial \theta}, 0, 0), \quad \phi_0 \text{ periodic} \tag{5.43}$$

$$- a \frac{\partial^2 \phi_1}{\partial \theta^2} - c \frac{\partial \phi_1}{\partial \theta} = H_\rho(x, Du, \theta, \frac{\partial \phi_0}{\partial \theta}, 0, 0) \tag{5.44}$$

$$+ \; H_\lambda(x \, , Du \, ,\theta, \frac{\partial \phi_0}{\partial \theta},0,0) \; \phi_1 \; + \; H_q \, (x \, ,Du \, ,\theta, \frac{\partial \phi_0}{\partial \theta}, \, 0,0) \, \frac{\partial \phi_1}{\partial \theta}$$

To solve (5.43) we apply ergodic control theory which determines $u$, $\phi_0$ simultaneously. One then solves (5.44) provided

$$H_\lambda(x \, ,p \, ,\theta,q \, ,0,0) \leq \; - \; \gamma \, , \gamma > 0 \tag{5.45}$$

which is true in the case (5.31). We can then obtain an estimate for the difference

$$\tilde{u}_\epsilon = u_\epsilon \; - \; u \; - \epsilon \; \phi_0 \; - \; \epsilon \, \rho \, \phi_1 \tag{5.46}$$

We can write the equation

$$- \; \Delta \tilde{u}_\epsilon \; - \; \frac{1}{\epsilon} \, \rho^2 \, b \, (\theta) \, \frac{\partial^2 \tilde{u}_\epsilon}{\partial \rho^2} \; - \; \frac{1}{\epsilon} \, a \, (\theta) \, \frac{\partial^2 \tilde{u}_\epsilon}{\partial \theta^2} \tag{5.47}$$

$$- \; \frac{\rho c}{\epsilon} \, \frac{\partial^2 \tilde{u}_\epsilon}{\partial \rho \partial \theta} \; + \; \beta \, \tilde{u}_\epsilon$$

$$= \epsilon \; \Delta \phi_0 \; + \; \epsilon \, \rho \, \Delta \phi_1 \; - \; \epsilon \, \beta \, \phi_0 \; - \; \epsilon \, \beta \, \rho \, \phi_1$$

$$+ \; H \, (x \, ,Du \; + \epsilon \, D \, \phi_0 \; + \epsilon \, \rho \, D \, \phi_1 \; + \; D\tilde{u}_\epsilon,\theta, \; \frac{\partial \phi_0}{\partial \theta} \; + \; \rho \, \frac{\partial \phi_1}{\partial \theta} \; + \; \frac{1}{\epsilon} \, \frac{\partial \tilde{u}_\epsilon}{\partial \theta}, \, \rho, \, \rho \, \phi_1 \; + \; \frac{\rho}{\epsilon} \, \frac{\partial \tilde{u}_\epsilon}{\partial \rho})$$

$$- \; H \, (x \, ,Du \, ,\theta, \, \frac{\partial \phi_0}{\partial \theta} \, ,0,0) \; - \; \rho \, H_\rho(x \, , \, Du \, ,\theta, \, \frac{\partial \phi_0}{\partial \theta},0,0)$$

$$- \; \rho \, \phi_1 \, H_\lambda(x \, , \, Du \, ,\theta, \, \frac{\partial \phi_0}{\partial \theta},0,0) \; - \; \rho \, \frac{\partial \phi_1}{\partial \theta} \, H_q \, (x \, , \, Du \, ,\theta, \, \frac{\partial \phi_0}{\partial \theta},0,0)$$

$$= H_p \, (\epsilon \, D \, \phi_0 \; + \; \epsilon \, \rho \, D \, \phi_1 \; + \; D\tilde{u}_\epsilon) \; + \; \frac{1}{\epsilon} \, H_q \, \frac{\partial \tilde{u}_\epsilon}{\partial \theta} \; + \; \frac{\rho}{\epsilon} \, H_\lambda \, \frac{\partial \tilde{u}_\epsilon}{\partial \rho}$$

$$+ \; H \, (x \, ,Du \, ,\theta, \, \frac{\partial \phi_0}{\partial \theta} \; + \; \rho \, \frac{\partial \phi_1}{\partial \theta}, \, \rho, \, \rho \, \phi_1) \; - \; H \, (x \, ,Du \, ,\theta, \, \frac{\partial \phi_0}{\partial \theta}, \, 0,0,)$$

$$- \; \rho \, H_\rho \; - \; \rho \, \phi_1 \, H_\lambda \; - \; \rho \, \frac{\partial \phi_1}{\partial \theta} \, H_q \; + \; \epsilon \, \Delta \phi_0 \; + \; \epsilon \, \rho \, \Delta \phi_1 \; - \; \epsilon \, \beta \, \phi_0 \; - \; \epsilon \, \rho \, \beta \, \phi_1$$

$$= H_p \; D\tilde{u}_\epsilon \; + \; \frac{1}{\epsilon} \, H_q \; \frac{\partial \tilde{u}_\epsilon}{\partial \theta} \; + \; \frac{\rho}{\epsilon} \, H_\lambda \; \frac{\partial \tilde{u}_\epsilon}{\partial \rho} \; + \; \chi_\epsilon(x \, ,\theta,\rho)$$

with the property

$$| \, \chi_\epsilon \, | \; \leq C \, \epsilon \; + \; C \, \epsilon \, \rho \; + \; C \, \rho^2 \tag{5.48}$$

We can then give the probabilistic interpretation of $u$. Considering a system

$$dx_\epsilon = H_p \;\; dt \;\; + \; \sqrt{2} \; dw$$

$$\epsilon \; d \, \theta_\epsilon = H_q \;\; dt \;\; + \; \sqrt{\epsilon} \sum_r \tilde{\xi} \, \sigma_r \;\, \xi \; db_r$$

$$\epsilon\, d\rho = \rho\, H_\lambda\, dt \;+\; \sqrt{\epsilon} \sum_r \xi\, \sigma_r\, \xi\, db_r$$

we have

$$\tilde{u}_\epsilon(x,\rho,\theta) = E \int_0^{\tau_\epsilon} e^{-\beta t}\, \chi_\epsilon(x_\epsilon(t),\, \theta_\epsilon(t),\, \rho_\epsilon(t))\, dt$$

$$+\; E\, \tilde{u}_\epsilon(x_\epsilon(\tau_\epsilon), \theta_\epsilon(\tau_\epsilon),\, \rho_\epsilon(\tau_\epsilon))\, e^{-\beta \tau_\epsilon}$$

and from (5.46), (5.48) it follows that

$$|\, \tilde{u}_\epsilon(x,\rho,\theta)\, | \;\leq\; C(\epsilon \;+\; \epsilon\, \rho \;+\; E \int_0^{\tau_\epsilon} e^{-\beta t}\, \rho_\epsilon(t)^2\, dt\, ).$$

Using the expression for $\rho_\epsilon$, and assuming that

$$-\; 2\gamma \;+\; \sum_r (\xi\, \sigma_r\, \xi)^2 \;\leq\; -\; \delta$$

we can conclude that

$$E\, \rho_\epsilon(t)^2 \leq \rho^2\, e^{-\delta \frac{t}{\epsilon}}$$

which implies

$$|\, \tilde{u}_\epsilon\, (x,\rho,\theta)\, | \;\leq\; C(\epsilon \;+\; \epsilon\, \rho \;+\; \epsilon\, \rho^2) \tag{5.49}$$

which justifies the convergence of $u_\epsilon$ to $u$.

## 6. Ergodic Control for diffusions in the whole space

### 6.1 Assumptions - Notation

Let us consider basically the situation of section 1, except that now the periodicity is left out. Of course, additional assumptions are necessary to recover the ergodicity. We consider

$$g\,(y\,,v\,)\colon \mathbb{R}^d \times U \to \mathbb{R}^d$$

$$l\,(y\,,v\,)\colon \mathbb{R}^d \times U \to \mathbb{R}^d \tag{6.1}$$

continuous and bounded

$$U_{ad} \text{ compact subset of } U \text{ a metric space.} \tag{6.2}$$

For a given feedback $v\,(y\,)$, which is a Borel function with values in $U_{ad}$, we shall solve in a weak sense the stochastic differential equation.

$$dy = (Fy + g(y, v(y)))\, dt + \sqrt{2}\, db_r(t), \quad y(0) = y. \tag{6.3}$$

The linear term $Fy$ will be useful to ensure an ergodicity property later on ($F$ a stable matrix). The Brownian motion $b_r$ is defined through a Girsanov transformation as explained in section 1. We can find a system $(\Omega, A, F^t, P_v^y)$ such that (6.3) holds.[4] We then consider the function

$$K_y^\alpha(v(\cdot)) = E_v^y \int_0^\infty e^{-\alpha t}\, l(y(t), v(t))\, dt \tag{6.4}$$

where

$$v(t) = v(y(t)), \text{ and we set}$$

$$\phi_\alpha(y) = \underset{v(\cdot)}{Inf}\, K_y^\alpha(v(\cdot)). \tag{6.5}$$

Setting $A = -\Delta - Fy \cdot D$, we can assert that $\phi_\alpha$ is the solution of

$$A\, \phi_\alpha + \alpha\, \phi_\alpha = H(y, D\, \phi_\alpha) \tag{6.6}$$

$$\phi_\alpha \text{ bounded}, \ \phi_\alpha \in W^{2,p,\mu}(\mathbb{R}^n),\ 2 \leq p < \infty$$

where $W^{2,p,\mu}(\mathbb{R}^n)$ denotes a Sobolev space with weight

$$\beta_\mu(y) = e^{-\mu(1 + |y|^2)^{1/2}} \tag{6.7}$$

and

$$L^{p,\mu} = \{z(y) \mid z\, \beta\mu \in L^p(\mathbb{R}^d)\}$$

$$W^{2,p,\mu} = \{ z \in L^{p,\mu} \mid \frac{\partial z}{\partial y_i}, \frac{\partial^2 z}{\partial y_i\, \partial y_j} \in L^{p,\mu} \}$$

## 6.2 Invariant measures

Since the diffusion $y(\cdot)$ does not lie in a compact set, some assumptions on the drift $g$ are necessary to ensure ergodicity. We shall mainly use the results of Khas'minskii [2] We make the following assumption:

*There exists a bounded smooth domain* D *and a function* $\psi$ *which is continuous and locally bounded on* $\mathbb{R}^d$ – D, $\geq 0$, $\psi \in W_{loc}^{2,p}(\mathbb{R}^d$ – D), *and*

[4]We limit ourselves to feedback controls, since only those will appear in the singular perturbation problem that we shall solve eventually. Of course, this is not at all necessary for the ergodic control itself.

$$A\psi - g(y,v)\cdot D\psi \geq l, \forall\, v,y \in \mathbb{R}^d \quad (6.8)$$

In general one can try to find $\psi$ of the form

$$\psi(y) = \operatorname{Log} Q(y) + k \qquad (6.9)$$

where

$$Q(y) = \frac{1}{2} My\cdot y + m\cdot y + \rho \qquad (6.10)$$

*$M$ symmetric positive definite $Q \geq 0$;*

**D** *is a region containing the zeros of $Q$.*

The following condition must be achieved to get (6.8)

$$\frac{|My + m|^2}{\frac{1}{2}My^2 + my + \rho} - \operatorname{tr} M - (Fy + g(y,v))\cdot(My + m) \qquad (6.11)$$

$$\geq \frac{1}{2}My^2 + my + \rho, \ \forall\, y \in \mathbb{R}^d - \mathbf{D}$$

for a convenient choice of $M$, $m$, and $\rho$. For instance if $d = 2$, we can take $M = I$, $m = 0$, and$\rho = 0$ and (6.9) is satisfied provided that for instance

$$F \leq (-\frac{1}{2} - \delta) I \qquad (6.12)$$

and **D** is a neighborhood of 0 sufficiently big. We now follow Khas'minskii [2] to introduce a compact set with an ergodic Markov chain as in section 1.4. We make the following assumption: There exists a bounded smooth domain **D** and a function $\psi \geq 0$ which is continuous and locally bounded on $\mathbb{R}^d - \mathbf{D}$ and

$$\psi \in W^{2,p}_{loc}(\mathbb{R}^d - \mathbf{D})$$

$$A\psi - g(y,v) D\psi \geq 1, \ \forall\, v, y \in \mathbb{R}^d - \mathbf{D} \qquad (6.8)$$

$$\psi > 0 < \ \psi \to \infty \text{ as } |y| \to \infty \text{ and } \frac{|D\psi|^2}{\psi} \text{ bounded}$$

In general, one can try to find $\psi$ of the form

$$\psi(y) = \log Q(y) + k \qquad (6.9)$$

where

$$Q(y) = \frac{1}{2} My \cdot y + m \cdot y + \rho \tag{6.10}$$

$M$ symmetric and positive definite and $Q > 0$; $\mathbf{D}$ is a region containing the zeros of $Q$.

The following condition must hold to have (6.8):

$$\frac{|My + m|^2}{\frac{1}{2}My^2 + m \cdot y + \rho} - \operatorname{tr} M - (Fy + g(y,v)) \cdot (My + m) \tag{6.11}$$

$$\geq \frac{1}{2} My \cdot y + m \cdot y + \rho, \quad \forall y \in \mathbb{R}^d - \mathbf{D};$$

for a convenient choice of $M$, $m$, and $\rho$. For instance, if $d = 2$, we can take $M = I$, $m = 0$, $\rho = 0$ and (6.9) is satisfied provided that, for instance

$$F \leq \left(-\frac{1}{2} - \delta\right) I \tag{6.12}$$

and $\mathbf{D}$ is a sufficiently large neighborhood of 0.

Consider a domain $\mathbf{D}_1$ such that $\overline{\mathbf{D}} \subset \mathbf{D}_1$, $\mathbf{D}_1$ smooth and bounded. Let $\Gamma$ and $\Gamma_1$ be the boundaries of $\mathbf{D}$, $\mathbf{D}_1$, respectively. We shall construct a Markov chain on $\Gamma_1$. Let $x \in \mathbb{R}^d$, we define

$$\theta'(x;\omega) = \operatorname{Inf}\{t \mid y_x(t) \in \mathbf{D}\} \tag{6.13}$$

$$\theta(x;\omega) = \operatorname{Inf}\{t \geq \theta'(x;\omega) \mid y_x(t) \notin \mathbf{D}_1\} \tag{6.14}$$

In (6.13), (6.14) $y_x(t)$ is the diffusion (6.3) with initial condition $x$. Using $\psi(x)$, we can write

$$E_v^x \theta'(x) \leq \psi(x). \tag{6.15}$$

This implies also that the exterior Dirichlet problem

$$A\eta - g(y, v(y)) \cdot D\eta = 0, \quad y \in \mathbb{R}^d - \mathbf{D} \tag{6.16}$$

$$\eta|_\Gamma = h, \quad h \in L^\infty(\Gamma)$$

has a bounded solution given explicitly by

$$\eta(x) = E_v^x h(y_x(\theta'(x))). \tag{6.17}$$

The Markov chain on $\Gamma_1$ is then constructed as follows. We define two sequences of stopping times (relative to $F^t$),

$$\tau_0, \tau_1, \tau_2, \ldots..,$$

$$\tau'_1, \tau'_2, \ldots.$$

such that

$$\tau_0 = 0$$

$$\tau_n = Inf \ \{ t > \tau'_n \mid y(t) \notin \mathbf{D}_1 \}, \ n \geq 1$$

$$\tau'_{n+1} = Inf \ \{ t \geq \tau_n \mid y(t) \in \mathbf{D} \}, \ n \geq 0$$

The process $y(t)$ in the brackets is the process defined by (6.3), i.e. with initial condition $y$. Let us set $Y_n = y(\tau_n), n \geq 1$. Then $Y_n \in \Gamma_1$ and is a Markov chain with transition probability defined by

$$E_v^y \left[ \phi(Y_{n+1}) \mid F^{\tau_n} \right] = E_v^x \phi(y_x(\theta(x))) \mid_{x = Y_n}. \tag{6.18}$$

We define the following operator on Borel bounded functions on $\Gamma_1$

$$\mathbf{P} \ \phi(x) = E_v^x \phi(y_x(\theta(x))) \tag{6.19}$$

We can give an analytic formula as follows. Consider the problem

$$A \varsigma - g(y, v(y)) \cdot D \varsigma = 0 \text{ in } \mathbf{D}_1, \ \varsigma \mid_{\Gamma_1} = \phi. \tag{6.20}$$

We first note that

$$E_v^x \phi(y_x(\theta(x))) = E_v^x \varsigma(y_x(\theta'(x))$$

therefore taking account of (6.17) , we have

$$\mathbf{P} \ \phi(x) = \eta(x) \tag{6.21}$$

where $\eta$ denotes the solution of (6.16) corresponding to the boundary condition $h = \varsigma$. Of course, in (6.21) $x \in \Gamma_1$ are the only relevant points. We then have

**Lemma 6.1.** *The operator* **P** *is ergodic.*

**Proof.** We proceed as in the proof of Lemma 3.1. Indeed, defining

$$\lambda_{xy}^v(B) = \mathbf{P} \ \chi_B(x) - \mathbf{P} \ \chi_B(y)$$

$$\forall \ x, y \in \Gamma_1, \ B \text{ Borel subset of } \Gamma_1$$

everything amounts to showing that

$$\sup_{v, x, y, B} \lambda_{xy}^v(B) < 1. \tag{6.22}$$

We have

$$\lambda_{xy}^{v}(B) = \eta(x) - \eta(y)$$

where

$$A\,\eta - g^{v} \cdot D\,\eta = 0 \text{ in } \mathbb{R}^{d} - \mathbf{D} \tag{6.23}$$

$$\eta \mid_{\Gamma} = \varsigma$$

$$A\,\eta - g^{v} \cdot D\,\varsigma = 0 \text{ in } \mathbf{D}_{1}$$

$$\varsigma \mid_{\Gamma_{1}} = \chi_{B}\,.$$

If (6.22) is false, there exists a sequence $(v_k, x_k, y_k, B_k)$ such that

$$\eta_k(x_k) \to 1, \quad \eta_k(y_k) \to 0 \tag{6.24}$$

where $\eta_k$ is defined by the set of relations (6.23) where we set $g_k = g^{v_k}$, $B$ changed in $B_k$. Note that $\varsigma_k$ remains bounded in $L^{\infty}(\mathbf{D}_1)$, and for any $\beta$ in $C_0^{\infty}(\mathbf{D}_1)$, $\beta\,\varsigma_k$ remains bounded in $W^{2,p}(\mathbf{D}_1)$. Moreover, $\eta_k$ remains bounded in $W^{2,p}(\mathbb{R}^d - \mathbf{D})$, and bounded in $L^{\infty}$. Assume $\beta = 1$ on $\overline{\mathbf{D}}$ and consider a subsequence such that

$$\beta\,\varsigma_k \to \beta\,\varsigma^{*} \text{ weakly in } W^{2,p}(\mathbf{D}_1)$$

$$\eta_k \to \eta^{*} \text{ weakly in } W^{2,p}(\mathbb{R}^d - \mathbf{D}).$$

In particular, $\varsigma_k \to \varsigma^{*}$ in $C^0(\Gamma)$, and we have

$$A\,\eta^{*} - g^{*} \cdot D\,\eta^{*} = 0 \text{ in } \mathbb{R}^{d} - \mathbf{D}$$

$$\eta^{*} \mid_{\Gamma} = \varsigma \mid_{\Gamma}$$

where $g^{*}$ denotes a weak limit of $g_k$. Since

$$\eta_k \to \eta^{*} \text{ in } C^0(\Gamma_1)$$

and $x_k \to x^{*}$, $y_k \to y^{*}$, we deduce from (6.24)

$$\eta^{*}(x^{*}) = 1,\ \eta^{*}(y^{*}) = 0,\ x^{*}, y^{*} \in \Gamma_1.$$

But $0 \leq \eta^{*} \leq 1$ and we get a contradiction with the maximum principle, since $\eta^{*}$ cannot be a constant. □

From ergodic theory, it follows (c.f. (1.30))

$$| \mathbf{P}^n \phi(y) - \int_{\Gamma_1} \phi(\eta) \pi(d\sigma) | \leq K \, || \phi || \, e^{-\rho n}, \quad x \in \Gamma_1 \tag{6.25}$$

where $K, \rho$ are uniform with respect to the feedback control $v(\cdot)$, and $\pi = \pi^v$ denotes the invariant probability on $\Gamma_1$.

It follows that, since

$$\mathbf{P}^n \phi(y) = E_v^y \phi(y(\tau_n))$$

we can write

$$| E_v^y \phi(y(\tau_n)) - \int_{\Gamma_1} \phi(\eta)\pi(d\sigma) | \leq K \, || \phi || \, e^{-\rho n}. \tag{6.26}$$

We can then define a probability on $\mathbb{R}^d$, by the formula

$$\int_{\mathbb{R}^d} \Lambda(y) \, d\mu(y) = \frac{\int_{\Gamma_1} [ E_v^\eta \int_0^{\theta(\eta)} \Lambda(y_\eta(t)) \, dt \, ] \, \pi(d\sigma)}{\int_{\Gamma_1} E_v^\eta \, \theta(\eta) \pi(d\sigma)} \tag{6.27}$$

$\forall$ $\Lambda$ Borel bounded in $\mathbb{R}^d$.

Following Khas'minskii, one can then prove that the invariant probability is unique, has a density with respect to Lebesgue measure, denoted by $m = m^v$ which is the solution of

$$A^* m + \text{div}\,(m g^v) = 0, \; m > 0, \tag{6.28}$$

$$\int_{\mathbb{R}^d} m(y) \, dy = 1.$$

where

$$A^* = -\Delta + \text{div}\,(Fy \, \cdot).$$

Consider now the Cauchy problem

$$\frac{\partial z}{\partial t} + A z - g^v Dz = 0 \tag{6.29}$$

$$z(y,0) = \phi(y)$$

We shall prove the following

**Lemma 6.2.** *We have*

$$z(y,1) \leq c \; | \phi |_{L_1} \tag{6.30}$$

**Proof.** We do not use the fundamental solution of (6.29), as in the periodic case, since the coefficients are unbounded. The method is also different from the Neumann case since the domain is unbounded. We first consider the adjoint Cauchy problem

$$-\frac{\partial u}{\partial t} - \Delta u + Fy \cdot Du + \operatorname{tr} Fu + \operatorname{div}(gu) = 0 \tag{6.31}$$

$$u(y,1) = 1$$

and we shall prove that

$$| u(y,t) | \leq C, \; t \in [0,1]. \tag{6.32}$$

For convenience we consider the forward adjoint. This is of course possible.

$$\frac{\partial u}{\partial t} - \Delta u + Fy \cdot Du + \operatorname{tr} Fu + \operatorname{div}(gu) = 0 \tag{6.33}$$

$$u(y,0) = 1.$$

Let $p \geq 1$ and $h(x,t)$ be a function which will be chosen later on. By integration by parts after multiplying (6.33) by $e^{2ph} \, |u|^{2p-2} u$, we deduce the energy equality

$$\frac{1}{2p} \frac{d}{dt} \int_{\mathbb{R}^d} e^{2ph(y,t)} u^{2p}(y,t) \, dt + (2p-1) \int_{\mathbb{R}^d} |Du|^2 e^{2ph} \, |u|^{2p-2} dy \tag{6.34}$$

$$- (2p-1) \int_{\mathbb{R}^d} e^{2ph} \, |u|^{2p-2} u \, Du \cdot g \, dy + \int_{\mathbb{R}^d} u^{2p} e^{2ph} \cdot$$

$$[ - h_t - \Delta h + \frac{2p-1}{2p} \operatorname{tr} F - Fy \cdot Dh - 2p \, |Dh|^2 - 2pg \cdot Dh \, ] \, dy = 0.$$

Let us define

$$v(y,t) = | u(y,t) |^p \, e^{ph(y,t)}$$

we have

$$| u |^{p-2} u \, Du \, e^{hp} = \frac{1}{p} Dv - v \, Dh.$$

Hence, we deduce from (6.34) that

$$\frac{1}{2p}\frac{d}{dt}\int_{\mathbb{R}^d} v^2(y,t)\,dy + \frac{2p-1}{p^2}\int_{\mathbb{R}^d} |Dv|^2\,dy + (2p-1)\int_{\mathbb{R}^d} v^2\,|Dh|^2\,dy$$

$$-\frac{2}{p}(2p-1)\int_{\mathbb{R}^d} v\,Dv\cdot Dh\,dy - (2p-1)\int_{\mathbb{R}^d} v\left(\frac{1}{p}\,g\,Dv - v\,g\,Dh\right)dy$$

$$+\int_{\mathbb{R}^d} v^2\left[-h_t - \Delta h + \frac{2p-1}{2p}\operatorname{tr}F - Fy\cdot Dh - 2\,p\,g\cdot Dh\right]dy = 0.$$

Simplifying, one obtains

$$\frac{1}{2p}\frac{d}{dt}\int_{\mathbb{R}^d} v^2(y,t)\,dy + \frac{2p-1}{p^2}\int_{\mathbb{R}^d} |Dv|^2\,dy - \frac{2p-1}{p}\int_{\mathbb{R}^d} v\,Dv\cdot g\,dy \tag{6.35}$$

$$+\int_{\mathbb{R}^d} v^2\left[-h_t + \frac{p-1}{p}\Delta h + \frac{2p-1}{2p}\operatorname{tr}F - Fy\cdot Dh - |Dh|^2 - g\,Dh\right]dy = 0$$

We now fix the function h, setting

$$h(y,t) = [-Q(t)(y-r(t))^2 - \rho(t)]$$

and choosing the functions $Q(t)$, $r(t)$, and $\rho(t)$ in order that

$$h_t - \Delta h + 2\,|Dh|^2 + Fy\cdot Dh = 0.$$

Performing the calculations we obtain

$$-\dot{Q} + 8\,Q^2 - F^*Q - FQ = 0$$

$$\dot{r} - F\,r = 0$$

$$-\dot{\rho} + 2\operatorname{tr}Q = 0.$$

We take

$$Q(1) = I,\ r(0) = \xi \text{ arbitrary},\ \rho(0) = 0.$$

It follows that $Q(t)$ is the solution of a Riccati equation and satisfies

$$q_0 I \le Q(t) \le q_1 I,\ q_0 \ge 0.$$

We deduce from (6.35)

$$\frac{1}{2}\frac{d}{dt}\int_{\mathbb{R}^d} v^2(y,t)\,dy + \frac{2p-1}{2p^2}\int_{\mathbb{R}^d} |Dv|^2\,dy \leq c\,p\int_{\mathbb{R}^d} v^2(y,t)\,dt.$$

Note that

$$\int_{\mathbb{R}^d} v^2(y,0)\,dy = \int_{\mathbb{R}^d} e^{ph(y,0)}\,dy = \int_{\mathbb{R}^d} e^{-pQ(0)(y-\xi)^2}\,dy$$

$$\leq \int_{\mathbb{R}^d} e^{-Q(0)(y-\xi)^2}\,dy \leq C.$$

Collecting results, we obtain finally

$$\int_{\mathbb{R}^d} v^2(y,t)\,dy + \int_0^1\int_{\mathbb{R}^d} |Dv|^2\,dy\,ds \tag{6.36}$$

$$\leq c\,p^2\left(\int_0^1\int_{\mathbb{R}^d} v^2(y,s)\,dy\,ds + 1\right)$$

where $c$ is a constant which does not depend on $p$ nor $\xi$. Note that (6.36) holds $p \geq 1$.

Using the norm

$$|||v|||^2 = \sup_{t\in[0,1]}\int_{\mathbb{R}^d} v^2(y,t)\,dy + \int_0^1\int_{\mathbb{R}^d} |Dv|^2\,dy\,dt$$

We can assert that

$$|||v||| \leq c\,p\,(|v|_{2,2} + 1) \tag{6.37}$$

recalling the notation

$$|v|_{q,r} = \left[\int_0^1\left(\int_{\mathbb{R}^d} v^q\,dy\right)^{\frac{r}{q}}dt\right]^{\frac{1}{r}}$$

We use the embedding (c.f. [10] p. 75 and Aronson [5] p. 624)

$$|v|_{q,r} \leq c\;|||v|||,\ \text{for}\ \frac{1}{r} + \frac{d}{2q} = \frac{d}{4},\quad q \geq 2$$

we deduce from (6.37)

$$|v|_{q,r} \leq c\,p\left(|v|_{2,2} + 1\right) \tag{6.38}$$

We choose $r > q$, $2 < q < 2 + \frac{4}{d}$. We have

$$| v |_{2,2} \leq c \; | v^{\frac{2}{q}} |^{\frac{q}{2}}_{q,r}.$$

Let us set $\frac{q}{2} = 1 + \alpha$, then we can assert that

$$| v |_{q,r} \leq c \; p \left( | v^{\frac{1}{1+\alpha}} |^{1+\alpha}_{q,r} + 1 \right)$$

We then take $p = (1 + \alpha)^k$, $k = 1,2...$ and note

$$\phi_k = | \; | u \; e^h |^{(1+\alpha)^k} |_{q,r}, \quad k \geq 1.$$

From (6.39) we deduce the induction relation

$$\phi_k \leq C_0 (1 + \alpha)^k \left( \phi_{k-1}^{1+\alpha} + 1 \right) \; k \geq 1. \tag{6.40}$$

Note that by (6.36)

$$\phi_1 \leq C \; ||| \; | u \; e^h |^{1+\alpha} ||| \; \leq C$$

(although we cannot assert that $\phi_0$ is finite).

From a Lemma in [10], p. 95, we deduce from (6.40) that (assuming $C_0 (1 + \alpha) \geq 1$),

$$\phi_k^{(1+\alpha)^{-k}} \leq (1 + \alpha)^{\beta_k} (2C_0)^{\frac{1-(1+\alpha)^{-k}}{\alpha}} (\phi_1 + 1)$$

with

$$\beta_k = \left[ 1 - (1 + \alpha)^{-k} \right] \left( \frac{2}{\alpha} + \alpha^2 \right) - \frac{k}{\alpha} (1 + \alpha)^{-k}$$

Therefore,

$$\lim_{k \to \infty} \phi_k^{(1+\alpha)^{-k}} \leq (1 + \alpha)^{\left( \frac{2}{\alpha} + \alpha^2 \right)} (2 \; C_0)^{\frac{1}{\alpha}} (\phi_1 + 1). \tag{6.41}$$

Since

$$\phi_k^{(1+\alpha)^{-k}} = | u \; e^h |_{q (1+\alpha)^k, \, r (2+\alpha)^k}$$

we get

$$| u(y,t)\, e^{h(y,t)} | \leq \lim_{k \to \infty} \phi_k^{(1+\alpha)^{-k}} \leq C$$

and the constant does not depend on $\xi$. Since for any $y$ and any $t$, we can pick $\xi$ so that $r(t) = y$, we deduce (6.32).

It follows from (6.32), going back to the solution $z$ of (6.29), that (6.42)

$$\int_{\mathbb{R}^d} z(y,t)\, dy = \int_{\mathbb{R}^d} \phi(y)\, u(y,0)\, dy = C\ |\phi|_{L^1}.$$

We consider the function

$$\varsigma(y,t) = e^{-k_0 t} z(y,t)$$

where $k_0$ is a constant which will be fixed later on. Note that $\varsigma$ is the solution of

$$\frac{\partial \varsigma}{\partial t} + A\varsigma - g^v \cdot D\varsigma - k_0 \varsigma = 0 \tag{6.,43}$$

$$\varsigma(y,0) = \phi(y).$$

Let us define

$$E_p(t) = \int_{\mathbb{R}^d} |\varsigma(y,t)|^p\, dy,\ p \geq 1.$$

From (6.42) we can assert that

$$E_1(t) \leq C\ |\phi|_{L^1}. \tag{6.44}$$

We compute

$$\frac{d}{dt} E_{2p}(t) = 2p \int_{\mathbb{R}^d} |\varsigma|^{2p-2} \varsigma \frac{\partial \varsigma}{\partial t}\, dt$$

$$= -2p(2p-1) \int_{\mathbb{R}^d} |\varsigma|^{2p-2} |D\varsigma|^2\, dy$$

$$+ 2p \int_{\mathbb{R}^d} |\varsigma|^{2p-2} \varsigma g \cdot D\varsigma\, dy + \int_{\mathbb{R}^d} |\varsigma|^{2p} (-2k_0 p - \operatorname{tr} F)\, dy.$$

Noting that

$$D |\varsigma|^p = p\ |\varsigma|^{p-2} \varsigma D\varsigma$$

We can write

$$\frac{d}{dt} E_{2p}(t) = \frac{-2p\,(2p-1)}{p^2} \int_{\mathbb{R}^d} |\, D \,|\, \varsigma \,|^p \,|^2\, dy$$

$$+ 2 \int_{\mathbb{R}^d} |\, \varsigma \,|^p\, g\, D \,|\, \varsigma \,|^p\, dy + \int_{\mathbb{R}^d} |\, \varsigma \,|^{2p}\, (-2k_0\, p - \operatorname{tr} F)\, dy.$$

Choosing $k_0$ so that

$$2\, k_0 \geq 2 + \|g\| - \operatorname{tr} F$$

we deduce

$$\frac{d}{dt} E_{2p}(t) \leq - \frac{2p-1}{p} \left[ \int_{\mathbb{R}^d} |\, D \,|\varsigma|^p \,|^2\, dy + \int_{\mathbb{R}^d} |\varsigma|^{2p}\, dy \right]. \tag{6.45}$$

Using the interpolation inequality

$$|\, u \,|_{L^2}^2 \leq C\, \|u\|_{H^1}^{\frac{2d}{d+2}}\, |\, u \,|_{L^1}^{\frac{4}{d+2}}$$

applied with $u = |\varsigma|^p$, we get

$$\|\, |\varsigma|^p \,\|_{H^1}^2 \geq C\, \frac{E_{2p}(t)^{\frac{d+2}{d}}}{E_p(t)^{\frac{4}{d}}}$$

hence, from (6.45) the inequality

$$\frac{d}{dt} (E_{2p}(t))^{-\frac{2}{d}} \geq C\, \frac{2p-1}{p}\, (E_p(t))^{-\frac{4}{d}}$$

or also

$$\frac{d}{dt} (E_{2^k}(t))^{-\frac{2}{d}} \geq C_0\, \frac{2^k - 1}{2^k}\, (E_{2^{k-1}}(t))^{-\frac{4}{d}}, \quad k \geq 1. \tag{6.46}$$

Applying (6.46) with $k = 1$, yields (taking account of (6.44))

$$\frac{d}{dt} (E_2(t))^{-\frac{2}{d}} \geq \frac{C_0}{2}\, C^{-\frac{4}{d}}\, |\, \phi \,|_{L^1}^{-\frac{4}{d}}$$

and integrating, one obtains

$$(E_2(t))^{-\frac{2}{d}} \geq \frac{C_0}{2}\, C^{-\frac{4}{d}}\, |\, \phi \,|_{L^1}^{-\frac{4}{d}}\, t = \frac{\chi}{2}\, t.$$

Following Besala [14] , we deduce by induction that

$$(E_{2^k}(t))^{\frac{1}{2^k}} \leq \chi^{-\frac{d}{2}(1-2^{-k})} \, 2^{\frac{d}{2}\sum_{l=1}^{k} l\, 2^{-k}} \, t^{-\frac{d}{2}(1-2^{-k})}$$

and letting $k$ tend to $\infty$, recalling the definition of $E_p(t)$, we deduce

$$|\varsigma(y,t)| \leq \left(\frac{\chi}{4}\, t\right)^{-\frac{d}{2}} \leq \frac{C}{t^{\frac{d}{2}}} \; |\phi|^2_{L^1}.$$

Therefore,

$$\sup_{|\phi|_{L^1}=1} |\varsigma(y,t)| \leq \frac{C}{t^{\frac{d}{2}}}$$

and by the linearity of equation (6.43), in fact,

$$|\varsigma(y,t)| \leq \frac{C}{t^{\frac{d}{2}}} \; |\phi|_{L^1}.$$

The same relation holds naturally for $z$, which completes the proof of (6.30). □

**Remark 6.1.** In the first part of the proof of Lemma 6.2, we have used a method due to Moser [15, 16] (c.f. also [10] p. 189). □

We deduce from Lemma 6.2 an estimate on the invariant probability solution of (6.28). Using

$$\int_{\mathbb{R}^d} m^v(y)\, z(y,1)\, dy = \int_{\mathbb{R}^d} m^v(y)\, \phi(y)\, dy$$

we deduce easily that

$$m^v(y) \leq \delta, \ \forall\, y, \ \forall\, v(\cdot) \tag{6.47}$$

It follows that $m^v$ is uniformly bounded in $L^p(\mathbb{R}^d)$, $\forall\, p,\ 1 \leq p < \infty$. Let $\theta$ be an element of $C_0^\infty(\mathbb{R}^d)$, we have

$$\begin{aligned} &-\Delta(m\,\theta) + \operatorname{div}(m\,\theta\, g) + Fy \cdot D(\theta m) \\ &= m\,(D\theta \cdot Fy - g\, \operatorname{div}\theta - \theta\, \operatorname{tr} F) = f \end{aligned} \tag{6.48}$$

and $f \in L^p(\mathbb{R}^d)$, $\forall\, p,\ 1 \leq p \leq \infty$.

From results on the Dirichlet problem, it follows that $m\ \theta$ belongs to $W^{1,p}(\mathbb{R}^d)$, $\forall$ $p$, $1 < p < \infty$. In particular, $m\ \theta$ is continuous. Therefore, we deduce that

$$m^v(y) \geq \delta_k > 0, \ \forall\ y \in K, \text{ compact} \tag{6.49}$$

where the constant $\delta_k$ does not depend on $v(\cdot)$.

**Remark 6.2.** The assumption (6.8) cannot be made without **D** nonempty. Otherwise (6.8) and (6.28) yield $\int m\ dy = 0$, which is impossible. □

We also shall consider the following approximation to $m$. Let $B_R$ be the ball of radius $R$, centered at 0. Let us consider $m_R$ defined by

$$A^* m_R + \operatorname{div}(m_R\ g^v) + \lambda\ m_R = \lambda\ \tau_R\ m \tag{6.50}$$

$$m_R \mid_{\partial B_R} = 0$$

$$m_R \in W_0^{1,p}(B_R)$$

in which $\lambda$ is sufficiently large so that

$$|\xi|^2 - \xi \cdot g\theta + (\lambda + \frac{1}{2} \operatorname{tr} F)\theta^2 \geq C(|\xi|^2 + \theta^2)$$

$$\forall\ \xi \in \mathbb{R}^d, \theta \in \mathbb{R}$$

Moreover, $\tau_R(y) = \tau(\frac{y}{R})$ where

$$\tau(y) \text{ is smooth } \tau(y) = 0 \text{ for } |y| \geq 1,$$

$$\tau(y) = 1, \text{ for } |y| \leq \frac{1}{2} \text{ and } 0 \leq \tau \leq 1.$$

We have

**Lemma 6.3.** *$\tilde{m}_R$ the extension of $m_R$ by 0 outside $B_R$, converges to $m$ in $H^1(\mathbb{R}^d)$ strongly and $\tilde{m}_q = \tilde{m}_{2^q}$ converges monotonically increasing to $m$.*

**Proof.** We compute

$$A^*(\tau_R\ m) + \operatorname{div}(\tau_R\ m\ g) = -\Delta\tau_R\ m - 2\,D\tau_R \cdot Dm$$

$$+ D\tau_R \cdot g\ m + D\tau_R \cdot Fy\ m$$

hence

$$A^*(m_R - \tau_R\, m) + \operatorname{div}((m_R - \tau_R\, m)\, g) + \lambda\,(m_R - \tau_R\, m) =$$

$$\Delta\tau_R\, m + 2D\,\tau_R \cdot Dm - D\,\tau_R \cdot (g + Fy)\, m$$

$$m_R - \tau_R\, m \,|_{\partial B_R} = 0$$

Therefore,

$$\int |\, D(\tilde{m}_R - \tau_R\, m)\,|^2\, dy - \int (\tilde{m}_R - \tau_R\, m)\, g \cdot D(\tilde{m}_R - \tau_R\, m)\, dy$$

$$+ (\lambda + \frac{1}{2}\operatorname{tr} F) \int (\tilde{m}_R - \tau_R\, m)^2\, dy$$

$$= \int (\tilde{m}_R - \tau_R\, m)\,(\Delta\tau_R\, m + 2\,D\,\tau_R \cdot Dm - D\,\tau_R \cdot (g + Fy)\, m)\, dy$$

But

$$|\, D\,\tau_R\,| \leq \frac{1}{R}, \quad |\,\Delta\tau_R\,| \leq \frac{1}{R^2} \text{ and is } 0 \text{ for } |\, y\,| \geq R$$

and

$$\int |\,D\,\tau_R\,|^2\, |\,Fy\,|^2\, m^2\, dy \to 0 \text{ as } R \to \infty$$

since the function $|\,D\,\tau_R\,|^2\, |\,Fy\,|^2\, m^2 \to 0$ pointwise and is bounded by $m^2$ which is integrable. It follows that

$$\|\, \tilde{m}_R - \tau_R\, m\,\|_{H^1} \to 0 \text{ as } R \to \infty.$$

Since $\|\, \tau_R\, m - m\,\|_{H^1} \to 0$, $\tilde{m}_R \to m$ in $H^1(\mathbb{R}^d)$.

Let us now prove the monotone convergence result. First of all we prove that

$$\tilde{m}_R \leq m.$$

We shall prove that $(m_R - m)^+ = 0$. But

$$A^*(m_R - m) + \operatorname{div}((m_R - m)\, g) + \lambda\,(m_R - \tau_R\, m) = 0 \text{ in } B_R$$

$$m_R - m = -m \leq 0 \text{ on } \partial B_R\, .$$

Therefore, $(m_R - m)^+ \in H_0^1(B_R)$. Multiplying by $(m_R - m)^+$ and integrating over $B_R$ we get

$$\int | D (m_R - m)^+ |^2 dy - \int_{B_R} (m_R - m)^+ g \cdot D (m_R - m)^+ dy$$

$$+ (\lambda + \frac{1}{2} \mathrm{tr}\, F) \int_{B_R} (m_R - m)^{+2} dy$$

$$+ \lambda \int_{B_R} (m - \tau_R m) (m_R - m)^+ dy = 0.$$

Hence, $(m_R - m)^+ = 0$ in $B_R$.

In a similar way we have

$$m_R \geq 0.$$

Indeed multiplying by $m_R^-$ and integrating yields

$$- \int_{B_R} | Dm_R^- |^2 dy + \int_{B_R} m_R^- g \cdot Dm_R^- dy$$

$$- (\lambda + \frac{1}{2} \mathrm{tr}\, F) \int_{B_R} (m_R^-)^2 dy = \lambda \int_{B_R} \tau_R m m_R^- dy \geq 0.$$

Hence, $m_R^- = 0$.

Let us next prove that

$$\tilde{m}_{q_1} \leq \tilde{m}_{q_2} \text{ if } q_1 \leq q_2.$$

Note that $\tau_{q_1}(y) = \tau_{2^{q_1}}(y) \leq \tau_{2^{q_2}}(y) = \tau_{q_2}(y)$.We have

$$A^* (m_{q_1} - m_{q_2}) + \mathrm{div}\, [ (m_{q_1} - m_{q_2}) g ] + \lambda (m_{q_1} - m_{q_2})$$

$$= m (\tau_{q_1} - \tau_{q_2})$$

$$m_{q_1} - m_{q_2} |_{\partial B_{2^{q_1}}} = - m_{q_2} |_{\partial B_{2^{q_1}}} \leqslant 0.$$

Multiplying by $(m_{q_1} - m_{q_2})^+$ and integrating, we deduce $(m_{q_1} - m_{q_2})^+ = 0$, which completes the proof. □

## 6.3 Hamilton-Jacobi-Bellman equation of ergodic control

We consider the following problem. To find a pair $\chi, \phi$ such that

$$\chi \text{ is a scalar } \phi \in W_{loc}^{2,p}(\mathbb{R}^d), \ \frac{\phi}{\psi} \text{ bounded at } \infty \tag{6.52}$$

$$A\,\phi + \chi = H(y, D\phi) \tag{6.53}$$

Our objective is to prove the following

**Theorem 6.1.** *We assume (6.1), (6.2), (6.8). Then there is one and only one* $\phi$ *(up to an additive constant) and a scalar* $\chi$ *such that (6.52), (6.53) hold.*

We begin with some preliminary steps. Let us consider a feedback $v_\alpha(\cdot)$ such that (c.f. (6.6)) we may write

$$A\,\phi_\alpha + \alpha\,\phi_\alpha = l(y, v_\alpha) + D\phi_\alpha \cdot g(y, v_\alpha). \tag{6.54}$$

Let then $m_\alpha$ be the invariant probability corresponding to the feedback $v_\alpha$ in equation (6.28). We then have

**Lemma 6.4.** *The following relation holds*

$$\int (\alpha\,\phi_\alpha - l(y, v_\alpha))\, m_\alpha\, dy = 0 \tag{6.55}$$

**Proof.** Consider the following approximation to $\phi_\alpha$, namely

$$A\,\phi_R + \alpha\,\phi_R = l(y, v_\alpha) + D\phi_R \cdot g(y, v_\alpha) \tag{6.56}$$

$$\phi_R \mid_{\partial B_R} = 0$$

Then $\phi_R$ is bounded by $\dfrac{||l||}{\alpha}$ and in $W_{\text{loc}}^{2,p}(\mathbb{R}^d)$. We have extended $\phi_R$ by 0 outside $B_R$, using the same notation.

It is easy to check that

$$\phi_R \to \phi_\alpha \text{ in } W_{\text{loc}}^{2,p} \text{ weakly and } L^\infty \text{ weak star.} \tag{6.57}$$

Multiplying (6.56) by $m_R$, yields

$$\int (\alpha\,\phi_R - l)\, \tilde{m}_R\, dy = \lambda \int (\tilde{m}_R - \tau_R\, m)\, \phi_R$$

and passing to the limit in $R$, we deduce (6.55). □

**Lemma 6.5.** *We have*

$$| \phi_\alpha(y) - \int_{\Gamma_1} \phi_\alpha(\eta)\, \pi_\alpha(d\sigma) | \leq \begin{cases} C\ \psi(y) \text{ in } \mathbb{R}^d - \mathbf{D} \\ C \text{ in} \mathbf{D} \end{cases} \tag{6.58}$$

*where the constant does not depend on* $\alpha$, *nor* $y$.

**Proof.** We write (6.54) as follows

$$A\ \phi_\alpha - D\phi_\alpha \cdot g\,(y\,, v_\alpha) = \overline{l}_\alpha(y). \tag{6.59}$$

$$| \phi_\alpha(y) - \int_{\Gamma_1} \phi_\alpha(\eta)\, \pi_\alpha(d\sigma) | \leq \begin{cases} C\ \psi(y) \text{ in } \mathbb{R}^d - \mathbf{D} \\ C \text{ in} \mathbf{D} \end{cases} \tag{6.58}$$

with

$$\overline{l}_\alpha(y) = l\,(y\,, v_\alpha) - \alpha\, \phi_\alpha(y).$$

From Lemma 6.4, we have

$$\int \overline{l}_\alpha(y)\, m_\alpha(y)\, dy = 0. \tag{6.60}$$

Moreover, clearly

$$|| \overline{l}_\alpha ||_{L^\infty} \leq C \tag{6.61}$$

Using the sequence of stopping times $\tau_n$ defined in (6.2), we can write

$$E^y_{v_\alpha}\ \phi_\alpha(y\,(\tau_N)) - \phi_\alpha(y) = E^y_{v_\alpha} \int_0^{\tau_N} \overline{l}_\alpha(y\,(t))\, dt. \tag{6.62}$$

Now from (6.26)

$$E^y_{v_\alpha}\ \phi_\alpha(y\, \tau_N)) \to \int_{\Gamma_1} \phi_\alpha(\eta)\, \pi_\alpha(d\sigma) \text{ as } N \to \infty. \tag{6.63}$$

Let us prove that

$$| E^y_{v_\alpha} \int_0^{\tau_N} \overline{l}_\alpha(y)\, dt | \leq C\ \psi(y)\ \forall\ \alpha, \forall\ N, \forall\ y \in \mathbb{R}^d - \mathbf{D}. \tag{6.64}$$

Indeed

$$E^y_{v_\alpha} \int_0^{\tau_N} \bar{l}_\alpha(y)\, dt = E^y_{v_\alpha} \int_0^{\tau_1} \bar{l}_\alpha(y)\, dt + \sum_{n=1}^{N-1} E^y_{v_\alpha} \int_{\tau_n}^{\tau_n+1} \bar{l}_\alpha(y)\, dt.$$

But let us define

$$\Lambda_\alpha(x) = E^x_{v_\alpha} \int_0^{\theta(x)} \bar{l}_\alpha(y(t))\, dt. \tag{6.65}$$

Note that

$$\int_{\Gamma_1} \Lambda_\alpha(\eta)\, \pi_\alpha(d\sigma) = \int_{\Gamma_1} \left[ E^\eta_{v_\alpha} \int_0^{\theta(\eta)} \bar{l}_\alpha(y(t))\, dt \right] \pi_\alpha(d\sigma) = 0. \tag{6.66}$$

by virtue of (6.60) and (6.27).

Moreover,

$$E^y_{v_\alpha} \int_{\tau_n}^{\tau_n+1} \bar{l}_\alpha(y(t))\, dt = E^y_{v_\alpha} \Lambda_\alpha(y(\tau_n)) \tag{6.67}$$

and from (6.26)

$$| E^y_{v_\alpha} \Lambda_\alpha(y(\tau_n)) | \leq K \, || \Lambda_\alpha || \, e^{-\rho n}$$

$$\leq || \bar{l}_\alpha || \, E^x_{v_\alpha} \theta(x)\, e^{-\rho n}.$$

Using the function $Z$ defined by

$$A Z - g_\alpha DZ = 1 \text{ in } \mathbf{D}_1, \quad Z \,|_{\Lambda_1} = 0$$

We have

$$E\, \theta(x) = EZ(y(\theta'))$$

Hence,

$$| E^y_{v_\alpha} \theta(x) | \leq C\, \psi(y) \quad x \in \mathbb{R}^d - \mathbf{D}.$$

Therefore,

$$| E^y_{v_\alpha} \Lambda_\alpha(y(\tau_n)) | \leq K_1\, e^{-\rho n}.$$

Therefore,

$$\sum_{n=1}^{N-1} \left| E^y_{v_\alpha} \int_{\tau_n}^{\tau_n+1} \bar{l}_\alpha(y)\, dt \right| \leq C_1 \text{ independent of } y, N.$$

Moreover,

$$\left| E^y_{v_\alpha} \int_0^{\tau_1} \bar{l}_\alpha(y)\, dt \right| \leq C \left| E^y_{v_\alpha} \theta(y) \right| \leq C\, \psi(y)$$

which implies (6.64) and the desired result. □

**Proof of Theorem 6.1.**

*Existence*

Let us set $\tilde{\phi}_\alpha = \phi_\alpha - \int_{\Lambda_1} \phi_\alpha(\eta)\, \pi_\alpha(d\sigma)$. Then $\left\| \frac{\tilde{\phi}_\alpha}{\psi} \right\|_{L^\infty} \leq C$. Moreover, from (6.6) we also have

$$A\, \tilde{\phi}_\alpha + \alpha\, \tilde{\phi}_\alpha + \chi_\alpha = H(y, D\tilde{\phi}_\alpha), \tag{6.68}$$

in which

$$\chi_\alpha = \alpha \int_{\Lambda_1} \phi_\alpha(\eta) \pi_\alpha(d\sigma).$$

It readily follows from (6.68) that

$$\frac{\tilde{\phi}_\alpha}{\psi} \text{ bounded in } W^{2,p,\mu}(\mathbb{R}^d),\quad 2 \leq p \leq \infty,\ \mu > 0.$$

We can extract a subsequence such that

$$\chi_\alpha \to \chi$$

$$\tilde{\phi}_\alpha \to \phi \text{ in } W^{2,p,\mu}(\mathbb{R}^d) \text{ weakly.}$$

We can assert that

$$\tilde{\phi}_\alpha,\ D\tilde{\phi}_\alpha \to \phi,\ D\phi \text{ pointwise,}$$

hence,

$$H(y, D\tilde{\phi}_\alpha) \to H(y, D\phi) \text{ pointwise,}$$

Noting that $H(y, D\tilde{\phi}_\alpha)$ is bounded in $L^{p,\mu}$, we can pass to the limit in (6.68), and the pair $\phi$, $\chi$ satisfies (6.52), (6.53).

*Uniqueness*

Let us prove first that

$$\chi = \underset{v(\cdot)}{Inf} \int l(y, v(y))\, m^v(y)\, dy. \tag{6.69}$$

Note that

$$\alpha \int \phi_\alpha\, m_\alpha\, dy = \alpha \int \tilde{\phi}_\alpha\, m_\alpha\, dy + \chi_\alpha.$$

Since $\int \tilde{\phi}_\alpha\, m_\alpha\, dy$ is bounded, it follows that for a subsequence

$$\alpha \int \phi_\alpha\, m_\alpha\, dy \rightarrow \chi.$$

But from (6.54)

$$\alpha \int \phi_\alpha\, m_\alpha\, dy \geq \chi^* = \underset{v(\cdot)}{Inf} \int l(y, v(y))\, m^v(y)\, dy$$

hence,

$$\chi \geq \chi^*.$$

On the other hand let us consider $\phi_R = \phi_{\alpha R}$ to be the solution of

$$A\,\phi_R + \alpha\,\phi_R = H(y, D\,\phi_R), \quad \phi_R \mid_{\partial B_R} = 0 \tag{6.70}$$

This function is of course different from the one defined in (6.56), although we have kept the same notation for simplicity. It is easy to check again that

$$\phi_R \rightarrow \phi_\alpha \text{ in } W^{2,p}_{\mathrm{loc}} \text{ weakly and } L^\infty \text{ weak star.} \tag{6.71}$$

Pick any control $v(\cdot)$. We deduce from (6.70)

$$A\,\phi_R + \alpha\,\phi_R \leq l(y, v(y)) + D\,\phi_R \cdot g(g, v(y)). \tag{6.72}$$

Let $m^v$ and $m_R^v$ correspond to $v(\cdot)$ by equation (6.28) and (6.50). Multiplying (6.72) by $m_R$ we get

$$\int (\alpha\,\phi_R - l)\, \tilde{m}_R\, dy \leq \lambda \int (\tilde{m}_R - \tau_R\, m)\, \phi_R$$

and letting $R$ tend to $+\infty$ we deduce

$$\int \alpha\, \phi_\alpha(y)\, m^v(y)\, dy \leq \int l(y, v(y))\, m^v(y)\, dy.$$

Hence, as $\alpha$ tends to 0,

$$\chi \leq \int l(y, v(y))\, m^v(y)\, dy$$

and since $v(\cdot)$ is arbitrary $\chi \leq \chi^*$. Therefore, (6.69) is proved.

Let $\hat{v}$ be a feedback associated to $\phi$, where $\phi$ is any solution of (6.53). Let us show that

$$\chi = \int l(y, \hat{v}(y))\, m^{\hat{v}}(y)\, dy. \tag{6.73}$$

Indeed call $\hat{\chi}$ the right hand side of (6.59). We have

$$A\,\phi - g^{\hat{v}} \cdot D\phi + \chi - \hat{\chi} = l(y, \hat{v}) - \hat{\chi} = \hat{l}(y)$$

and

$$\int \hat{l}(y)\, m^{\hat{v}}(y)\, dy = 0.$$

We deduce as in (6.62)

$$E_{\hat{v}}^{y}\, \phi(y(\tau_N)) - \phi(y) + (\chi - \hat{\chi})\, E_{\hat{v}}^{y}\, \tau_N$$

$$= - E_{\hat{v}}^{y} \int_0^{\tau_N} \hat{l}(y(t))\, dt$$

hence,

$$(\chi - \hat{\chi})\, E_{\hat{v}}^{y}\, \tau_N \quad \text{bounded in } N.$$

However,

$$E_{\hat{v}}^{y}\, \tau_N \to +\infty \text{ as } N \to \infty,$$

since

$$E_{\hat{v}}^{y}\, (\tau_{n+1} - \tau_n) = E_{\hat{v}}^{x}\, \theta(x)\, \big|_{x = y(\tau_n)}$$

$$= E_{\hat{v}}^{x}\, Z(y_x(\theta'(x))\, \big|_{x = y(\tau_n)}$$

and

$$Z(y_x(\theta'(x))) \Longrightarrow \underset{\xi \in \Gamma}{Inf}\ Z(\xi) > 0.$$

Therefore necessarily, $\chi = \hat{\chi}$, which is (6.73).

Let now $\phi^1$, $\phi^2$ be solutions of (6.52), (6.,53). We define $\phi_R^1$, $\phi_R^2$ as follows:

$$A\,\phi_R^1 + \lambda\,\phi_R^1 + \chi = H(y, D\,\phi_R^1) + \lambda\phi^1 \tag{6.74}$$

$$\phi_R^1 \mid_{\partial B_R} = 0$$

$$A\,\phi_R^2 + \phi_R^2 + \chi = l(y, \hat{v}_2(y)) + D\,\phi_R^2 \cdot g(y, \hat{v}_2(y)) + \lambda\phi^2 \tag{6.75}$$

$$\phi_R^2 \mid_{\partial B_R} = 0.$$

In (6.75) $\hat{v}_2(\cdot)$ represents a feedback such that

$$H(y, D\,\phi^2(y)) = l(y, \hat{v}_2(y)) + D\,\phi^2 \cdot g(y, \hat{v}_2(y)).$$

Note that $\phi_R^1$, $\phi_R^2$ are not defined in a symmetric way. We take $\lambda$ sufficiently large. It is clear that

$$\phi_R^1 \to \phi^1,\ \phi_R^2 \to \phi^2 \text{ in } W_{\text{loc}}^{2,p} \text{ weakly}$$

$$\text{and in } L^\infty \text{ weak star.}$$

We deduce from (6.74), (6.75) that

$$A\,(\phi_R^1 - \phi_R^2) + \lambda(\phi_R^1 - \phi_R^2) \tag{6.76}$$

$$\leq D\,(\phi_R^1 - \phi_R^2) \cdot g(y, \hat{v}_2(y)) + \lambda(\phi^1 - \phi^2).$$

Consider also $m_R^{\hat{v}_2}$. Multiplying (6.76) by $m_R^{\hat{v}_2}(\phi_R^1 - \phi_R^2)^+$ we deduce

$$\int_{B_R} m_R^{\hat{v}_2} \mid D\,(\phi_R^1 - \phi_R^2)^+ \mid^2 dy + \lambda \int_{B_R} m_R^{\hat{v}_2}(\phi_R^1 - \phi_R^2)^{+2}\,dy$$

$$\leq \lambda \int_{B_R} m_R^{\hat{v}_2}(\phi^1 - \phi^2)\,(\phi_R^1 - \phi_R^2)^+\,dy.$$

Letting $R$ tend to $+\infty$, and using Fatou's Lemma, one obtains

$$\int_{\mathbb{R}^d} m^{\hat{v}_2} \mid D\,(\phi^1 - \phi^2)^+ \mid^2 dy = 0.$$

Therefore,

$$D\ (\phi^1 - \phi^2)^+ = 0.$$

interchanging $\phi^1$ and $\phi^2$, leads finally to $D\ (\phi^1 - \phi^2) = 0$, and this completes the proof of the uniqueness. □

## 7. Singular perturbations with diffusions in the whole space

### 7.1 Setting of the problem

Again we basically consider the setting of section 2, in which we shall drop the assumptions of periodicity as far as the fast system is concerned. We consider

$$f\,(x\,,y\,,v\,) : \mathbb{R}^n \times \mathbb{R}^d \times U \to \mathbb{R}^n \tag{7.1}$$

$$g\,(x\,,y\,,v\,) : \mathbb{R}^n \times \mathbb{R}^d \times U \to \mathbb{R}^d$$

$$l\,(x\,,y\,,v\,) : \mathbb{R}^n \times \mathbb{R}^d \times U \to \mathbb{R}$$

continuous bounded

$$U_{ad} \text{ compact of } U \text{ (metric space).} \tag{7.2}$$

On a convenient set $(\Omega, A\,, F^t\,, P^\epsilon)$ (c.f. section 2.1), we define a dynamic system, composed of a slow and a fast system described by the equations (2.7), with $g$ replaced by $Fy + g\,(x\,,y\,,v\,)$. The cost function is defined by (2.8), and we are interested in the behavior of the value function $u_\epsilon(x\,,y\,)$. It is given as the solution of the H.J.B. equation (noting $A_y = -\Delta_y - Fy \cdot D$)

$$-\Delta_x\, u_\epsilon - \frac{1}{\epsilon} A_y\, u_\epsilon + \beta\, u_\epsilon = H(x\,, D_x\, u_\epsilon, y\,, \frac{1}{\epsilon} D_y\, u_\epsilon) \tag{7.3}$$

$$u_\epsilon = 0 \text{ for } x \in \Gamma, \forall\, y$$

$$u_\epsilon \in W^{2,p,\mu}(\mathcal{O} \times \mathbb{R}^d),\ 2 \leq p < \infty$$

By $W^{2,p,\mu}(\mathcal{O} \times \mathbb{R}^d)$ we mean in fact, (since $\mathcal{O}$ is bounded) the set of functions $z$ such that $z\ \beta_\mu(y\,)$ belongs to $W^{2,p}(\mathcal{O} \times \mathbb{R}^d)$.

We shall denote by $v_\epsilon(x\,,y\,)$ the optimal feedback attached to (2.3), as defined in (2.13). The assumption (6.8) is replaced by

$$A\psi - g(x,y,v)\cdot D\psi \geq 1, \forall\ x,v,y \text{ in } \mathbb{R}^d - \mathbf{D} \tag{7.4}$$

$\mathbf{D}$, $\psi$ have the same properties as in (6.8).

## 7.2 Approximation to the invariant measure

We shall consider the following invariant measures. For a feedback $v(y)$, consider $m^v(x,y)$ which is the solution of

$$A_y^* m + \text{div}_y (m\ g^v) = 0 \tag{7.5}$$

$$m > 0,\ \int_{\mathbb{R}^d} m(x,y)\, dy = 1,\ m \in H^1(\mathbb{R}^d),\ \forall\ x.$$

For a feedback $v(x,y)$ we shall consider $m_\epsilon^v(x,y)$ which is the solution of

$$-\epsilon\Delta_x m_\epsilon + A_y^* m_\epsilon + \text{div}_y (m\ g^v) = 0 \tag{7.6}$$

$$\frac{\partial m_\epsilon}{\partial \nu}\Big|_\Gamma = 0,\quad m_\epsilon \in H^1(\mathcal{O}\times\mathbb{R}^d)$$

$$m_\epsilon > 0,\ \int_{\mathbb{R}^d} m_\epsilon(x,y)\, dy = 1\ \forall\ x.$$

In particular, we shall call $m_\epsilon$ the solution of (7.6) corresponding to the feedback $v_\epsilon(x,y)$ as defined in the preceding paragraph. The construction of the invariant probability $m_\epsilon$ is done in a way similar to that of $m$. Let us consider $\mathbf{D}$, $\mathbf{D}_1$ as in (6.8). To avoid confusion in the notation, let us call $\gamma$, $\gamma_1$ the respective boundaries of $\mathbf{D}$, $\mathbf{D}_1$ (instead of $\Gamma$, $\Gamma_1$, since now $\Gamma$ denotes the boundary of $\mathcal{O}$). We consider the stochastic processes

$$dx = \sqrt{\epsilon}\, dw - \chi_\Gamma(x,t)\,\nu\, d\xi,\quad x(0) = x$$

$$dy = Fy + g(x,y,v(x,y))\, dt + \sqrt{2}\, db_v(t),\quad y(0) = y$$

which are defined on a system $(\Omega, A, F^t, P_v^{x,y})$ and $w, b$ are independent standard Wiener processes.

We define

$$\theta'(x,y;\omega) = Inf\ \{t \mid y(t) \in \mathbf{D}\}$$

$$\theta(x,y;\omega) = Inf\ \{t \geq \theta' \mid y(t) \notin \mathbf{D}_1\}$$

and we have (c.f. (6.15))[5]

$$E\ \theta'(x,y) \leq \psi(y).$$

Define the sequence of stopping times $\tau_0 = 0,\ \tau_n,\ \tau'_{n+1}$ as in section 6.2, and the Markov chain $X_n = x(\tau_n),\ Y_n = y(\tau_n)$ which is a Markov chain on $O \times \gamma_1$. We then define the linear operator on Borel bounded functions on $O \times \gamma_1$ by the relation

$$P^\epsilon\ \phi(x,y) = E_v^{xy}\ \phi(x(\theta), y(\theta)). \tag{7.7}$$

We deduce the analytic formula (c.f. (6.21))

$$P^\epsilon\ \phi(x,y) = \eta_\epsilon(x,y) \tag{7.8}$$

where

$$-\epsilon\,\Delta_x\,\eta + A_y\,\eta - g^v \cdot D_y\,\eta = 0, \text{ on } O \times (\mathbb{R}^d - D) \tag{7.9}$$

$$\eta\,|_\gamma = \varsigma, \quad \frac{\partial \eta}{\partial \nu}\,|_\Gamma = 0$$

$$-\epsilon\,\Delta_x\,\varsigma + A_y\,\varsigma - g^v \cdot D_y\,\varsigma = 0 \text{ on } O \times D_1 \tag{7.10}$$

$$\varsigma\,|_{\gamma_1} = \phi, \quad \frac{\partial \varsigma}{\partial \nu}\,|_\Gamma = 0$$

The ergodicity of $P^\epsilon$ is proved like that of $P$ (c.f. Lemma 6.1). Let $\pi^\epsilon(dx, d\sigma)$ be the corresponding invariant probability on $O \times \gamma_1$. We then define the probability $\mu^\epsilon(dx, dy)$ *on* $O \times \mathbb{R}^d$ by the formula

$$\int\!\!\int_{O\,\mathbb{R}^d} \Lambda(x,y)\, d\mu^\epsilon(x,y) = \frac{\int\!\!\int_{O\,\gamma_1} [\, E_v^{\xi\eta} \int_0^{\theta(\xi,\eta)} \Lambda(x(t), y(t))\, dt\, ]\ \pi^\epsilon(d\xi, d\eta)}{\int\!\!\int_{O\,\gamma_1} E_v^{\xi\eta}\ \theta(\xi,\eta)\ \pi^\epsilon(d\xi, d\eta)} \tag{7.11}$$

for any $\Lambda$ Borel bounded on $O \times \mathbb{R}^d$

Let us note that we can also give an analytic formula for the quantity

$$\alpha^\epsilon(x,y) = E_v^{xy} \int_0^{\theta(x,y)} \Lambda(x(t), y(t))\, dt$$

namely

[5] Here $E = E_v^{x,y}$ for short.

$$- \epsilon \Delta_x \alpha + A_y \alpha - g^v \cdot D_y \alpha = \Lambda \text{ in } O \times (\mathbb{R}^d - D)$$

$$\alpha \mid_\gamma = \beta, \quad \frac{\partial \alpha}{\partial \nu} \mid_\Gamma = 0$$

$$- \epsilon \Delta_x \beta + A_y \beta - g^v \cdot D_y \beta = \Lambda \text{ in } D_1$$

$$\beta \mid_{\gamma_1} = 0, \quad \frac{\partial \beta}{\partial \nu} \mid_\Gamma = 0.$$

We have

$$d \mu^\epsilon(x, y) = m_\epsilon(x, y) \, dx dy. \tag{7.13}$$

Moreover, considering the Cauchy problem

$$\frac{\partial z}{\partial t} - \epsilon \Delta_x z + A_y z - g^v \cdot D_y z = 0 \tag{7.14}$$

$$\frac{\partial z}{\partial \nu} \mid_\Gamma = 0, \quad z(x, y, 0) = \Lambda(x, y)$$

we have

$$\int_O \int_{\mathbb{R}^d} \Lambda(x, y) \, m^\epsilon(x, y) \, dx dy = \int_O \int_{\mathbb{R}^d} m^\epsilon(x, y) \, z(x, y, t) \, dx dy,$$

$$\forall \; t > 0$$

and reasoning as in Lemma 2.1, we deduce

$$0 < \delta_K < m_\epsilon(x, y) < \delta_1, \; \forall \; x \in O, \; \forall \; y \in K, \text{ compact of } \mathbb{R}^d \tag{7.15}$$

with constants uniform with respect to $v(\cdot)$, the left constant (but not the right) depending on the compact $K$.

To proceed we shall slightly reinforce the assumption (7.14) as follows

$$A \psi - k_0 \mid D \psi \mid \geq 1, \; \forall \; y \in \mathbb{R}^d - D \tag{7.16}$$

and $D$, $\psi$ have the same properties as in (6.8). In (7.16) $k_0$ is a constant such that

$$\mid g(x, y, v) \mid \leq k_0 \tag{7.17}$$

Note that (7.16) is satisfied in the example (6.12).

Let us prove the following.

**Lemma 7.1.** *Let* $B_\rho$ *be the ball of radius* $\rho$ *in* $\mathbb{R}^d$, and $\bar{B}_\rho = \mathbb{R}^d - B_\rho$. *Then*

$$\int\int_{O\bar{B}_\rho} m_\epsilon(x,y)\, dx\, dy \leq \delta(\rho) \tag{7.18}$$

*where* $\delta(\rho) \to 0$ *as* $\rho \to \infty$.

**Proof.** Consider in (7.12), $\Lambda(x,y) = \chi_{\bar{B}_\rho}(y)$. Assuming $\rho$ sufficiently large so that $\bar{B}_\rho \cap \mathbf{D}_1 = \phi$, we deduce that $\beta = 0$ and thus $\alpha = \alpha^\epsilon$ is a solution of

$$- \epsilon\Delta_x \alpha + A_y \alpha - g^v \cdot Dy\, \alpha = \chi_{\bar{B}_\rho} \text{ in } \mathbf{O} \times (\mathbb{R}^d - \mathbf{D}) \tag{7.19}$$

$$\alpha \mid_\gamma = 0, \quad \frac{\partial\alpha}{\partial\nu} \mid_\Gamma = 0.$$

Let us consider the solution $u = u_\rho(y)$ of

$$A_y u - k_0 \mid D_y u \mid = \chi_{\bar{B}_\rho} \tag{7.20}$$

$$u \mid_\gamma = 0.$$

We have

$$\alpha^\epsilon(x,y) \leq u_\rho(y) \tag{7.21}$$

since clearly

$$- \epsilon \Delta_x (\alpha - u) + A_y (\alpha - u) - g^v \cdot D(\alpha - u)$$

$$= g^v Du - k_0 \mid Du \mid \leq 0 \text{ in } \mathbf{O} \times (\mathbb{R}^d - \mathbf{D}),$$

$$(\alpha - u) \mid_\gamma = 0, \quad \frac{\partial(\alpha-u)}{\partial\nu} \mid_\Gamma = 0$$

Note also that

$$u_\rho(y) \leq \psi(y) \tag{7.22}$$

since

$$A_y (u - \psi) - k_0 \frac{Du}{\mid Du \mid} (Du - D\psi) \leq 0 \text{ in } \mathbb{R}^d - \mathbf{D}$$

$$(u - \psi) \mid_\gamma \leq 0.$$

As $\rho$ tends to $+\infty$, $u_\rho$ remains bounded in $W^{2,p,\mu}(\mathbb{R}^d - \mathbf{D})$.

Since $\chi_{\bar{B}_\rho} \to 0$ in $L^{p,\mu}(\mathbb{R}^d - \mathbf{D})$, clearly $u_\rho \to 0$ in $W^{2,p,\mu}(\mathbb{R}^d - \mathbf{D})$ weakly. In particular,

$$\alpha^\epsilon(x, y) \leq u_\rho(y) \leq \delta_0(\rho) \to 0 \text{ as } \rho \to \infty, \tag{7.23}$$

for any $y$ in $\gamma_1$ and $x \in \mathbf{O}$. Therefore,

$$\iint_{O\gamma_1} [\, E_v^{\xi\eta} \int_0^{\theta(\xi,\eta)} \chi_{\bar{B}_\rho}(y(t))\, dt \,]\, \pi^\epsilon(d\,\xi, d\,\eta) \tag{7.24}$$

$$= \iint_{O\gamma_1} \alpha^\epsilon(\xi,\eta)\, \pi^\epsilon(d\,\xi, d\,\eta)$$

$$\leq \delta_0(\rho)$$

Consider next in (7.12) $\Lambda = 1$. Let $\beta_0$ be the solution of

$$A_y\, \beta_0 + k_0 \,|\, Dy\, \beta_0 \,| = 1 \text{ in } \mathbf{D}_1$$

$$\beta_0 \,|_{\gamma_1} = 0$$

then one has

$$\beta^\epsilon(x, y) \geq \beta_0(y)$$

Therefore,

$$\beta^\epsilon(y) \,|_\gamma \geq c_0 > 0$$

and

$$\alpha^\epsilon(x, y) \geq c_0$$

which implies

$$\iint_{O\gamma_1} [\, E_v^{\xi,\eta}\, \theta(\xi,\eta) \,]\; \pi^\epsilon\, (d\,\xi, d\,\eta) \geq c_0.$$

This estimate, together with (7.24) and formula (7.11) implies the desired result (7.18).

Consider also as in (6.50) the solution $m_{\epsilon R}$ of

$$-\,\epsilon\, \Delta_x\, m_{\epsilon R} + A_y^*\, m_{\epsilon R} + \mathrm{div}_y\, (m_{\epsilon R}\, g^v) + \lambda\, m_{\epsilon R} = \lambda\, \tau_R\, m_{\epsilon R} \tag{7.25}$$

$$\frac{\partial m_{\epsilon R}}{\partial \nu} \,|_\Gamma, \quad m_{\epsilon R} \,|_{\partial B_R} = 0$$

then we have

$$m_{\epsilon R} \to m_\epsilon \text{ in } L^1 \cap H^1 \text{ as } R \to \infty. \tag{7.26}$$

### 7.3 *A priori* estimate

We shall need the approximation of $u_\epsilon$ given by

$$- \Delta_x u_{\epsilon R} - \frac{1}{\epsilon} A_y u_{\epsilon R} + \beta u_{\epsilon R} = H(x, D_x u_{\epsilon R}, y, \frac{1}{\epsilon} D_y u_{\epsilon R}) \tag{7.27}$$

$$u_\epsilon = 0 \text{ on } \partial(O \times B_R)$$

and

$$u_{\epsilon R} \to u_\epsilon \text{ in } W^{2,p}_{\text{loc}} \text{ weakly and in } L^\infty \text{ weak star} \tag{7.28}$$

where *loc* is meant only for the $y$ variable. We shall need also a similar approximation in the case of explicit feedbacks; in particular $v_\epsilon$

**Lemma 7.1.** *The following estimates hold*

$$| D_x u_\epsilon |^2_{L^2_{\text{loc}}} \leq C, \quad | u_\epsilon |_{L^\infty} \leq C \tag{7.29}$$

$$| D_y u_\epsilon |^2_{L^2_{\text{loc}}} \leq C_\epsilon$$

**Proof.** Using the feedback $v_\epsilon$, equation (7.3) reads

$$- \Delta_x u_\epsilon - \frac{1}{\epsilon} A_y u_\epsilon + \beta u_\epsilon = l(x, y, v_\epsilon) + Du_\epsilon \cdot f(x, y, v_\epsilon) + \tag{7.30}$$

$$+ \frac{1}{\epsilon} D_y u_\epsilon \cdot g(x, y, v_\epsilon)$$

$$u_\epsilon |_\Gamma = 0$$

Similarly, define $u_{\epsilon R}$ corresponding to (7.30)

$$- \Delta_x u_{\epsilon R} - \frac{1}{\epsilon} A_y u_{\epsilon R} + \beta u_{\epsilon R} = l(x, y, v_\epsilon) + D u_{\epsilon R} \cdot f(x, y, v_\epsilon) \tag{7.31}$$

$$+ \frac{1}{\epsilon} D_y u_{\epsilon R} \cdot g(x, y, v_\epsilon)$$

$$u_\epsilon = 0 \text{ on } \partial(O \times B_R).$$

Consider similarly $m_\epsilon$ and $m_{\epsilon R}$. Multiplying (7.31) by $m_{\epsilon R} u_{\epsilon R}$ and integrating, one obtains

$$\iint m_{\epsilon R} \mid D_x u_{\epsilon R} \mid^2 + \frac{1}{\epsilon}\iint m_{\epsilon R} \mid D_y u_{\epsilon R} \mid^2 + \beta \iint m_{\epsilon R} u_{\epsilon R}^2 \tag{7.32}$$

$$= \iint (l_\epsilon + D_x u_{\epsilon R} \cdot f_\epsilon) m_{\epsilon R} u_{\epsilon R} + \iint \lambda (m_{\epsilon R} - \tau_{\epsilon R} m_\epsilon) \frac{u_{\epsilon R}^2}{2} .$$

Letting R tend to $\infty$ after majoring and using Fatou's Lemma yields

$$\iint m_\epsilon \mid D_x u_\epsilon \mid^s + \frac{1}{\epsilon} \iint m_\epsilon \mid D_y u_\epsilon \mid^2 + \beta \iint m_\epsilon u_\epsilon^2 \leq C.$$

Using the estimates (7.15) it follows that

$$\int\int_{O\times K} \mid D_x u_\epsilon \mid^2 + \frac{1}{\epsilon}\int\int_{O\times K} \mid D_y u_\epsilon \mid^2 \leq C_K \tag{7.33}$$

which implies (7.29), noting that $\mid u_\epsilon \mid_{L^\infty} \leq \frac{1}{\beta} \mid l \mid_{L^\infty}$. Let us note also that since from (7.32) $\tilde{m}_{\epsilon R}^{1/2} D_x u_{\epsilon R}$ remains bounded in $L^2(O \times \mathbb{R}^d)$, it is possible to pass to the limit in (7.32) as $R \to \infty$, and obtain the inequality

$$\iint m_\epsilon \mid D_x u_\epsilon \mid^2 dx\, dy + \frac{1}{\epsilon} \iint m_\epsilon \mid D_y u_\epsilon \mid^2 dx\, dy + \beta \iint m_\epsilon u_\epsilon^2 dx\, dy \tag{7.34}$$

$$\leq \iint (l_\epsilon + D_x u_\epsilon \cdot f_\epsilon) m_\epsilon u_\epsilon\, dx\, dy$$

**Lemma 7.2.** *Let $\phi(x) \in H_0^1(O) \cap H^2(O)$, then we have the inequality*

$$\iint m_\epsilon \mid D_x (u_\epsilon - \phi) \mid^2 dx\, dy + \frac{1}{\epsilon} \iint m_\epsilon \mid D_y u_\epsilon \mid^2 dx\, dy \tag{7.35}$$

$$+ \iint \beta m_\epsilon (u_\epsilon - \phi)^2 dx\, dy \leq \iint m_\epsilon u_\epsilon (\Delta\phi - \beta\phi)\, dx\, dy$$

$$+ \int ( \mid D_x \phi \mid^2 + \beta \phi^2 )\, dx$$

**Proof.** Using $u_{\epsilon R}$ and $m_{\epsilon R}$ as intermediaries like in Lemma 7.1, and making calculations similar to those of Lemma 2.3, one obtains

$$2 \iint D_x u_\epsilon D_x \phi\, m_\epsilon\, dx\, dy + \iint m_\epsilon u_\epsilon \Delta\phi\, dx\, dy \tag{7.36}$$

$$+ \beta \iint m_\epsilon u_\epsilon \phi\, dx\, dy = \iint (l_\epsilon + D_x u_\epsilon \cdot f_\epsilon) m_\epsilon \phi\, dx\, dy ,$$

and we have equality instead of an inequality in (7.34). We can then complete the proof of (7.35) as in Lemma 2.3. □

### 7.4 Convergence

We have

**Lemma 7.3.** *Let us consider a subsequence of* $u_\epsilon$ *such that*

$$u_\epsilon \to u \quad \text{in } H^2_{\text{loc}}(O \times \mathbb{R}^d) \text{ weakly.} \tag{7.37}$$

*Then* $u$ *is a function of* $x$ *only, belongs to* $H^1_0(O)$, *and the convergence (7.37) is strong.*

**Proof.** Setting

$$X_\epsilon = \int\int m_\epsilon \mid D_x (u_\epsilon - u) \mid^2 dx\, dy + \frac{1}{\epsilon} \int\int m_\epsilon \mid D_y u_\epsilon \mid^2 dx\, dy$$
$$+ \beta \int\int m_\epsilon (u_\epsilon - u)^2 dx\, dy$$

and making use of Lemma 7.1, one can prove as in Lemma 2.4, that

$$\overline{\lim_{\epsilon \to 0}} X_\epsilon = 0$$

and using the estimates (7.15) we deduce

$$\int_{O \times K}\int \mid D_x (u_\epsilon - u) \mid^2 dx\, dy + \frac{1}{\epsilon} \int_{O \times K}\int \mid Dy\, u_\epsilon \mid^2 dx\, dy +$$
$$+ \beta \int_{O \times K}\int (u_\epsilon - u)^2 dx\, dy \to 0,$$

for any $K$ compact subset of $\mathbb{R}^d$. This proves the desired result. □

We now identify the limit. Let us recall the definition of $m^v$ given in (7.5). Define $\chi(x, p)$ by the formula

$$\chi(x, p) = \underset{v(\cdot)}{Inf} \int_{\mathbb{R}^d} m^{v(x,y)} (l(x, y, v(y)) + p \cdot f(x, y, v(y))) \, dy \tag{7.38}$$

and consider the Dirichlet problem

$$-\Delta u + \beta u = \chi(x, Du), \quad u \mid_\Gamma = 0, \quad u \in W^{2,p}(O) \tag{7.39}$$

We can then state the following

**Theorem 7.1.** *We assume (7.1), (7.2) and (7.16). Then we have*

$$u_\epsilon \to u \quad \text{in } H^2_{\text{loc}}(O \times \mathbb{R}^d) \text{ strongly} \tag{7.40}$$

The proof is similar to that of Theorem 2.1, making use in particular of Lemma 7.1.

## References


[1] J.L. Doob , *Stochastic Processes,* J. Wiley, New York (1967).

[2] R.Z. Khas'minskii, *Stochastic Stability of Differential Equations,* Sijthoff and Noordhoof, Alphan aan den Rijn, The Netherlands (1980).

[3] R. Jensen and P.L. Lions, *Some asymptotic problems in fully nonlinear elliptic equations and stochastic control.* to appear.

[4] M. Robin, "Long - term average cost control problems for continuous time Markov processes: A survey," *Acta Applicandae Mathematicae* vol. 1, pp. 281-299. (1983).

[5] D.G. Aronson, "Nonnegative solutions of linear parabolic equations," *Annali della Scuol. Norma Sup. Pisa* vol. XXII, no. Fasc. IV, (1968).

[6] A. Bensoussan, J.L. Lions, and G.C. Papanicolaou, *Asymptotic Analysis of Periodic Structures,* North-Holland, Amsterdam (1978).

[7] A. Bensoussan , *Methodes de Perturbations en Controle Optimal,* Dunod, Paris (to be published).

[8] J. Chow and P. Kokotovic, "Near optimal feedback stabilization of a class of nonlinear singularly perturbed systems," *SIAM J. Control* vol. 16, (1978).

[9] Y.A. Kogan, "On optimal control of a non-terminating diffusion process with reflection," *Theory of Prob. and Appl.* vol. 14, pp. 496-502 (1969).

[10] O.A. Ladyzenskaya, V.A. Solonnikov, and N.N. Ural'tseva, "Linear and Quasi Linear Equations of Parabolic Type," *Translations of Math. Monographs* vol. 33, American Mathematics Society, (1968).

[11] D. Kinderlehrer and G. Stampacchia, *An Introduction to Variational Inequalities and their Applications,* Academic Press, New York (1980).

[12] J.L. Lions and E. Magenes, "Problemes aux limits non homogenes (V)," *Ann. Sc. Normale Sup. Pisa,* (1962).

[13] W. Fleming and R. Rishel, *Optimal Deterministic and Stochastic Control,* Springer - Verlag, New York (1975).

[14] P. Besala, "On the existence of a fundamental solution for a parabolic differential equation with unbounded coefficients," *Ann. Polonici Math.* vol. XXIX, pp. 403-409 (1975).

[15] J. Moser , "A new proof of de Giorgi's theorem concerning the regularity problem for elliptic differential equations," *Comm. Pure Appl. Math.* vol. 13, pp. 457-468 (1960).

[16] J. Moser , "A Harnack inequality for parabolic differential equations," *Comm. Pure Appl. Math.* vol. 17, pp. 101-134 (1964).

# Part II: LARGE SCALE SYSTEMS

# SINGULAR PERTURBATION OF MARKOV CHAINS

F. Delebecque, O. Muron, J.P. Quadrat

INRIA
Domaine de Voluceau
B.P. 105
78150 LE CHESNAY, FRANCE

**ABSTRACT.**

This paper studies some aspects of perturbation theory applied to Markov chains. In the first part we introduce the notion of agregated chain and show how these chains arise in the context of perturbation and time scales. In the second part, we study some applications of perturbed Markov chains to the Reliability of large scale repairable systems. In the third part we give some applications to optimal control.

# 1. PERTURBATION AND AGGREGATION OF MARKOV CHAINS

## 1.1. Review on Markov chains and notations

A discrete time Markov $(X_n)$ is characterized by its transition probability matrix M. M(x,y) is the conditional probability $P_r(X_{n+1} = y \mid X_n = x)$ where x and y are elements of a state space E. We will assume that E is finite with $m_o$ elements.

The matrix M defines an operator acting on "functions" (column-vectors) or "measures" (row-vectors) with the following interpretation :

$$M^n f(x) = \mathbf{E}_x f(X_n).$$

Here $\mathbf{E}_x$ denotes the expectation operator associated with the probability law $\mathbf{P}_x$ of the chain starting at x : $\mathbf{P}_x(X_o = x) = 1$. The measure $mM^n$ is the probability law of $X_n$ when the probability law of $X_o$ is m.

A discrete time Markov chain is also characterized by its "generator" $A = M - I$. One can write :

$$Af(x) = q(x) \{ \sum_{y \neq x} p(x,y) f(y) - f(x)\}$$

It is easy to check that $T_1$, the date of the first jump of $(X_n)$ is a geometric r.v. such that

$$\mathbf{E}_x T_1 = \frac{1}{q(x)} .$$

For continuous time Markov chains one generally starts with a generator A of the same type, $T_1$ is an exponential r.v. with mean $\frac{1}{q(x)}$ and q(x,y) is the transition probability matrix of the chain seen at instants of jumps.

One has $\mathbf{E}_x f(X_t) = e^{At} f(x)$.

Properties of A :

Clearly the off-diagonal elements of A are non negative. Also $A1 = 0$ i.e. zero is an eigenvalue of A with eigenvector 1. In general however there are several eigenvectors associated with the zero eigenvalue of A. After reordering the elements of E one can get the following figure for A :

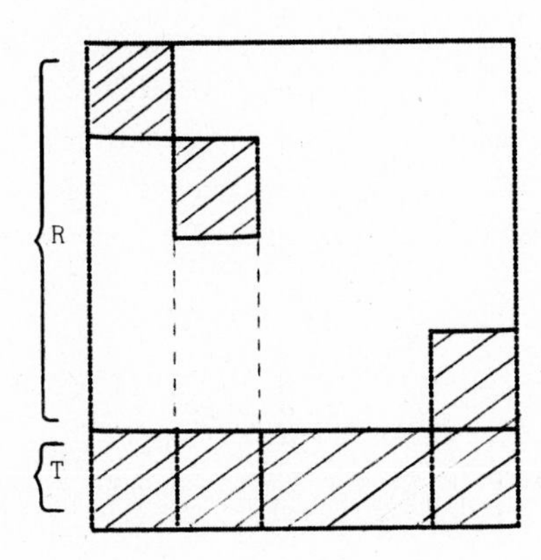


A state x is either recurrent ($x \in R$) or transient ($x \in T$). If x is recurrent we will denote by $\bar{x}$ its recurrent class (independent block in the preceding figure).

We will denote by $m_1$ the number of such blocks. The events $A_{\bar{x}}$ : "$X_n$ is absorbed in the class $\bar{x}$ " make a partition of E. We will denote by $q_{\bar{x}}$ the function defined by $q_{\bar{x}}(x) = \mathbf{P}_x(A_{\bar{x}})$. Clearly

$$\sum_{\bar{x}} q_{\bar{x}}(x) = 1.$$

These $m_1$ functions are right eigenvectors associated with the zero eigenvalue of A. $Aq_{\bar{x}} = 0, \forall \bar{x}$ .

If $X_o \in \bar{x}$ , then the probability law of $X_n$ tends to $m_{\bar{x}}$ , the invariant probability measure with support $\bar{x}$ which satisfies $m_{\bar{x}} A = 0$, $m_{\bar{x}} 1 = 1$. So these $m_1$ measures are left eigenvectors for the eigenvalue zero of A. It turns out that the $q_{\bar{x}}$ (resp. $m_{\bar{x}}$) vectors make a basis of Ker A (resp. Ker $A^T$). Although it is quite natural to consider such bases because of their probabilistic interpretation it is easy to see that these eigenvectoirs are not "robust" to perturbation : indeed if some of the structural zero elements of the matrix A are perturbed by some small quantity a recurrent state may become transient or a transient state may become recurrent. It is therefore important to know the splitting of the eigenvalue zero of A under perturbation : indeed the multiplicity of this eigenvalue decrease whenever recurrent classes are linked together by the perturbation. The most important case for the applications is the case $T = \emptyset$ where the multiplicity moves from $m_1$ (number of blocks along the diagonal) to one, Courtois [4].

Resolvent :

The resolvent of the Markov chain is the matrix $R(\lambda) = (\lambda - A)^{-1}$. One has the following probabilistic interpretation for $\lambda R(\lambda)$

$$\lambda R(\lambda) f(x) = E_x f(X_T)$$

where T is an exponential (or geometric in the discrete time case) random variable with mean $\frac{1}{\lambda}$ . In particular one has $R(\lambda) f \geqq 0$ if $f \geqq 0$ and $R(\lambda) f(x) > 0$ if in addition $f(x) > 0$. This property is basic for the control of Markov chains.

The resolvent $R(\lambda)$ gives the Cauchy's formula :

$$p(A) = \frac{1}{2i\pi} \int_G p(\lambda) R(\lambda) d\lambda.$$

Here G is a closed contour enclosing all the eigenvalues of A.

In particular, we will denote :

$$P_i = \frac{1}{2i\pi} \int_{(\lambda_i)} R(\lambda)d\lambda = \text{Residue of } R(\lambda) \text{ at } \lambda = \lambda_i .$$

$$D_i = \frac{1}{2i\pi} \int_{(\lambda_i)} \lambda R(\lambda)d\lambda - \lambda_i P_i = (A - \lambda_i)P_i .$$

For a Markov generator A we have $D_o = 0$. This means that the eigenvalue 0 of A is diagonalizable (no Jordan blocks).

Moreover one knows an explicit full rank factorisation of P (mean ergodic theorem) :

$$P = \lim_{\lambda \to 0} \lambda R(\lambda) = \sum_{\bar{x}} (m_{\bar{x}} \otimes q_{\bar{x}}).$$

So the $m_1$ right eigenvectors $q_{\bar{x}}$ make a basis of Ker A = $\mathbb{R}(P_o)$ and the left eigenvectors make the dual basis :

$$(m_{\bar{x}}, q_{\bar{x}}) = 1 , \quad (m_{\bar{y}}, q_{\bar{x}}) = 0 \qquad \bar{y} \neq \bar{x}$$

We have the direct sum decomposition

$$\mathbb{R}^n = \text{Ker } A + \mathbb{R}(A) = \mathbb{R}(P) + \text{Ker } P.$$

With respect to any basis compatible with this decomposition the operator (P - A) is represented by the matrix :

| | Ker A | R(A) |
|---|---|---|
| Ker A | 1 ... 1 | 0 |
| R(A) | 0 | $-\bar{A}$ |

(P - A)

Here the matrix $\bar{A}$ is invertible so $(P-A)$ is invertible and we denote $S = (P - A)^{-1} - P$ the "reduced resolvent of a at zero" : it is the inverse of $-A$ restricted to $\mathbb{R}(I-P)$. The preceding direct sum decomposition may be written

(1) $$f = Pf + S(-A)f = Pf + (-A)Sf.$$

In the litterature $-S$ is also sometimes called the "group-inverse" or Drazin inverse of A. [15].

1.2. Aggregation

Let us consider two Markov generators $A_o$ and $A_1$ let $P = Q.M$ be the full rank factorization of the spectral projection associated with the eigenvalue zero of $A_o$ considered above.

The operator $PA_1P$ defines a new Markov chain. Indeed, seen as mapping $\mathbb{R}(P)$ to $\mathbb{R}(P)$, is represented in the base $\{q_{\bar{x}}\}$ by the matrix $\bar{A}_1 = M A_1 Q$ i.e. $\bar{A}_1(\bar{x},\bar{y}) = m_{\bar{x}} A_1 q_{\bar{y}}$ and one can easily check that $\bar{A}_1$ is the generator of an $m_1$ states Markov chain. Its states are the recurrent classes of the umperturbed chain $A_o$.

Example [13] : Let B be a subset of the state space E and let us consider the instants at which the chain return in B, (assuming $X_o \in B$) namely $T_o = 0$, $T_1 = \inf\{n \geq 0, X_n \in B\}, \ldots$

The chain induced on B is by definition the process $Y_n^B$ such that $Y_o^B = X_o, Y_1^B = X_{T_1}, \ldots$

$(Y_n^B)$ is a new Markov chain with generator $P^B\ A\ P^B / \mathbb{R}(P^B)$ where $P^B$ is the projection abtained by making the states of B absorbing i.e. the spectral projection associated with the transition probability matrix $I_B + I_{E\setminus B}M$ (where $I_B$ is the diagonal matrix $I_B\ f(x) = f(x)$ if $x \in B$, $I_B\ f(x) = 0$ if $x \in E\setminus B$). So the above aggregation procedure leads to the chain induced on B.

Iterated aggregation :

Let us set $A_o^1 = P A_1 P$.

Then the resolvent of $A_o^1$ may be written :

$$(\lambda - A_o^1)^{-1} = (\lambda - A_o^1)^{-1} P + \frac{1}{\lambda}(I - P) = Q(\lambda - \bar{A}_1)^{-1} M + \frac{1}{\lambda}(I - P)$$

We deduce that the spectral projection R for the eigenvalue 0 of $P A_1 P$ may be written :

$$R = P^1 + (I - P)$$

where $P^1 = Q^1 . M^1$, $Q^1 = Q\bar{Q}$, $M^1 = \bar{M} M$ and $\bar{Q}$ and $\bar{M}$ are obtained from the agregate $\bar{A}_1 = M A_1 Q$ in the same manner as Q and M were from $A_o$ : the spectral projection $\bar{P}$ for $\bar{A}_1$ admits the factorization $\bar{P}_o = \bar{Q}.\bar{M}$.

$P^1$ is a subprojection of P : $P^1 P = P P^1 = P^1$ and rank $(P^1) = m_2$ is the number of recurrent classes of the agregate $\bar{A}_1$. The reduced resolvent of $A_o^1$ is :

$$S^1 = (R - A_o^1)^{-1} - R = Q \bar{S}^1 M$$

where $\bar{S}^1$ is the reduced resolvent at zero of $\bar{A}_1$.

One has $S^1 P = P S^1 = S^1$, $S S^1 = S^1 S = S^1 P^1 = 0$ and R(P) may be decomposed into a direct sum :

$$P = P^1 + S^1(-A_o^1) = P^1 + (-A_o^1)S^1. \tag{2}$$

This decomposition is similar to (1). Since $P^1$ may be interpreted as the spectral projection for the zero eigenvalue of a Markov chain it may be used to aggregate a new generator $A_2$ as done above with P. In other words $P^1 A_2 P^1 / \mathbb{R}(P^1)$ may be interpreted as the generator of a new chain $\bar{A}_2 = \bar{M} M A_2 Q\bar{Q}$. The process may be iterated. Let us examine for instance the previous example.

Example : Let C be a subset of B. We have :

$$P^C P^B = P^B P^C = P^C$$

and the chain induced on C is also the chain induced on C by the induced chain on B. The computations involve the solution of two linear systems of size $|E\backslash B|$ and $|B\backslash C|$ instead of the solution of a linear system of size $|E\backslash C|$. This result may be useful to compute the restriction of the invariant measure to C which is known to be also the invariant measure of the trace chain on C.

## 1.3. Perturbation

In this section we show how the above aggregation process arises when perturbed Markov chain are studied with particular time scales. We will consider first the computation of a discounted cost with small interest rate for a perturbed Markov chain which will be useful for control purpose, then we will give some properties of perturbed eigenvalues which are useful in reliability theory.

### 1.3.1. Computation of discounted cost.

Let us consider a perturbed Markov chain with generator $A_o + \varepsilon A_1$ and the discounted cost :

$$V_\varepsilon(x) = \mathbf{E}_x \int_o^\infty e^{-\varepsilon^k \lambda s} \,.\, \varepsilon^k f(X_s)\, ds$$

where $\mathbf{E}_x$ stands for the mathematical expectation of the perturbed chain starting at x and k is a positive integer. One has also the interpretation $V(x) = \mathbf{E}_x f(X_T)$ where the perturbed chain is stopped at a random time with mean $\frac{1}{\varepsilon^k}$ so the "weak interactions" cannot be neglected. Let us show how the computation can be done for k=1 and k=2 taking into account the particular structure of $A_o$ and $A_1$. One looks for a solution $V = V_o + \varepsilon V_1 + \ldots$. By identifying the factors of $\varepsilon$ in the equation verifed by V i.e.

$$- \varepsilon\lambda V_\varepsilon + (A_0 + \varepsilon A_1)V_\varepsilon + \varepsilon f = 0$$

we obtain the following system of equations :

i) $\quad A_o V_o = 0$

ii) $\quad -\lambda V_o + A_o V_1 + A_1 V_o + f = 0$

$\quad A_o V_2 + A_1 V_1 = 0$ ... etc...

This system may be solved easily thanks to the decomposition (1) : $V_i = PV_i + S(-A_o)V_i$. For instance the first equation gives $(I - P) V_o = 0$ and the second one implies

$$P(ii) : -\lambda PV_o + PA_1 P V_o + P_o f = 0$$

We obtain the aggregate Markov chain $PA_1P$ of the preceding section and P(ii) uniquely defines

$$V_o = PV_o = (\lambda - PA_1P)^{-1} P f = Q(\lambda - \bar{A}_1)^{-1} M f$$

ii) also determines the part of $V_1$ in $\mathbb{R}(I-P)$ : $(I - P)V_1 = S(-A_o)V_1 = SA_1 V_o + S f.$

The above process is interesting because the computation of $(V_o, V_1, \ldots)$ is made in a decentralized-aggregated way : first we compute $\bar{A}_1$ and then we solve the aggregate system.

More generally the computation of $\varepsilon^k (\varepsilon^k \lambda - (A_o + \varepsilon A_1))^{-1}$ may be done by recurrence on k. Let us consider the case k=2. To compute $V_o$ one has to solve the system :

(i) $\quad A_o V_o = 0$

(ii) $\quad A_o V_1 + A_1 V_o = 0$

(iii) $\quad -\lambda V_o + A_o V_1 + A_1 V_1 + f = 0$

Using (i) in (ii) we obtain :

P(ii) : $PA_1P\,V_o = 0$ and $(I - P)V_1 = SA_1PV_o$

Then we obtain :

P(iii) : $-\mu PV_o + PA_1\,PV_o + (PA_1SA_1P)V_1 + Pf = 0$

We remark that the system P(ii), P(iii) is analogous to the system (i), (ii) that has been solved for k=1 with the change of variable :

$$A_o \leftarrow A_o^1 = PA_1P \quad , \quad A_1 \qquad A_1^1 = PA_1SA_1\,P$$

Since $A_o^1$ is a Markov generator the system P(i), P(ii) is well-posed and we have :

$$V_o = (\mu - P^1A_1SA_1P^1)^{-1}\,P^1f = \bar{Q}Q\,(\mu - \bar{A}_2)\bar{M}Mf = Q^1(\mu - \bar{A}_2)M^1\,f$$

and

$$\bar{A}_2 = \bar{M}\,M\,A_1^1\,\bar{Q}Q = M^1\,A_1^1\,Q^1.$$

Passing to the limit we obtain that $\bar{A}_2$ is a Markov generator.

In the general case we can define a finite sequence of aggregated chains $\bar{A}_1$, $\bar{A}_2, \ldots$ by the same process as described in the preceding section. Each generator $\bar{A}_k$ is associated with a time scale and allows to define a projector $P^{(k)} = Q^{(k)}\,.\,M^{(k)}$ which is used to define $\bar{A}_{k+1}$.

### 1.3.2. Perturbed eigenvalues.

We have already seen that 0 is an eigenvalue of $A_o$ with multiplicity $m_1$ and a basis of right eigenvectors is given by $\{q_{\bar{x}}\ \ \bar{x} \in (1,\ldots,m_1)\}$.

Assume that the eigenvalues of $A_o + zA_1$ in the neighborhood of 0 are of the form $z\,\lambda_{\bar{x}}^1 + o(z)$ , $\bar{x} \in 1,\ldots,m_1$, $\lambda_{\bar{x}}^1 \neq 0$ and that the corresponding eigenvectors are $e_{\bar{x}} + zq^1_{\bar{x}} + o(z)$ $(\bar{x} = 1,\ldots,m_1)$.

We will denote by E the matrix $E = [e_1| \dots | e_{m_1}]$.

The eigenvalue equation may be written :

$$(A_o + z A_1)(E + z Q^1 + O(z)) = z(E + z Q^1 + O(z))\Lambda^1$$

where $\Lambda^1 = \mathrm{diag}(\lambda_1^1, \dots, \lambda_{\bar{x}}^1)$.

By identification of the factors of z, we obtain :

$$A_o E = 0$$

$$A_o Q^1 + A_1 E = E\Lambda^1$$

The first equation is trivially satisifed if E is obtained from Q by a change a variables E = QT with T invertible. If we multiply from the left the second equation by M, we obtain :

$$M A_1 Q T = M Q T \Lambda^1 = T \Lambda^1$$

and so if we choose T the matrix which diagonalize (if possible) the aggregate operator $M A_1 Q$ that is $T^{-1}(M A_1 Q)T = \Lambda^1$ we see that the z-term vanish. These formal calculations show the following important fact well known in perturbation theory : if the eigenvalues split in the form described above then they are solution of an aggregate eigenvalue problem and also a basis QT of eigenvectors of $A_o$ may be constructed made of limit of eigenvectors for the perturbed eigenvalue problem (the first order eigenvalues fix the zero order eigenvectors).

For perturbed eigenvalues of Markov chains one is mainly interested in computing the limit of $m_\varepsilon$ where $m_\varepsilon$ is the invariant probability measure (left-eigenvector) corresponding to the zero eigenvalue of $A_o + \varepsilon A_1$. The most common situation is the case of a generator $A_o$ with a block-diagonal structure with weak interactions between the blocks. In eg. [5] it is shown that $\lim m_\varepsilon$ is associated with the eigenvalue zero of the aggregate $M A_1 Q$ that is $m_1 - 1$ perturbed eigenvalues split and $\lim_{\varepsilon \to o} m_\varepsilon = \bar{m}$ where $\bar{m}$ is the invariant probability measure of $MA_1 Q$.

The general situation (with transient states) is more complicated but it can be shown that the perturbed eigenvalues are all of the form $\varepsilon^k \lambda_i + o(\varepsilon^k)$ with k integer (so the Newton diagram is made of lines with slopes 1/k, k integer). Let us now briefly describe the "reduction process" following Kato [9]. One introduces the "total projection"

$$(3) \qquad Q_i(z) = (2i\pi)^{-1} \int_{(\lambda_i)} (\lambda - A(z))^{-1} \, d\lambda$$

where $(\lambda_i)$ is a fixed contour enclosing the unperturbed eigenvalue $(\lambda_i)$.

The $Q_i(z)$ are projections which reduce $A(z)$ i.e. $A(z)P_i(z) = P_i(z)\, A(z)$ and are holomorphic in $|z| < \delta_i$.

The restriction of $A(z)$ to Im $Q_i(z)$ admits as eigenvalues the set of perturbed eigenvalues which tends to $\lambda_i$ (the"$\lambda_i$-group"). The $Q_i(z)$ define a decomposition of the space into a direct sum :

$$(4) \qquad I = Q_1(z) + \ldots + Q_r(z) \qquad |z| < \delta.$$

This decomposition is the spectral decomposition of $A_o$ in the case $z = 0$.

Let $A^1(z) = z^{-1}(A(z) - \lambda_i)\, Q(z)$.

When $\lambda = \lambda_i$ is a diagonalizable eigenvalue one has $D = (A_o - \lambda)Q(0) = 0$ and so $A^1(z)$ is holomorphic in a neighborhood of 0 (0 included). Moreover if $\lambda^1(z)$ is an eigenvalue of $A^1(z)$ then $z^{-1}(\lambda(z) - \lambda) = \lambda^1(z)$ for an eigenvalue $\lambda^1(z)$ of the $\lambda$-group, and $\lambda(z)$ can be written under the form

$$(5) \qquad \lambda(z) = \lambda + z\lambda^1(z).$$

The operator $A^1(z)$ induces a decomposition of Im $Q(z)$ similar to (4) :

$$(6) \qquad Q(z) = \sum_{i=1}^{r_1} Q_i^1(z) \quad \text{where } Q_i(z) = \frac{1}{2i\pi} \int_{(\lambda_i^1)} (\lambda - A^1(z))^{-1} \, d\lambda$$

Here $Q_i^1$ is the total projection associated with the unperturbed eigenvalue $\lambda_i^1$ of $A^1(0) = PA_1P$ for the series $A^1(z)$.

(When 0 is an eigenvalue of $A^1(z)/\text{Im } Q(z)$ then one has to define

$$Q_o^1(z) \text{ by } \frac{1}{2i\pi} Q_o(z) \, . \, (\int_{(o)} (\lambda - A^1(z))^{-1} d\lambda) \, )$$

The operators $Q_i^1(z)$ are holomorphic subprojections of $Q(z)$ which reduce $A(z)$ and the characteristic polynomial of $A(z)$, (which is $p(\lambda,A(z)) = \prod_{i=1}^{r_o} p(\lambda,A(z)/\text{Im}Q_i(z))$ thanks to 4) may be factorized once more : $p(\lambda,A(z) / \text{Im } Q_i(z)) = \prod_{i=1}^{r_1} p(\lambda,A(z)/Q_i^1(z))$.

This "process of reduction" (Kato [9]) may be pursued whenever an eigenvalue of the first term of the aggregate series $A^1(z)$, $A^2(z)$,... is diagonalizable.

In [2], [5], it is shown that this is always true for the eigenvalue 0 of a perturbed Markov chain. In particular, if $\bar{A}_n$ the n-th aggregate chain admits $\lambda \neq 0$ as eigenvalue with multiplicity m then there exist m repeated eigenvalues of $A(\varepsilon)$ of the form $\varepsilon^n\lambda + o(\varepsilon^n)$.

In the next paragraph we will derive a simpler way to compute a perturbed eigenvalue, for systems with particular structure.

## 2. APPLICATION OF PERTURBED MARKOV CHAINS TO THE RELIABILITY OF LARGE SCALE REPAIRABLE SYSTEMS

### 2.1. Introduction

Let us consider a physical system having two components. The components are supposed to be independent from the reliability point of view. Moreover one assumes as often in reliability theory (for electronic parts for example) that the random variable modelling the time before failure of each component follows an exponential distribution. The parameter $\lambda$ of the law is called "failure rate" of the component.

When a component has failed it is repaired ; the repair time is also assumed to follow an exponential distribution : the parameter $\mu$ is called "repair rate" of the component.

Supose one repairman only is available for the physical system : if the second component fails when the first is being repaired one awaits the end of the first repair to begin the second.

This simple system can be modelled by a continuous time Markov chain $X_t$. $X_t$ can take three values :

$X_t$ = 2 if the two components are working
$X_t$ = 1 if one is working and the other is being repaired
$X_t$ = 0 if one is being repaired and the other is awaiting repair.

The transition rates of the Markov chain are the following :

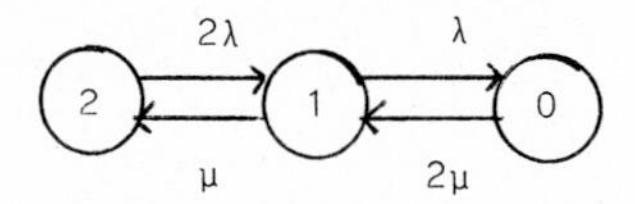


Moreover the mean times between failures (MTBF) are much longer (many thousand hours) than the mean times to repair (a few hours). One can therefore consider that it is a perturbed Markov chain.

Assume the components are in "parallel" mode that is the system fails only when the two components are failed. The reliability engineers are interested in three parameters :

- Availability = P (the system works at time t) = $1-P(X_t = 0)$
- Reliability = P (the system works between 0 and t) = $P(X_s \neq 0, \quad s \leq t)$
- MTTF = Mean time before the first failure of the system (the initial state being the state where both components works).

When the system is large (the number of components is more than 10) an exact computation of these parameters is not possible ; a perturbation method will provide approximations for then.

## 2.2. Model of the system

The large scale repairable systems involving a repair policy with the constitution of queues can in the exponential case be modelled by a Markov chain $X_t$ with the following properties :

There exists a partition of the set E of states in subsets $G_i$ :

$$E = G_o \cup G_1 \dots \cup G_k$$

and the following hypothesis are made :

H1 : the only nonzero transitions occurs between neighbooring subsets (from $G_i$ to $G_{i-1}$ or $G_{i+1}$)

H2 : from any state e in $G_i$ $(i > 0)$ there exists an nonzero transition to the subset $G_{i-1}$.

H3 : the largest of the transition rates $G_i \rightarrow G_{i+1}$ is much smaller than the smallest of the transition rates $G_i \rightarrow G_{i-1}$.

H4 : Any state e in $G_i$ $(i > 0)$ can be reached from at least one state in $G_{i-1}$.

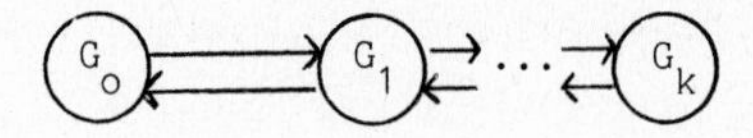


Figure 1

From these hypothesis one deducts that the transition matrix $\Lambda$ of the Markov chain is block tridiagonal. In Figure 2 the submatrices marked "$\varepsilon$" contains the rates that are assumed to be small by hypothesis H3.

=

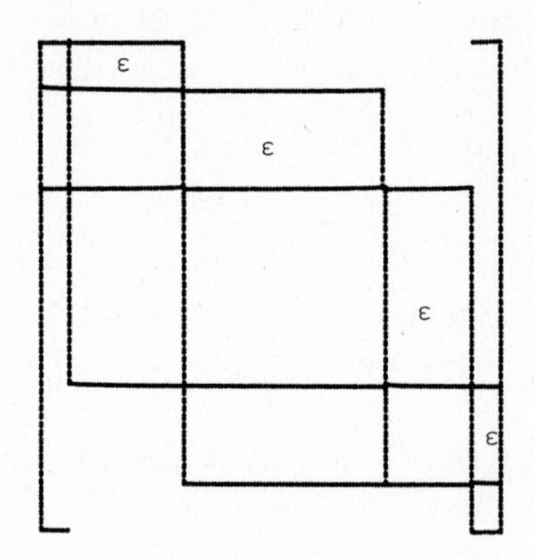


Figure 2

If the physical system contains k components then $G_i$ is the subset of states where i components are not working (either in repair or a waiting repair). The subset $G_o$ is constituted of one state : the initial state where all components are working.

Let $n_i$ = card $(G_i)$, $n_o = 1$. Let A be the transposed matrix of $\Lambda$. A has the following form :

$$
\begin{bmatrix}
-D_o(\varepsilon), B_1 & & & & \\
\varepsilon A_o, -D_1(\varepsilon), B_2 & & & & \\
 & \cdots & & & \\
 & & \varepsilon A_{i-1}, -D_i(\varepsilon), B_{i+1} & & \\
 & & & \cdots & \\
 & & & & \varepsilon A_{k-2}, -D_{k-1}(\varepsilon), B_k \\
 & & & & \varepsilon A_{k-1}, \quad -D_k(\varepsilon)
\end{bmatrix}
$$

Figure 3

The submatrices constituting A have the following properties :

a) $A_i$ $(i = 0, k-1)$ are of dimension $(n_{i+1}, n_i)$ ; their elements are between 0 and 1.

b) $B_i$ $(i = 1, k)$ are of dimension $(n_{i-1}, n_i)$ ; their elements are greater than one ; moreover the sums of each column is strictly positive by H2.

c) The diagonal elements are such that the sum of the columns is zero. $D_i(\varepsilon)$ $(i = 0,k)$ can therefore be expressed as a function of $A_i$ and $B_i$.

Let $D_i = D_i(0)$ $(i = 0,k)$

d) The sum of each line of $A_i$ $(i = 0,k-1)$ is non zero by H4.

## 2.3. Approximates for availabilities

Availabilities can be expressed as function of the stationary probabilities of the Markov chain. The first result gives approximates for them when $\varepsilon$ is small.

**Theorem 1**

Let $P_i(\varepsilon)$ $(i = 0,k)$ be the column vector of the stationary probabilities for the states in $G_i$

$$P(\varepsilon) = \begin{pmatrix} P_o(\varepsilon) \\ \vdots \\ P_k(\varepsilon) \end{pmatrix}$$

and let $Q_i(\varepsilon)$ $(i = 0,k)$ be the columns vectors defined by :

$$Q_o(\varepsilon) = 1,$$

$$Q_i(\varepsilon) = \varepsilon^i \prod_{i=1}^{j=1} (D_j^{-1} A_{j-1}) \quad (i = 1,k)$$

and

$$Q(\varepsilon) = \begin{pmatrix} Q_o(\varepsilon) \\ \\ Q_k(\varepsilon) \end{pmatrix}$$

The ratio of the ith component of P to the ith component of Q goes to 1 as $\varepsilon$ goes to zero.

Pratical use

The $Q_i$ will provide approximates for the availabilites ; let us consider the reduced graph obtained by keeping only the downward transitions. And the modified graph where the $\alpha$ rate from state f to f' is replaced by $\alpha/\beta$ where $\beta$ is the sum of the upward transition rates leaving f' that were supressed during the first step. ($\beta$ is non zero by H2).

The probability Q corresponding to the state f is then obtained by identifying the paths leading from the initial state $f_o$ to f, multiplying along them the rates and adding the products.

Proof

Let $P_t(\varepsilon)$ be the vector of the probabilities of the n states at time t ; by definition of A, $P_t(\varepsilon)$ is given by

$$\frac{d}{dt} P_t(\varepsilon) = A\, P_t(\varepsilon)$$

and $P(\varepsilon) = \lim_{t\to\infty} P_t(\varepsilon)$ will be obtained by solving :

$$(1, 1, \ldots, 1)\ P(\varepsilon) = 0,$$
$$A\ \ P(\varepsilon) = 0.$$

Solving the system formally by Gauss elimination method leads to :

$$- D_o(\varepsilon) P_o(\varepsilon) + B_1 P_1(\varepsilon) = 0,$$

$$\varepsilon A_i P_i(\varepsilon) - D_{i+1}(\varepsilon) P_{i+1}(\varepsilon) + B_{i+2} P_{i+2}(\varepsilon) = 0$$

$(i = 0, k-2)$

$$\varepsilon A_{k-1} P_{k-1}(\varepsilon) - D_k(\varepsilon) P_k(\varepsilon) = 0.$$

$D_i(\varepsilon)$ is inversible ; moreover

$$D_i(\varepsilon) \to D_i \text{ as } \varepsilon \to 0.$$

From the last equation one can deduct :

$$P_{k-1}(\varepsilon) = D^{-1}_{k-1}(\varepsilon) A_{k-2} P_{k-2}(\varepsilon) ;$$

substituting in the preceeding one :

$$\varepsilon A_{k-3} P_{k-3}(\varepsilon) = [D_{k-2}(\varepsilon) - B_{k-1} D^{-1}_{k-1}(\varepsilon) A_{k-2}] P_{k-2}(\varepsilon).$$

Let

$$D'_{k-2}(\varepsilon) = D_{k-2}(\varepsilon) - \varepsilon B_{k-1} D^{-1}_{k-1}(\varepsilon) A_{k-2}.$$

$$D'_{k-2}(\varepsilon) \to D_{k-2} \text{ as } \varepsilon \to 0.$$

By induction one gets then :

$$\varepsilon A_i P_i(\varepsilon) = D'_{i+1}(\varepsilon) P_{i+1}(\varepsilon) \quad (i = 0, k-1).$$

An explicit formula for $P_i(\varepsilon)$ is then :

$$P_i(\varepsilon) = \varepsilon^i \left[ \prod_{j=i}^{j=1} (D'^{-1}_j(\varepsilon) A_{j-1}) \right] P_o(\varepsilon) \quad (i = 0, k).$$

Let $C_i(\varepsilon)$ be the vector whose expression is between the brackets

$$C_i(\varepsilon) \to C_i \text{ as } \varepsilon \to 0$$

where

$$C_i = \prod_{j=i}^{1} (D_j^{-1} A_{j-1})$$

The value of the scalar $P_o(\varepsilon)$ can be obtained by substituting the $P_i$ in the equation

$$(1, 1, \ldots, 1)\, P(\varepsilon) = 1.$$

One deducts then

$$P_o(\varepsilon) \to 1 \text{ as } \varepsilon \to 0.$$

Let then

$$Q_o = 1,\ Q_i(\varepsilon) = \varepsilon^i\, C_i \quad (i = 1, k)$$

and

$$Q(\varepsilon) = \begin{pmatrix} Q_o \\ Q_1(\varepsilon) \\ Q_k(\varepsilon) \end{pmatrix}$$

The jth component of P is then equivalent to the jth component of Q as goes to zero (j = 1,n).

## 2.4. Approximations for the mean time to failure and the reliability

To compute these parameters one must first define the subset P of E consisting of the states where the system is failed.

Let A' be the submatrix of A corresponding to $P^c$ the complementary subset of P in E. The structure of A is similar to that of A :

$$\begin{bmatrix} -D''_0(\varepsilon), & B'_1 & & \\ \varepsilon A'_{0'} & -D''_1(\varepsilon), & B'_2 & \\ & & \cdots & \\ & & \varepsilon A'_{k'-1}, & -D''_k(\varepsilon) \end{bmatrix}$$

Let $Q_t(\varepsilon)$ be the vector whose components are the mean time spent in the states of $P^c$ between 0 and t, $Q_t(\varepsilon)$ is solution of :

$$A' \; Q_t(\varepsilon) + P_o(\varepsilon) = P_t(\varepsilon)$$

Let

$$T = \lim_{t \to \infty} \; Q_t(\varepsilon), \; T = \begin{pmatrix} T_0 \\ T_1 \\ \\ T_{k'} \end{pmatrix}$$

T is solution of :

$$A'T = \begin{pmatrix} -1 \\ 0 \\ \\ 0 \end{pmatrix}$$

$T_i$ (i = 0, k') is a column vector of dimension $n'_i$. Solving the system as in 2, one gets :

$$T_i = \varepsilon^i \; C'_i(\varepsilon) \; T_0 \qquad (i = 1, k')$$

with

$$C'_i(\varepsilon) = D''^{-1}_i(\varepsilon) \; A'_{i-1} \; C'_{i-1} \qquad (i = 1, k') \; ; \; C'_o = 1 \; ;$$

Moreover

$$C'_i(\varepsilon) \to C'_i \text{ as } \varepsilon \to 0$$

where $C'_i = D''^{-1}_i \; A'_{i-1} \; C'_{i-1}$ $(i = 1,k')$, $C'_o = 1$, and all components of $C'_i$ are strictly positive.

Let $i_o$ be the smallest i such that $\Lambda_i$ is non zero $(i = 0, k')$ ; one gets, then :

$$T_o \sim \frac{1}{\varepsilon^{i_o+1} \; \Lambda_{i_o} \; C'_{i_o}} \quad \frac{-1}{\varepsilon^{i_o+1} \; \Lambda_{i_o} \; C'_{i_o}} \quad \text{as } \varepsilon \to 0$$

and

$$E(T_1) = \text{MTTF} \sim T_o \text{ as } \varepsilon \to 0.$$

Let us define the "equivalent failure rate" $\Lambda_e(\varepsilon)$ by

$$\Lambda_e(\varepsilon) = - \varepsilon^{i_o+1} \; \Lambda_{i_o} \; C'_{i_o} .$$

The following result gives approximates for the MTTF and the reability.

**Theorem 2**

a) $E(T_p) \quad \frac{1}{\Lambda_e(\varepsilon)}$ as $\varepsilon \to 0$ ;

b) The random variable $\Lambda_e(\varepsilon) \; T_p$ converges in law to an exponential variable of parameter 1 as $\varepsilon$ goes to 0.

The approximates obtained in Theorems 1 and 2 are very useful for large systems. The reliability, parameters would otherwise be obtained by solving a very large linear system. We will now give an example.

## 2.5. Example

The system considered contains n independent components. The failure and repair rates of the ith component are $\lambda_i$ and $\mu_i$. Suppose that the number of repairmen available is $r \leq n$ ; when all repairmen are busy a queue is formed on a first come first serve basis.

Let $G_i$ be the subset of states where i components are not working (either awaiting repair or in repair). To define a state in $G_i$ $i \leq r$, it is enough to give the list $j_1 < j_2 < \ldots < j_i$ of components in repair. The number of such states is $\binom{n}{i}$.

In the subsets $G_i$ $i > r$, it is necessary to know the components in repair and the order of the queue. There are

$$\binom{n}{r} A_{i-r}^{n-r} \quad \text{such states.}$$

The corresponding process is easily seen to satisfy assumptions $H_1$, $H_2$, $H_3$ and $H_4$ and approximates can therefore be found for the availability of the system.

Let us get for example the probability that the subset $P = [l_1 < l_2 < \ldots < l_i]$ is not available. If $i > r$ a state e of P is defined by $R = [j_1 < j_2 < \ldots < j_r]$ the list of the components being repaired and by the order of the queue for the components awaiting repair.

Applying Theorem 1 to each of these states and summing one gets the following approximation for the unavailability of P :

$$Q = (i - r)! \prod_{j \in P} \lambda_j \sum_{R \subset P} \frac{1}{\prod_{j \in R} \mu_j \cdot \left( \sum_{j \in R} \mu_j \right)^{i-r}}$$

Similar results can be obtained for various repair policies (see [14] for details).

Approximation for the reliability parameters of a system can therefore easily be obtained for large scale systems by a simple inspectrion of the transition graph of the associated Markov Chain.

REFERENCES

[ 1 ] P. BERNHARD. On singular implicit linear dynamical systems, Siam J of Control and Opt. vol 20 n° 5 sept 82.

[ 2 ] M. CODERCH, A.S. WILLSKI, S.S. SASTRY, D.A. CASTANON. Hierarchical aggregation of linear systems with multiple time scales, MIT Report LIDS-P-1187, mars 1982.

[ 3 ] M. CODERCH, A.S. WILLSKY, S.S. SASTRY. Hierarchical aggregation of singulary perturbed finite state Markov chains submitted to stochastics.

[ 4 ] P.J. COURTOIS. Decomposability, ACM Monograph Series, Academic Press, 1977.

[ 5 ] F. DELEBECQUE. A reduction process for pertubed Markov chains, a paraître SIAM J. of applied math. to appear.

[ 6 ] F. DELEBECQUE, J.P. QUADRAT. Optimal control of Markov chains admitting strong and weak interactions, Automatica, Vol. 17, n° 2, pp. 281-296, 1981.

[ 7 ] F. DELEBECQUE, J.P. QUADRAT. The optimal cost expansion of finite controls finite states Markov chains with weak and strong interactions. Analysis and optimization of systems, Lecture Notes an control and Inf. Science 28 Springer Verlag, 1980.

[ 8 ] A.A. PERVOZVANSKII, A.V. GAITSGORI. Decomposition aggregation and approximate optimization en Russe, Nauka, Moscou, 1979.

[ 9 ] T. KATO. Perturbation theory for linear operator, Springer Verlag, 1976.

[10] B.L. MILLER, A.F. VEINOTT. Discrete dynamic programming with small interest rate. An. math. stat. 40, 1969, pp. 366-370.

[11] R. PHILIPS, P. KOKOTOVIC. A singular perturbation approach to modelling and control of Markov, chains IEEE A.C. Bellman issue, 1981.

[12] H. SIMON, A. ANDO. Aggregation of variables in dynamic systems, Econometrica, 29, 111-139, 1961.

[13] J. KEMENY, L. SNELL. Finite Markov chains, Van Nostrand, 1960.

[14] O. MURON. Evaluation de politiques de maintenance pour un système complexe, RIRO, vol. 14, n° 3, pp. 265-282, 1980.

[15] S.L. CAMBELL, C.D. MEYER jr. Generalized inverses of linear transformations. Pitman, London, 1979.

[16] TKIOUAT. Thèse Rabat à paraitre.

[17] J.P. QUADRAT. Commande optimale de chaines de Markov perturbées Outils et Modèles Math. pour l'automatique... t3 edition CNRS 1983.

[18] J.P. QUADRAT Optimal control of perturbed, Markov chain the multitime scale case. Singular pertubation in systems and control. CISM courses and lectures n° 280, Springer Verlag 82.

[19] F. DELEBECQUE, J.P. QUADRAT. Contribution of stochastic control, team theory and singular perturbation to an example of large scale systems : Management of hydropower production. IEEE AC avril 1978.

# OPTIMAL CONTROL OF PERTURBED MARKOV CHAINS

*J.P. Quadrat*[†]

Given a controlled perturbed Markov chain of transition matrix $m^u(\varepsilon)$, where $\varepsilon$ is the perturbation scale and u the control, we study the solution expansion in $\varepsilon$ , $w^\varepsilon$ , of the dynamic programming equation :

$$\min_u \; [m^u(\varepsilon)\, w^\varepsilon + c^u(\varepsilon)] = (1 + \lambda(\varepsilon))\, w^\varepsilon .$$

$m^u(\varepsilon)$, $c^u(\varepsilon)$, $\lambda(\varepsilon)$ are polynomials in $\lambda$ . The case $\lambda(\varepsilon) = \varepsilon^\ell$ leads to study Markov chains on a time scale of order $1/\varepsilon^\ell$. The state space and the control set are finite.

†INRIA, Domaine de Voluceau, Rocquencourt, B.P. 105, 78150 LE CHESNAY CEDEX, France.

PLAN



## 1 - INTRODUCTION

Stochastic or deterministic control problems can be reduced after discretization to the control of Markov chains. This approach leads to control of Markov chains which have a large number of states. An attempt to solve this difficulty is to see the initial Markov chains as the perturbation of a simpler one, and to design algorithms which use the hierarchical structure of more and more aggregated models, described in the previous paper of Delebecque, to increase the computation speed of the optimal control.

The two time scale control problem (actualization rate of order $\varepsilon$) is solved in Delebecque-Quadrat [6] , [7]. The ergodic control problem when the unperturbed chain has no transient classes has been studied in Philips-Kokotovic [19]. In this paper we give the construction of the complete expansion of the optimal cost of the control problem in the general multi-time scale situation. This presentation is a very little improved version of Quadrat [17] , [18]. For that, we use three kinds of results :

- the Delebecque's result discussed in the previous paper.
- the realization theory of implicit systems developed by Bernhard [1]. Indeed this method gives a recursive mean of computing the complete cost expansion in the uncontrolled case.
- the Mille -Veinott [10] way of constructing the optimal cost expansion of an unperturbed Markov chain having a small actualization rate.

## 2 - NOTATIONS AND STATEMENT OF THE PROBLEM

We study the evaluation of a cost associated to the trajectory of a discrete Markov chain in the controlled perturbed case. For that let us introduce a coherent system of notation which allows discussion on the simplest structure that we need for each concept that we have to study.

2.1 - $(T, \mathcal{X}, m, c, \lambda)$ is associated to the unperturbed uncontrolled case and shall be called the Markov chain n-uple.

- T is the time set isomorphic to $\mathbb{N}$ ;
- $\mathcal{X}$ is the state space of the Markov chain. It is a finite discrete space. $|\mathcal{X}|$ denotes card ( $\mathcal{X}$) that is the number of elements of $\mathcal{X}$ . x will be the generic element of $\mathcal{X}$ ;
- m denotes the transition matrix of the Markov chain, that is a $(|\mathcal{X}|, |\mathcal{X}|)$ - matrix with positive entries such that $\sum_{x' \in \mathcal{X}} m_{x\,x'} = 1$ ;
- c is the instantaneous cost that is a $|\mathcal{X}|$- vector with positive entries ;
- $\lambda$ is an actualization rate that is, $\lambda \in R$ and $\lambda > 0$.

The set of possible trajectories is denoted by $\Omega = \mathcal{X}^T$, a trajectory by $\omega \in \Omega$, the position of the process at time t if the trajectory is $\omega$ by $X(t,\omega)$. The conditional probability of the cylinder :

$$B = \{\omega : X_t(\omega) = x_t,\ t = 0,1,\dots,n\}$$

knowing $X(0, \omega)$ is :

$$P^{X_0}(B) = \prod_{t=0}^{n-1} m_{x_t\, x_{t+1}}$$

To the trajectory $\omega$ is associated the cost :

$$j(\omega) = \sum_{t=0}^{+\infty} \frac{1}{(1+\lambda)^{t+1}}\, c_{X(t,\omega)} \tag{2.1}$$

The conditional expected cost knowing $X(0,\omega)$ is a $|\mathcal{X}|$- vector denoted w defined by :

$$w_x := E[j(\omega) \mid X(0,\omega) = x], \quad \forall x \in \mathcal{X} \qquad (2.2)$$

The Hamiltonian is the operator :

$$h : \mathbb{R}^{|\mathcal{X}|} \to \mathbb{R}^{|\mathcal{X}|}$$
$$w \qquad [m - (1 + \lambda)i]\, w + c \qquad (2.3)$$

where i denotes the identify of the $(|\mathcal{X}|, |\mathcal{X}|)$ - matrices set.

Then w defined by (2.2) is the unique solution of the Kolmogorov equation :

$$h(w) = 0 \qquad (2.4)$$

2.2 - In the perturbed situation the n-tuple defining the perturbed Markov chain is :

$$(T, \mathcal{X}, \mathcal{E}, m(\varepsilon), c(\varepsilon), \lambda(\varepsilon))$$

- $\mathcal{E}$ is now the space of the perturbations ; in all the following it is $\mathbb{R}^+$ ;
- $m(\varepsilon)$, $c(\varepsilon)$, $\lambda(\varepsilon)$ have the same definition as previously but depends on the parameter $\varepsilon \in \mathcal{E}$ , and we suppose that they are polynomials in this variable.

We denote by d the degree of a polynomial and by v its valuation (the smallest non zero power of the polynominal). In the following $d(m) = 1$, $v(m) = 0$, $v(\lambda) = v(c) = d(\lambda) = \ell$ . From this particular case the general case can be understood.

The Hamiltonian of the perturbed problem is denoted by :

$$h(w,\varepsilon) = [m(\varepsilon) - (1 + \lambda(\varepsilon))\, i]\, W + c(\varepsilon) \qquad (2.5)$$

The expected conditional cost is denoted $w^{\varepsilon}$ and is solution of the Kolmogorov equation :

$$h(w^{\varepsilon},\varepsilon) = 0 \tag{2.6}$$

We shall prove that $w^{\varepsilon}$ admits an expansion in $\varepsilon$ that we shall denote by $W(\varepsilon) = \sum_{n=0}^{\infty} \varepsilon^n w_n$ where $W_i$ are $|\mathcal{X}|$-vectors. Then we have :

$$m(\varepsilon)\, W(\varepsilon) = \sum_{n=0}^{\infty} \varepsilon^n (MW)_n \tag{2.7}$$

with :

$$M = \begin{bmatrix} m_0 & & & 0 \\ m_1 & m_0 & & \\ 0 & m_1 & m_0 & \\ | & \ddots & \ddots & \ddots \end{bmatrix} \tag{2.8}$$

an infinite block matrix.

For the Hamiltonian we can introduce the same notation :

$$h(W(\varepsilon),\, \varepsilon) = \sum_{n}^{\infty} \varepsilon^n H_n(W) \tag{2.9}$$

where $H_n(W)$ are the $|\mathcal{X}|$- vectors defined in (2.9) by identification of the $\varepsilon^i$ terms , that is :

$$\begin{cases} H_0(W) = (m_0 - i)w_0 \\ H_1(W) = m_1\, w_0 + (m_0 - i)w_1 \\ \vdots \\ H_\ell\,(W) = -\,\lambda_\ell\, w_o + m_1\, w_{\ell-1} + (m_0 - i)\, w_\ell + c_\ell \\ \vdots \end{cases} \tag{2.10}$$

Then with :

$$H(W) \equiv [M - (I + \Lambda)]\, W + C, \qquad (2.11)$$

where :

$C = (x_n,\ n \in \mathbb{N},\ c_n$ are $|\mathcal{X}|$-vectors)

$I$ : the identity operator $\begin{bmatrix} i & 0 & 0 & 0 & \cdots \\ 0 & i & 0 & 0 & \cdots \\ 0 & 0 & i & & \\ \vdots & \vdots & & \ddots & \end{bmatrix}$

$\Lambda$ : the operator $\ell^{th}$ $|\mathcal{X}|$-block $\begin{bmatrix} \diagdown & & 0 \\ & i\lambda_\ell & \\ 0 & & \diagdown \end{bmatrix}$

an expansion of the cost is obtained by solving :

$$H(W) = 0 \qquad (2.12)$$

Morever the sequence $(W_i,\ i \in \mathbb{N})$ can be computed recursively. These two results will be shown in part 4.

2.3 - For the control problem we need the introduction of the n-tuple :

$$(T, \mathcal{X}, \mathcal{U}, m^u, c^u, \lambda)$$

- $\mathcal{U}$ is the set of control which is here a finite set. $|\mathcal{U}|$ denotes the cardinal of $\mathcal{U}$ . Its generic element is denoted by u.
- m denotes the $(|\mathcal{U}|, |\mathcal{X}|, |\mathcal{X}|)$ tensor of entries $m^u_{xx'}$, the probability to go in x', starting from x, the control being u.
- c denotes the $(|\mathcal{U}|, |\mathcal{X}|)$ matrix of entries $c^u_x$, the cost to be in x, the control being u.

A policy is an application :

$$s : \mathcal{X} \rightarrow \mathcal{U}$$

The set of policies is $\mathcal{P} := \mathcal{U}^{\mathcal{X}}$

For a policy s, mos denotes the $(|\mathcal{X}|, |\mathcal{X}|)$ transition matrix of entries :

$$(mos)_{xx'} = m_{xx'}^{s_x} : \qquad (2.13)$$

cos denotes the $|\mathcal{X}|$-vector ;

$$(cos)_x = c_x^{s_x}. \qquad (2.14)$$

we associate to a policy $s \in \mathcal{P}$ and a trajectory $\omega$ , the cost

$$j^s(\omega) = \sum_{t=0}^{+\infty} \frac{1}{(1+\lambda)^{t+1}} (cos)_{X(t,\omega)} \qquad (2.15)$$

and the optimal conditional expected cost knowing the initial condition is :

$$w_x^* = \operatorname*{Min}_{s \in \mathcal{P}} \mathbb{E}(j^s(\omega) \mid X(0,\omega) = x) \qquad (2.16)$$

The Hamiltonian is defined as the operator :

$$h : \mathcal{U} \times \mathbb{R}^{\mathcal{X}} \rightarrow \mathbb{R}^{\mathcal{X}}$$

$$(u,w) \qquad h^u(w) = [m^u - (1+\lambda)i]\, w + c^u. \qquad (2.17)$$

The notation $(hos)_x$ for $h_x^{s_x}$ will be used.

Then the optimal Hamiltonian is the operator :

$$h^* : \mathbb{R}^{\mathcal{X}} \to \mathbb{R}^{\mathcal{X}}$$

$$w \qquad h^*_x(w) = \min_u h^u_x(w), \ \forall x \in \mathcal{X} \tag{2.18}$$

The optimal expected cost $w^*$ is the unique solution of the dynamic programming equation :

$$h^*(w^*) = 0 \tag{2.19}$$

An optimal policy is given by :

$$s^* : \mathcal{X} \to \mathcal{U}$$

$$x \qquad s^*_x \in \operatorname{argmin} h^u_x(w^*), \ \forall x \in \mathcal{X}$$

2.4 - The perturbed control problem is defined by the n-tuple :

$$(T, \mathcal{X}, \mathcal{U}, \mathcal{S}, m^u(\varepsilon), \lambda(\varepsilon)).$$

Its interpretation is clear from the previous paragraphs.

By analogy the notation $H^u(w,\varepsilon)$, $h^*(w,\varepsilon)$, $w^{*\varepsilon}$ , $H^u(W)$ are clear, but we need a definition of $H^*(W)$. For that let us introduce the lexicographic order, $\geq$ , for sequences of real numbers, that is :

$$\begin{aligned} &(y_0, y_1, \ldots) \geq (y'_0, y'_1, y'_2, \ldots) \text{ is true } \Leftrightarrow \\ &(\text{if } y_n = y'_n, \ \forall n < m \text{ then } y_m \geq y'_m) \ \forall m \in \mathbb{N}. \end{aligned} \tag{2.20}$$

We denote by $\overrightarrow{\min}$ the minimum for this order. Then we define $H^*$ by :

$$H^*_x(W) = \overrightarrow{\min} \ H^u_{.x}(W) \tag{2.21}$$

(indeed $H^u_{.x}(W)$ is a sequence of real numbers the coefficients of $h^u_x(W(\varepsilon),\varepsilon)$ in its expansion in $\varepsilon$).

We shall prove that $w^{*\varepsilon}$ admits an expansion in $\varepsilon$ denoted by $W^*(\varepsilon)$ which satisfies :

$$H^*(W^*) = 0 \qquad (2.22)$$

The purpose of this paper is to prove this last result and to show that $W^*$ can be computed recursively. By this way we can design faster algorithm than the ones obtained by a direct solution of $h^*(w^{*\varepsilon},\varepsilon) = 0$.

## 3 - PERTURBED MARKOV CHAIN

We give some algebraîc complement to the previous study of the perturbed Markov chain $(T, \mathcal{X}, \mathcal{S}, m(\varepsilon), c(\varepsilon), \lambda(\varepsilon))$. For that we study the transfer function $\varepsilon^\ell \mu\ (\varepsilon^\ell \mu + i - m(\varepsilon))^{-1}$ in $\varepsilon$, where $1/\varepsilon$ denotes the avance operator. This interpretation gives a general way to find a finite memory algorithm to compute the expansion of $w^\varepsilon$.

We have seen in (2.11) that when the conditional expected cost $w^\varepsilon$ admits an expansion, $W(\varepsilon)$, in $\varepsilon$ this expansion satisfies :

$$H(W) \equiv (M - I - \Lambda)\ W + C = 0 \qquad (3.1)$$

(3.1) is an infinite set of linear equations. Conversely if a solution of (3.1) exists with for example $(W_i,\ i \in \mathbb{N})$ bounded the $W(\varepsilon)$ converges, for $\varepsilon < 1$, and is a solution of :

$$h(w,\varepsilon) = 0 \qquad (3.2)$$

Let us show now that (3.1) can be computed recursively.

For that we build the implicit realization of $W$ :

$$\begin{cases} E\ y_{n+1} = F\ y_n - G\ C_{n+\ell+1} \\ W_{n+1} = J\ y_{n+1} \qquad y_{-1} = 0 \end{cases} \qquad (3.3)$$

with :

$$a_0 = m_0 - i \tag{3.4}$$

$$E = \begin{bmatrix} a_0 & & 0 \\ m_1 & \ddots & \\ 0 \ddots & \ddots & \\ -\mu & m_1 & a_0 \end{bmatrix} \Big\} (\ell+1) \text{ blocks} \tag{3.5}$$

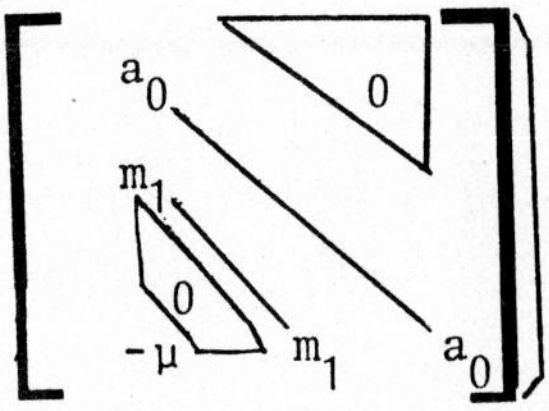

$$F = \begin{bmatrix} 0 & a_0 & & 0 \\ | & m_1 & \ddots & \\ | & 0 & m_1 & a_0 \\ 0 & — & — & 0 \end{bmatrix} \Big\} (\ell+1) \text{ blocks} \tag{3.6}$$

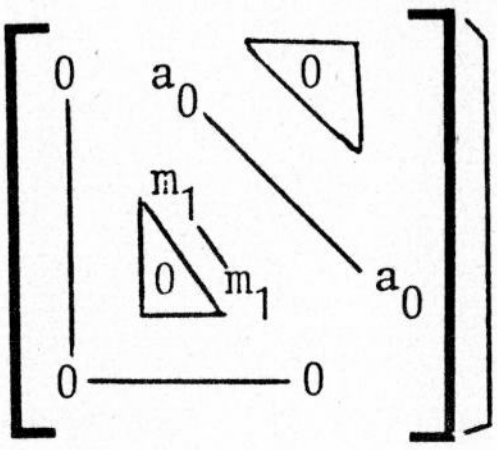

$$G = \begin{bmatrix} 0 \\ | \\ 0 \\ i \end{bmatrix} \Big\} (\ell+1) \text{ blocks} \tag{3.7}$$

$$J = \underbrace{[i \;\; 0 —— 0]}_{(\ell+1) \text{ blocks}} \tag{3.8}$$

Indeed if $W$ is a solution of (3.1) :

$$y_n = (W_n, W_{n+1}, \ldots, W_{n+\ell})$$

is a solution of (3.3).

Conversely if $W$ is a solution of (3.3), by elimination of the variables $y$ we see that $W$ satisfies (3.1).
Let us denote by $\mathcal{L}$ the space $\mathbb{R}^{|\mathcal{X}|} \times (\ell+1)$ in which lives $y$.

Following Bernhard [1], to prove the existence of a solution of (3.3), we have to show that there exists $\mathcal{Z} \subset \mathcal{L}$ which satisfies :

$$F\mathcal{Z} \subset E\mathcal{Z} \tag{3.9}$$

$$G \subset E\mathcal{Z}. \tag{3.10}$$

We can take $\mathcal{Z} = \mathcal{L}$ . Indeed (3.9) is equivalent to finding a $z$ such that :

$$Ez = Fy \ , \ \forall y \in \mathcal{L}$$

But by the change of variables $z'^{k} = z^{k+1}$, $k = 1,..\ell$ $z'_{\ell} = z_{\ell}$ (3.11) becomes :

$$Ez' = Gc \text{ with } c = -\mu\, y^{2} + m_{1}\, y^{\ell} \in \mathbb{R}^{|\mathcal{X}|} \tag{3.12}$$

with is a relation of (3.10) kind.

Delebecque has proved that (3.12) has a solution, and that the $|\mathcal{X}|$-first entries of z' are uniquely defined.

Morever Bernhard [1] Th.3 has described the non-unicity space of (3.3). It is the smallest space satisfying :

$$F\mathcal{V} \subset E\mathcal{V}$$

$$\mathcal{V} \supset \mathcal{N}(E)$$

Let us show that $\mathcal{V} = \mathcal{N}(E)$ is a solution. For that we have only to verify :

$$F\mathcal{N}(E) = E\mathcal{N}(E) = 0$$

but $\quad x \in \mathcal{N}(E) \quad$ implies $\quad Ex = 0$

Thus we have to prove $\quad (Ex = 0 \Longrightarrow Fx = 0)$

But the result of Delebecque shows that $Ex = 0 \Longrightarrow |\mathcal{X}|$- first entries of x are 0 which implies $Fx = 0$.

Now the fact that $\mathcal{N}(J) \supset \mathcal{N}(E)$ implies that the sequence $W_n$ is uniquely defined.

We have proved the :

Theorem 1 : The solution $w^{\varepsilon}$ of :

$$h(W,\varepsilon) := (m(\varepsilon) - i - \lambda(\varepsilon)W + c(\varepsilon) = 0 \qquad (3.13)$$

admits an expansion $W(\varepsilon)$ which is the unique solution of :

$$H(W): = (M-I-\Lambda)W + C = 0 \qquad (3.14)$$

Moreover W can be computed recursively by constructing the implicit system realization of

$$\begin{cases} Ey_{n+1} = Fy_n - GC_{n+\ell+1},\ y_{-1} = 0 \\ W_{n+1} = Jy_{n+1}, \end{cases} \qquad (3.15)$$

where E, F, G, H are defined in (3.5) to (3.8).

This implicit system has an output uniquely defined and it admits a strictly causal realization.
A specific algorithm is given in Tkiovat [16]

## 4 - REVIEW OF CONTROLLED MARKOV CHAINS

Given the controlled Markov chain n-tuple : $(T,\mathcal{X},\mathcal{U}, m^u, c^u, \lambda)$. The optimal conditional expected $w^*$ cost is the unique solution in w of the dynamic programming equation :

$$h^*_x(w) \ \min_u [(m^u - 1 - \lambda)w + c^u]_x = 0,\ \forall x \in \mathcal{X}. \qquad (4.1)$$

This result can be proved using the Howard algorithm :

Step 1 : Given a policy $s \in \mathcal{U}^{\mathcal{X}}$, let us compute w, solving, in w, the linear equation :

$$hos(w) = 0 \tag{4.2}$$

Step 2 : Given a conditional expected cost w, let us improve the policy by computing :

$$\min_u h_x^u(w), \forall x \in \mathcal{X} \tag{4.3}$$

We change $s(x)$ only if $h_x^u(w) < 0$. Then we return to step 1.

By this way we generate a sequence :

$$((s^n, w^n) \; ; \; n \in \mathbb{N})$$

which converges after a finite number of steps. The sequence $(w^n, n \in \mathbb{N})$ is decreasing.

Indeed :

$$hos^n(w^n) = 0 \tag{4.4}$$

$$hos^{n+1}(w^{n+1}) = 0 \tag{4.5}$$

Then (4.4)-(4.5) gives :

$$(mos^{n+1} - 1 - \lambda) \; (w^n - w^{n+1}) + hos^n(w^n) - hos^{n+1}(w^n) = 0 \tag{4.6}$$

But by (4.3) we have :

$$hos^n(w^n) - hos^{n+1}(w^n) \geq 0 \tag{4.7}$$

Then (4.6) and (4.7) proves that :

$$w_n - w_{n+1} \geq 0 \tag{4.8}$$

Indeed, (4.6) can be seen as a Kolmogorov equation in $(w_n - w_{n+1})$, with a positive instantaneous cost.

The existence and the uniqueness of a solution in w of (4.1) follows easily from this result.

## 5 - CONTROL OF PERTURBED MARKOV CHAINS

Given the perturbed controlled Markov chain n-tuple $(T, \mathcal{X}, \mathcal{U}, \mathcal{B}, m^u(\lambda), c^u(\varepsilon), \lambda(\varepsilon))$. The optimal cost is the unique solution in w of the dynamic programming equation :

$$h^*_x(w,\varepsilon) \equiv \min_u [(m^u(\varepsilon) - 1 - \lambda(\varepsilon))w + c^u(\varepsilon)]_x = 0, \forall x \in \mathcal{X} \qquad (5.1)$$

We have the :

Theorem 2 : The solution of (5.1) denoted by $w^{*\varepsilon}$ admits an expansion in $\varepsilon$ denoted by $W^*(\varepsilon)$ which is the unique solution in $\mathcal{W}$ of the vectorial dynamic programming equation :

$$H^*_{.x}(W) \equiv \overrightarrow{\min_u} [(M^u - I - \Lambda)W + C^u]_{.x} = 0, \forall x \in \mathcal{X} \qquad (5.2)$$

Let us remember that $\overrightarrow{\min}$ means the minimum for the lexicographic order on the sequence of real numbers.

The solution $W^*$ can be computed by the vectoriel Howard algorithm :

Step 1 : Given a policy $s \in \mathcal{U}^{\mathcal{X}}$, let us compute W using the results of part 4 :

$$Hos(w) = 0 \qquad (5.3)$$

Step 2 : Given a conditional expected cost W, let us improve the policy by computing :

$$\overrightarrow{\min_u} H^u_{.x}(W) \qquad (5.4)$$

We change $s(x)$ only if $H^u_{.x}(W) < 0$. Then we return to step 1.

By this way we generate a sequence :

$$((s^n, W^n) \; ; \; n \in \mathbb{N})$$

which converges after a finite number of steps. The sequence $(W^n, n \in \mathbb{N})$ is decreasing for the lexicographich order >.

This decreasing property can be proved easily using the following equivalence :

$$h_x^u(W(\varepsilon),\varepsilon) \geq h_x^{u'}(W(\varepsilon)\ \varepsilon) \iff H^u_{.x}(W) \geq H^{u'}_{.x}(W), \tag{5.5}$$

From this property the theorem can be proved easily.

A priori it is not clear if we may restrict the minimization to finite part of the infinite sequence.

The following result shows that this is possible and gives an estimate on the length of the sequence part on which we have to apply the lexicographic order minimization.

Theorem 3 : The vectoriel minimization (in 2.4) may be applied only on the 2 first terms of the sequence, without changing the convergence to the solution of (5.1), with $\eta = (d(c) + (v(\lambda) + 2)\ |\mathcal{X}|)$

Proof : Let us show that :

$$H_n^u(W) = H_n^{u'}(W),\ \forall i = d\ (c) + 1,\ldots,\eta \implies H_n^u(W) = H_n^{u'}(W),\ \forall n > d(c) \tag{5.6}$$

By theorem 1, W admits a strictly causal realization that is there exists E', F', G' such that :

$$\begin{cases} z_{n+1} = E'z_n + F'C_{n+\ell+1} \\ W_{n+1} = J'z_{n+1} \end{cases} \tag{5.7}$$

By (3.15) we know that the order of the matrix E is smaller than $(v(\lambda)+1)|\mathcal{X}|$.
The entry $C_{n+\ell+1}$ is equal to zero for $n \geq d(c) - \ell - 1$

We add $|\mathcal{X}|$ new states to z, denoted by z' with :

$$z'_{n+1} = W_n. \tag{5.8}$$

With the new states $z'' = (z,z')$ the second part of (6.6) can be written :

$$([a_0^u - a_0^{u'}][J',0] + [m_1^u - m_1^{u'}][0,i])\ z''_n = 0 \tag{5.9}$$

(5.9) has the form :

$$J'z''_n = 0 \tag{5.10}$$

with J' an observation matrix of the dynamical system of state $z''_n$. It follows by the Cayley-Hamilton theorem that if (5.10) is true $\forall n$ : $\eta \geq n > d(c)$ then (5.10) is true $\forall n > d(c)$. The theorem 3 is deduced easily from this result.

Remark :

In Tkiouat [16] a method is given to compute for each state x a bound $q(x)$ on the size of vectors on which we have to make the vectorial minimization.

## 6-Example and application

Let us show on a trivial 2 time scale example how these results can be useful to design fast algorithm to solve stochastic control problem.

Let us take the most simple example :

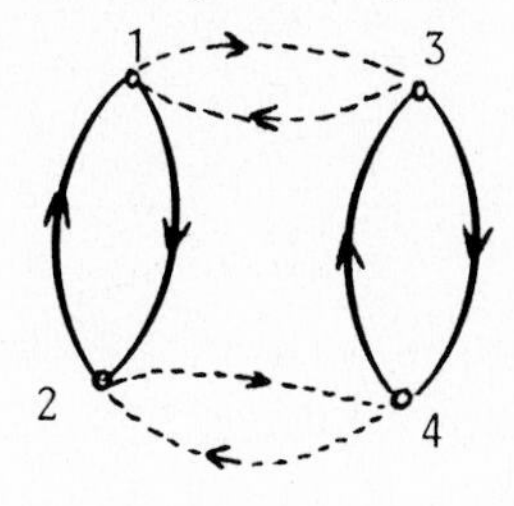

The dotted line denotes $\varepsilon$ -order probability transition, the others 1-order ones. All these transition depends of a control u which can take a finite number of values.

In this case the transition matrix has the following structure :

$$m^u(\varepsilon) = \left[\begin{array}{c|c} B_1 & \varepsilon A_2 \\ \hline \varepsilon A_1 & B_2 \end{array}\right]$$

Then,

$$m_0 = \left[\begin{array}{c|c} B_1 & 0 \\ \hline 0 & B_2 \end{array}\right]$$

$$m_1 = \left[\begin{array}{c|c} d_1 & A_1 \\ \hline A_2 & d_2 \end{array}\right]$$

Where $d_1$, $d_2$ are diagonal matrices.

The Howard algorithm becomes :

1 - given a policy that is the vector of decision $s = (u_1, u_2, u_3, u_4)$ the entry i gives the decision $u_i$ when the state is i.

2 - we solve

$$\begin{cases} m_0 W = 0 \\ (a_1 - \mu)W + m_0 W + c = 0 \end{cases}$$

using the particular structure of W, and the results explained in Delebecque, that is we compute :

$p_1$ solution of:

$$\begin{cases} p_1 B_1 = 0 \\ p_1 \begin{bmatrix} 1 \\ 1 \end{bmatrix} = 1 \end{cases}$$

and $p_2$ solution of :

$$\begin{cases} p_2 B_2 = 0 \\ p_2 \begin{bmatrix} 1 \\ 1 \end{bmatrix} = 1 \end{cases}$$

Then we compute the aggregate whain transition matrix :

$$\bar{A} = \begin{array}{|c|c|} \hline P_1 & 0 \quad 0 \\ \hline 0 \quad 0 & P_2 \\ \hline \end{array} \qquad A \begin{array}{|c|c|} \hline 1 & 0 \\ 1 & 0 \\ \hline 0 & 1 \\ 0 & 1 \\ \hline \end{array}$$

and the aggregated cost :

$$\bar{C}_1 = p_1 \begin{bmatrix} c_1 \\ c_2 \end{bmatrix}$$

$$\bar{C}_2 = p_2 \begin{bmatrix} c_3 \\ c_4 \end{bmatrix}$$

Then we solve

$$(\bar{A} - \mu)\, \bar{W} + \bar{C} = 0$$

which defines completely $\bar{W}$

Then we can compute $W_1$ solution of

$$\begin{cases} B_1 \; V_1 + [d_1 \mid A_1] \begin{bmatrix} 1 & 0 \\ 1 & 0 \\ \hline 0 & 1 \\ 0 & 1 \end{bmatrix} \bar{W} + \begin{bmatrix} c_1 \\ c_2 \end{bmatrix} = 0 \\ B_1 \; V_1 = 0 \end{cases}$$

and $w_2$ solution of

$$\begin{cases} B_2 \; V_2 + [d_2, \; A_2] \begin{bmatrix} 1 & 0 \\ 1 & 0 \\ \hline 0 & 1 \\ 0 & 1 \end{bmatrix} \bar{W} + \begin{bmatrix} c_3 \\ c_4 \end{bmatrix} = 0 \\ B_2 \; V_2 = 0 \end{cases}$$

Now we know

$$W_0 = \begin{bmatrix} 1 & 0 \\ 1 & 0 \\ \hline 0 & 1 \\ 0 & 1 \end{bmatrix} \bar{W}$$

and

$$W_1 = \begin{bmatrix} V_1 \\ \hline V_2 \end{bmatrix}$$

Then it is possible to improve the strategy by minimizing in u all the entries of the 4 - vector

$$(m_1(u)-\mu)\, \dot{W}_0 + (m_0(u) - i)W_1 + c(u)$$

the result is a new strategy $(u_1,u_2,u_3,u_4)$.

We have seen that in this process we have never solved a linear of size 4 but three systems of size 2, and 4 minimizations.

More generaly when the matrice $m_0$ has a block diagonal structure, this perturbation method improve the speed of the Howard algorithm. In the best situation we can obtain a $|\mathcal{X}|^2$ algorithm to solve the problem.

In this discussion we have only compute the first term of the expansion where the vectorial minimization defines completely the control after computing only the two first terms of the expansion $W_0$ and $W_1$.
The general algorithm is complicated to be implemented.
It is done in Tkiouat [16].

This method can be applied for discrete version of the following diffusion process :

$$dX_t = b_1(X_t,Y_t,U_t)\, dt + \sigma_1(X_t,Y_t,\, U_t)\, dW_t^1$$

$$dy_t = \frac{1}{\varepsilon}\, b_2(X_t,Y_y,U_t)\, dt + \frac{1}{\sqrt{\varepsilon}}\, \sigma_2(X_t,Y_t,U_t)\, dW_t^2$$

that is diffusion process having two time scales. Some dam management problems can be described in this formalism see Delebecque-Quadrat [19]

## REFERENCES

[1] P. BERNHARD. On singular implicit linear dynamical systems, Siam J of Control and Opt. vol 20 n° 5 sept 82.

[2] M. CODERCH, A.S. WILLSKI, S.S. SASTRY, D.A. CASTANON. Hierarchical aggregation of linear systems with multiple time scales, MIT Report LIDS-P-1187, mars 1982.

[3] M. CODERCH, A.S. WILLSKY, S.S. SASTRY. Hierarchical aggregation of singulary perturbed finite state Markov chains submitted to stochastics.

[4] P.J. COURTOIS. Decomposability, ACM Monograph Series, Academic Press, 1977.

[5] F. DELEBECQUE. A reduction process for pertubed Markov chains, a paraître SIAM J. of applied math. to appear.

[6] F. DELEBECQUE, J.P. QUADRAT. Optimal control of Markov chains admitting strong and weak interactions, Automatica, Vol. 17, n° 2, pp. 281-296, 1981.

[7] F. DELEBECQUE, J.P. QUADRAT. The optimal cost expansion of finite controls finite states Markov chains with weak and strong interactions. Analysis and optimization of systems, Lecture Notes an control and Inf. Science 28 Springer Verlag, 1980.

[8] A.A. PERVOZVANSKII, A.V. GAITSGORI. Decomposition aggregation and approximate optimization en Russe, Nauka, Moscou, 1979.

[9] T. KATO. Perturbation theory for linear operator, Springer Verlag, 1976.

[10] B.L. MILLER, A.F. VEINOTT. Discrete dynamic programming with small interest rate. An. math. stat. 40, 1969, pp. 366-370.

[11] R. PHILIPS, P. KOKOTOVIC. A singular perturbation approach to modelling and control of Markov, chains IEEE A.C. Bellman issue, 1981.

[12] H. SIMON, A. ANDO. Aggregation of variables in dynamic systems, Econometrica, 29, 111-139, 1961.

[13] J. KEMENY, L. SNELL. Finite Markov chains, Van Nostrand, 1960.

[14] O. MURON. Evaluation de politiques de maintenance pour un système complexe, RIRO, vol. 14, n° 3, pp. 265-282, 1980.

[15] S.L. CAMBELL, C.D. MEYER jr. Generalized inverses of linear transformations. Pitman, London, 1979.

[16] TKIOUAT. Thèse Rabat à paraitre.

[17] J.P. QUADRAT. Commande optimale de chaines de Markov perturbées Outils et Modèles Math. pour l'automatique... t3 edition CNRS 1983.

[18] J.P. QUADRAT Optimal control of perturbed, Markov chain the multitime scale case. Singular pertubation in systems and control. CISM courses and lectures n° 280, Springer Verlag 82.

[19] F. DELEBECQUE, J.P. QUADRAT. Contribution of stochastic control, team theory and singular perturbation to an example of large scale systems : Management of hydropower production. IEEE AC avril 1978.

# TIME SCALE MODELING OF DYNAMIC NETWORKS WITH SPARSE AND WEAK CONNECTIONS

*J.H. Chow*[†] *P.V. Kokotovic*[††]

## 1. INTRODUCTION

Many properties of more detailed models of large scale dynamic systems can be deduced from their simplified representations as networks with a storage element at each node and non-storage branches connecting the nodes. Depending on a particular application, each storage is described by one or two state variables and hence the network model is described by a system of first or second order differential equations. An example of such a network model is a power system in which each synchronous machine is modeled as a rotating mass and the transmission line transients are neglected. Mass-spring networks are also used as models of large space structures and are mathematically similar to physically different Markov chain models of queueing systems. To stress the dynamic aspect of such network models, we call them dynamic networks.

Several recent publications [1-5] investigate time-scale properties of dynamic networks with weak connections and develop a decomposition-aggregation methodology. They construct a slow aggregate model as a "long-term equivalent" of the network with each aggregate state representing a subsystem or an "area" of the network. The short term model of the same network consists of the decoupled area models. This methodology, which has been applied to power systems [1] and Markov chain models [3,4], points out that a distinction should be made between the terms weak connections and weak dynamic coupling. While the term "connection" refers to the strength of a branch, that is the value of a matrix entry, the notion of "coupling" is deeper. Two subsystems are weakly coupled if the dynamic properties of the overall system, such as its eigenvalues, can be determined, at least approximately, from the subsystems treated separately. It has been shown that power systems and Markov chains with weak connections are weakly coupled in a fast time-scale,

[†]Electric Utility Systems Engineering Department, General Electric Company, Schenectady, NY 12345.

[††]Coordinated Sciences Laboratory and Electrical Engineering Department, University of Illinois, 1101 W. Springfield Avenue, Urbana, IL 61801.

but they may be strongly coupled in a slow time-scale. In other words, weak connections determine the long-term dynamic behavior of the network. The aggregate model captures this long-term behavior by assuming that the strength of internal connections in each area forces its nodes to lose their identity and act as a single "aggregate" node connected to other "aggregate" nodes through weak external connections. The same weak external connections have a negligible effect on the short-term behavior of the network which is modeled with the areas disconnected from each other.

The outlined methodology is a result of a singular perturbation analysis with respect to a scalar parameter $\varepsilon$ representing the weak connections [1]. This asymptotic analysis investigates the limit as $\varepsilon \to 0$ in two time-scales, $t$ and $\tau = t/\varepsilon$. A singularly perturbed model of the network is obtained by transforming the original state variables into a set of aggregate (slow) and local (fast) variables. The transformation itself is found automatically by a computer algorithm which groups the nodes into areas. The details of this grouping algorithm and its applications to large power systems can be found in [6]. A summary of the algorithm is given in Appendix C.

A limitation of the methodology presented in [1] is that each individual external connection is assumed to be weak, that is, proportional to the small parameter $\varepsilon$. This formulation excludes more common situations in which the weakness of external connections is due to their sparsity. In many applications, individual external connections are as strong as, but much sparser than, the internal connections. In this paper, we allow strong but sparse external connections. Relating the sparsity pattern of a network with its time scale properties, we extend the asymptotic analysis of [1] to a larger class of networks. From the graph theory point of view [7], our analysis also contains a new result on the dependence of the eigenvalues of a graph on its sparsity pattern.

We first characterize the network topology by two parameters $d$ and $\delta$, and show that the sparse external connections induce the time-scale properties similar to those of dynamic networks with weak connections. Although the aggregate model depends on two parameters, a singular perturbation model is still obtained by the same decomposition-aggregation transformation.

An application of these results is to find the unknown sparsity pattern of a large network. We show that this can be determined by our grouping algorithm. The network considered represents the power system of the Western U.S., but with connections treated as 1 or 0. In this case, the algorithm produces a grouping into nine areas quite close to an earlier grouping obtained for the same system with actual values of the connections [6]. This dominance of the topological factor over the actual numerical values of the connections is of interest not only for networks, but also for more general large scale systems, such as those discussed in [17,18]. An alternative approach to the determination of time-scales can be based on the generalized Gerschgorin Circle results in [19].

The sparse connection results are then combined with the weak connection results to analyze practical networks which, in general, contain both sparse and weak connections. We derive time-scale results for the special case when the sparse connections are also weak. The same Western U.S. power system is used to illustrate a procedure using the grouping algorithm to find the sparse and weak connections in practical networks.

## 2. CHARACTERIZATION OF SPARSITY

Suppose that an n-node network has r internally dense, sparsely connected areas, which are unknown to us. To find the unknown partitioning, we first characterize the sparsity pattern of such a network and then, in Section 3, we determine its time scale properties, which lead to the desired decomposition-aggregation result in Section 4.

### 2.1 Node and Area Parameters

In our terminology, an _external_ connection is any connection between two nodes from different areas and an _internal_ connection is any connection between two nodes from the same area. The i-th node in area $\alpha$ has _dense_ internal connections and _sparse_ external connections if

$$c^E_{\alpha i}/c^I_{\alpha i} \ll 1 \qquad (2.1)$$

where

$c^E_{\alpha i}$ = the number of external connections of node i in area $\alpha$,

$c^I_{\alpha i}$ = the number of internal connections of node i in area $\alpha$.

From the point of view of a decomposition along the sparse area boundaries, the least favorable nodes are those with the sparsest internal connections $\underline{c}^I$, or with the densest external connections $\overline{c}^E$, where

$$\underline{c}^I = \min_{\alpha,i} \{c^I_{\alpha i}\}, \qquad \bar{c}^E = \max_{\alpha,i} \{c^E_{\alpha i}\}. \tag{2.2}$$

It is desired that a partitioning into areas be such that even the worst node has more internal than external connections, that is, the node parameter

$$d = \bar{c}^E/\underline{c}^I \tag{2.3}$$

should be small, $d<<1$. In a large network, a small number of nodes may violate this requirement. Then the average values of $c^I_{\alpha i}$ and $c^E_{\alpha i}$ can be used to define the node parameter d. This will be further discussed in conjunction with the large power network example in Section 5.

The $\alpha$-th area has sparse external and dense internal connections if

$$\gamma^E_\alpha/\gamma^I_\alpha<<1, \tag{2.4}$$

where

$\gamma^E_\alpha$ = the total number of external connections in area $\alpha$,

$\gamma^I_\alpha$ = the total number of internal connections in area $\alpha$.

The least favorable areas are those with the sparsest internal connections $\underline{\gamma}^I$, and the densest external connections $\bar{\gamma}^E$, where

$$\underline{\gamma}^I = \min_{\alpha} \{\gamma^I_\alpha\}, \qquad \bar{\gamma}^E = \max_{\alpha} \{\gamma^E_\alpha\}. \tag{2.5}$$

Clearly, $\underline{\gamma}^I$ is bounded below by $\underline{m}\, \underline{c}^I$, that is,

$$\underline{\gamma}^I \geq \underline{m}\, \underline{c}^I, \tag{2.6}$$

where

$\underline{m} = \min_{\alpha} \{m_\alpha\}$,

$m_\alpha$ = the number of nodes in area $\alpha$.

Our goal is to find a partitioning in which even the worst area has more internal than external connections, that is, the "area parameter"

$$\delta = \bar{\gamma}^E / (\underline{m}\ \underline{c}^I) \qquad (2.7)$$

is small, $\delta \ll 1$. As we shall see, this requirement is more critical than $d \ll 1$.

If a partitioning of a network into r areas can be found such that the nodes satisfy $d \ll 1$ and the areas satisfy $\delta \ll 1$, then the network possesses r internally dense, sparsely connected areas. Strictly speaking, the methodology developed in this paper applies to such networks only. However, for large networks with internally dense areas which at a few nodes fail to satisfy the requirement $d \ll 1$, then requirement $\delta \ll 1$ is usually satisfied because $\underline{m}$ is large. Then the decomposition-aggregation methodology still provides approximate results.

## 2.2 Examples

Let us illustrate these two situations by the networks in Figures 2.1 and 2.2. When the 10-node network in Figure 2.1 is partitioned into two areas as indicated, we have

$$\underline{m} = 5, \quad \bar{\gamma}^E = 2, \quad \underline{c}^I = 3, \quad \bar{c}^E = 1.$$

Thus, the node parameter and the area parameter

$$d = \bar{c}^E / \underline{c}^I = 0.333,$$

$$\delta = \bar{\gamma}^E / (\underline{m}\ \underline{c}^I) = 0.133$$

both satisfy our requirements and the network has two internally dense, sparsely connected areas. For this simple network, it is obvious that no other partition can yield a smaller d or $\delta$.

A network in which our requirement $d \ll 1$ is violated is the 8-node longitudinal network in Figure 2.2. For the given 2-area partition, we have

$$\underline{m} = 4, \quad \bar{\gamma}^E = 1, \quad \underline{c}^I = 1, \quad \bar{c}^E = 1,$$

and, hence, $\delta$ is small, but d is not:

$$\delta = 0.25, \quad d = 1.$$

No other partitioning into two or more areas yields a smaller d or $\delta$. In Section 3, we show that the decomposition-aggregation method

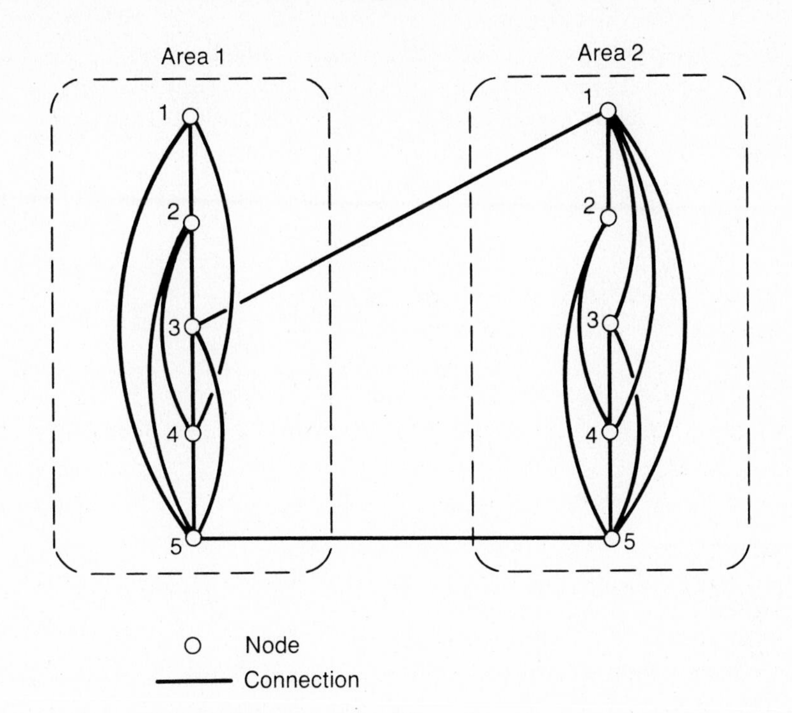


Figure 2.1 A 10-Node, 2-Area Network

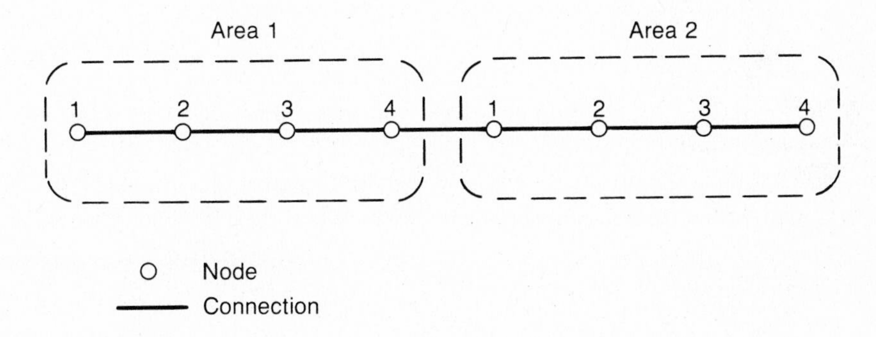


Figure 2.2 An 8-Node, 2-Area Longitudinal Network

is still applicable, but with less accuracy. Additional examples of sparse networks are given in [8].

2.3 Dynamic Networks

Although our analysis applies to other types of networks, for clarity we present it in the context of electromechanical networks. To stress the structural aspect of the problem, we make the simplifying assumption that all the connections (resistances, impedances) normalized with respect to the storages (capacitances, inertias) are equal to unity. Under this assumption, if the "potentials" of the storage elements are used as the state variables, the network dynamic equations are

$$\dot{x}_i^\alpha = \sum_{\beta=1}^{r} \sum_{j=1}^{m_\beta} k_{ij}^{\alpha\beta}(x_j^\beta - x_i^\alpha), \text{ for } i = 1,\dots,m_\alpha,$$

$$\alpha = 1,2,\dots,r, \text{ and } j \neq i \text{ when } \beta = \alpha, \tag{2.8}†$$

$$k_{ij}^{\alpha\beta} = 1, \text{ if node } i \text{ in area } \alpha \text{ and node } j \text{ in area } \beta \text{ are connected,}$$

$$k_{ij}^{\alpha\beta} = 0, \text{ otherwise,}$$

where $x_i^\alpha$ is the potential of the storage element of the i-th node in area $\alpha$.

Considering $x_i^\alpha$ as the i-th entry of the $m_\alpha$-vector $x^\alpha$, which is the $\alpha$-th subvector of the n-vector x, we rewrite (2.8) as

$$\dot{x} = Kx = (K^I + K^E)x, \tag{2.9}$$

where $K^I$ is the internal connection matrix whose entries are $k_{ij}^{\alpha\alpha}$ and and $K^E$ is the external connection matrix whose entries are $k_{ij}^{\alpha\beta}$, $a \neq \beta$. It is important to note that a property of the network matrix K is that its diagonal entries are the negative of the sums of the other entries in the same row. This is in contrast to the graph adjacency matrix whose diagonal entries are zero [7].

The connections in the dynamic network model are bidirectional, and hence the matrices K, $K^I$, and $K^E$ are symmetric. To make our results applicable to a larger class of models such as Markov chains, we do not assume this symmetry property.

By our definition of the internal connections, the matrix $K^I$ is block-diagonal

$$K^I = \text{diag}(K_1^I, K_2^I, \dots, K_r^I), \tag{2.10}$$

---

† For mass-spring and power system models where $\ddot{x}_i^\alpha$ appears instead of $\dot{x}_i^\alpha$, our analysis is modified in an obvious way.

where $K^I_\alpha$ is the $m_\alpha x m_\alpha$ internal connection matrix of area $\alpha$. Since an area itself is a dynamic network, the i-th diagonal entry of $K^I_\alpha$ is $-c^I_{\alpha i}$, that is the sum of the other entries in row i. The sum of the off-diagonal entries of $K^I_\alpha$ is $2\gamma^I_\alpha$. Although the diagonal entries of $K^I_\alpha$ are much larger in magnitude than any of the off-diagonal entries, $K^I_\alpha$ is not diagonally dominant because its rank is $m_\alpha-1$, due to a zero eigenvalue with the eigenvector

$$u_\alpha = [1 \quad 1 \quad \dots \quad 1]^T. \tag{2.11}$$

This zero mode is the equilibrium manifold of the area $\alpha$ when disconnected from the rest of the network. It expresses the fact that, when isolated, area $\alpha$ is at equilibrium, $\dot{x}^\alpha = 0$, whenever all its storage potentials are the same, $x^\alpha_i = x^\alpha_j$, $i, j = 1,2,\dots,m_\alpha$.

The corresponding partition of $K^E$ into $r^2$ block matrices $K^E_{\alpha\beta}$, $\alpha$, $\beta = 1,2,\dots,r$, is such that $K^E_{\alpha\beta}$ contains the connections between areas $\alpha$ and $\beta$. We note that $K^E_{\alpha\alpha}$ is a diagonal matrix whose i-th diagonal entry is $-c^E_{\alpha i}$, the negative of the sum of the other entries in row i. It is also helpful to observe that the sum of the entries of $K^E_{\alpha\beta}$ is equal to the number of external connections between areas $\alpha$ and $\beta$, and hence $d<<1$ and $\delta<<1$ imply that $K^I_\alpha$ is dense and $K^E_{\alpha\beta}$ is sparse, for all $\alpha$ and $\beta$.

## 3. TIME SCALE SEPARATION

We now demonstrate how the equilibrium manifold property of the areas and their sparse external connections induce a two-time-scale behavior of the overall dynamic network. The network motion decomposes into the fast local motions of the storage elements within the same area and the network-wide slow motion of the areas. During this slow motion all the nodes in the same area are "coherent," that is, the areas move as "rigid bodies." Intuitively, the local motions are fast because the dense internal connections allow the node potentials in the same area to rapidly equalize, that is, to reach an area equilibrium. During the fast motion, the exchange with other areas is negligible due to the sparsity of external connections. This exchange becomes significant over a longer period, that is, the motion of the areas as heavier rigid bodies is slow. This two-time-scale behavior is made apparent by a new choice of state variables.

### 3.1 Slow and Fast Variables

To describe the slow motion, we define for each area an aggregate variable

$$y_\alpha = \sum_{i=1}^{m_\alpha} x_i^\alpha / m_\alpha = (1/m_\alpha)\, u_\alpha^T x^\alpha, \qquad \alpha = 1,\dots,r. \tag{3.1}$$

In mass-spring networks this variable is the familiar area "center of inertia" [1], while in Markov chains it represents the probability for a Markov process to be in a group of the states [3]. Denoting by $y$ the r-vector whose $\alpha$-th entry is $y_\alpha$, the matrix form of (3.1) is

$$y = Cx = M_a^{-1} U^T x, \tag{3.2}$$

where

$$M_a = \text{diag}\,(m_1, m_2, \dots, m_r) \tag{3.3}$$

is the rxr aggregate "inertia" matrix, and

$$U = \text{diag}\,(u_1, u_2, \dots, u_r) \tag{3.4}$$

is the nxr matrix whose diagonal $m_\alpha$x1 blocks are the vectors (2.11).

For the fast motion, we select in each area a reference node, say the first node, and define the motions of the other nodes in the same area relative to this reference node by the local variables

$$z_{i-1}^\alpha = x_i^\alpha - x_1^\alpha, \qquad i = 2,3,\dots,m_\alpha, \qquad \alpha = 1,\dots,r. \tag{3.5}$$

Denoting by $z^\alpha$ the $(m_\alpha-1)$-vector whose i-th entry is $z_i^\alpha$ and considering $z^\alpha$ as the $\alpha$-th subvector of the (n-r)-vector z, we rewrite (3.5) as

$$z = Gx = \text{diag}(G_1, G_2, \dots, G_r)x, \tag{3.6}$$

where $G_\alpha$ is the $(m_\alpha-1)$x$m_\alpha$ matrix

$$G_\alpha = \begin{bmatrix} -1 & 1 & 0 & . & 0 \\ -1 & 0 & 1 & . & 0 \\ . & . & . & . & . \\ -1 & 0 & 0 & . & 1 \end{bmatrix}. \tag{3.7}$$

We have thus defined a transformation of the original state x into the aggregate and local variables

$$\begin{bmatrix} y \\ z \end{bmatrix} = \begin{bmatrix} C \\ G \end{bmatrix} x . \tag{3.8}$$

The inverse of this transformation is explicitly known

$$x = [U \quad G^+] \begin{bmatrix} y \\ z \end{bmatrix} , \tag{3.9}$$

where $G^+$ is block diagonal

$$G^+ = G^T(GG^T)^{-1} = \mathrm{diag}(G_1^+, G_2^+, \ldots, G_r^+) , \tag{3.10}$$

and $G_\alpha^+$ is the $m_\alpha x(m_\alpha - 1)$ matrix

$$G_\alpha^+ = \frac{1}{m_\alpha} \begin{bmatrix} -1 & -1 & . & -1 \\ m_\alpha - 1 & -1 & . & -1 \\ -1 & m_\alpha - 1 & . & -1 \\ . & . & . & . \\ -1 & -1 & . & m_\alpha - 1 \end{bmatrix} . \tag{3.11}$$

In the new variables y and z, the network model (2.9) becomes

$$\begin{bmatrix} \dot{y} \\ \dot{z} \end{bmatrix} = \begin{bmatrix} \bar{A}_{11} & \bar{A}_{12} \\ \bar{A}_{21} & \bar{A}_{22} \end{bmatrix} \begin{bmatrix} y \\ z \end{bmatrix} , \tag{3.12}$$

where

$$\begin{aligned} \bar{A}_{11} &= CK^E U \quad , \quad \bar{A}_{12} = CK^E G^+ \\ \bar{A}_{21} &= GK^E U \quad , \quad \bar{A}_{22} = G(K^I + K^E)G^+ . \end{aligned} \tag{3.13}$$

An important property of this model is that the matrix $K^I$ does not appear in $\bar{A}_{11}$, $\bar{A}_{12}$ and $\bar{A}_{21}$ because C and U defined by (3.2) and (3.3) span the left and the right null spaces of $K^I$, respectively. For the same reason, $GK^IG^+$ is nonsingular, namely,

$$CK^I = 0 \quad , \quad K^IU = 0 \quad , \quad \det (GK^IG^+) \neq 0. \tag{3.14}$$

If each node is connected to at least 3/4 of all the other nodes in the same area, then the following row norm ($\|\bullet\|_\infty$) bounds are derived in Appendix A:

$$\|GK^IG^+\|_\infty \geq \min_{\alpha,i} \{c^I_{\alpha i}/3\} \geq \underline{c}^I/3, \tag{3.15}$$

$$\|CK^EU\|_\infty = \|\bar{A}_{11}\|_\infty = \max_{\alpha} \{2\gamma^E_\alpha/m_\alpha\} \leq 2\bar{\gamma}^E/\underline{m} = 2\underline{c}^I\delta, \tag{3.16}$$

$$\|CK^EG^+\|_\infty = \|\bar{A}_{12}\|_\infty = \max_{\alpha} \{4\gamma^E_\alpha/m_\alpha\} \leq 4\bar{\gamma}^E/\underline{m} = 4\underline{c}^I\delta, \tag{3.17}$$

$$\|GK^EU\|_\infty = \|\bar{A}_{21}\|_\infty = \max_{\alpha,i} \{2c^E_{\alpha i} + 2c^E_{\alpha 1}\} \leq 4\bar{c}^E = 4\underline{c}^Id, \tag{3.18}$$

$$\|GK^EG^+\|_\infty \leq \max_{\alpha,i} \{3c^E_{\alpha 1} + 4c^E_{\alpha i}\} \leq 7\bar{c}^E = 7\underline{c}^Id. \tag{3.19}$$

Similar bounds can be obtained using the column norm ($\|\bullet\|_1$), but the same norm has to be used throughout the analysis since, in general, $\|A\|_\infty$ is not equal to $\|A\|_1 = \|A^T\|_\infty$.

### 3.2 Time Scale Properties

For $\delta$ and d sufficiently small, the norms of $\bar{A}_{11}$, $\bar{A}_{12}$ and $\bar{A}_{21}$ are much smaller than the norm of $\bar{A}_{22}$ which is of order $\underline{c}^I$, that is $O(\underline{c}^I)$. Furthermore, for d sufficiently small, $\bar{A}_{22}$ is nonsingular since $GK^IG^+$ is nonsingular by (3.14). This suggests that an appropriate fast time-scale is

$$t_f = \underline{c}^I t. \tag{3.20}$$

We also rescale the matrices in (3.12) as

$$A_{11} = \bar{A}_{11}/(\underline{c}^I\delta), \qquad A_{12} = \bar{A}_{12}/(\underline{c}^I\delta),$$
$$A_{21} = \bar{A}_{21}/(\underline{c}^I d), \qquad A_{22} = \bar{A}_{22}/\underline{c}^I, \tag{3.21}$$

so that the norms of $A_{11}$, $A_{12}$, $A_{21}$ and $A_{22}$ are all O(1). This scaling reveals that the model (3.12) is singularly perturbed by the parameter $\delta$, that is,

$$dy/dt_f = \delta A_{11}y + \delta A_{12}z,$$
$$dz/dt_f = dA_{21}y + A_{22}z. \tag{3.22}$$

In the slow time-scale

$$t_s = \delta t_f \tag{3.23}$$

the same model has the so-called explicit singular perturbation form

$$dy/dt_s = A_{11}y + A_{12}z,$$
$$\delta dz/dt_s = dA_{21}y + A_{22}z. \tag{3.24}$$

This well-known model allows us to make use of many existing results. One of them is the following Theorem [1,9,20] about the time-scale properties of (3.24).

<u>Theorem 3.1</u>. There exist $\delta^*$ and $d^*$ such that for all $0 < \delta \leq \delta^*$, $0 < d \leq d^*$, the system (3.24) has r slow eigenvalues

$$\lambda_s = \lambda[(A_{11} - dA_{12}A_{22}^{-1}A_{21}) + O(\delta d)] \tag{3.25}$$

and n-r fast eigenvalues

$$\lambda_f = \lambda[A_{22} + O(\delta d)]/\delta \tag{3.26}$$

and, hence, $|\lambda_s|/|\lambda_f| = O(\delta)$. Furthermore, the slow and fast subsystems of (3.24) are

$$dy_s/dt_s = (A_{11} - dA_{12}A_{22}^{-1}A_{21})\, y_s = A_0 y_s, \quad y_s(0) = y(0), \tag{3.27}$$

$$dz_f/dt_f = A_{22}z_f, \quad z_f(0) = z(0) + dA_{22}^{-1}A_{21}y(0), \tag{3.28}$$

respectively, and their solutions approximate the solution y,z of (3.24) as

$$y = y_s(t_s) + O(\delta d), \tag{3.29}$$

$$z = -dA_{22}^{-1}A_{21}y_s(t_s) + z_f(t_f) + O(\delta d). \tag{3.30}$$

Proof: See Appendix B.

Theorem 3.1 shows that the area parameter $\delta$ is a singular perturbation parameter whose smallness guarantees the existence of two time-scales. The node parameter d is a regular perturbation parameter whose smallness guarantees that the slow variables y are weakly coupled to the fast variables z. In the next section, we show that the smallness of d also implies that the fast local models are weakly coupled and the contribution of the internal connections to the aggregate model is small. We recall that in the networks with $\varepsilon$-connections a single parameter performs both the time-scale separation and the decoupling of the fast dynamics [1]. In sparse networks, these two roles are played by d and $\delta$. In all other respects, the time-scale separation results for spare networks are analogous to those for networks with $\varepsilon$-connections. In particular, the notion of "slow coherency" [1] remains the same, that is, the states in the same area are coherent with respect to the slow modes. A practically important consequence is that the grouping algorithm [1] and its sparsity-based version [6] can now be used for aggregate modeling of sparse networks.

Furthermore, if the slow subsystem (3.27) also has sparsely connected areas, a nested application of the same decomposition and aggregation approach, starting from the fastest time-scale, will result in a multi-time-scale hierarchy of aggregate models. The technique is essentially similar to the multi-time-scale methodology developed by Peponides [10], Delebecque [11], and Coderch, et. al. [12] for networks with $\varepsilon$-connections.

The counterpart of Theorem 3.1 for second order systems

$$\begin{aligned} d^2y/dt_f^2 &= \delta A_{11}y + \delta A_{12}z \\ d^2z/dt_f^2 &= dA_{21}y + A_{22}z \end{aligned} \tag{3.31}$$

is based on Theorem 1 in [21]. In this case, the slow-fast time relation (3.23) is $t_s = \sqrt{\delta}\, t_f$, the eigenvalue approximations (3.25), (3.26) are only $O(\sqrt{\delta d})$. The state approximations (3.27)-(3.30) are also $O(\sqrt{\delta d})$, but because of the oscillatory character, are valid only for a finite time interval.

Our development thus far has demonstrated a relationship between sparsity and time-scale properties of dynamic networks. In the remainder of the paper we show how these properties are used in a decomposition-aggregation procedure and for a grouping into areas when the partitioning is unknown.

## 4. DECOMPOSITION AND AGGREGATION

The presence of the node parameter d in $A_0$ and $A_{22}$ suggests the construction of several approximate models for the slow and fast subsystems.

### 4.1 Aggregate Models

Since the states of the slow subsystem (3.27) are the area centers of inertias, we treat (3.27) as an aggregate model. To simplify (3.27), we first express the terms in $A_0$ in a clearer form:

$$(\underline{c}^I\delta)A_{11} = \bar{A}_{11} = M_a^{-1}\,(U^T K^E U) = M_a^{-1}K_a, \tag{4.1}$$

which is derived in [1], and

$$(\underline{c}^I\delta)dA_{12}A_{22}^{-1}A_{21} = \bar{A}_{12}\bar{A}_{22}^{-1}\bar{A}_{21} = CK^E G^+(GKG^+)^{-1}GK^E U = M_a^{-1}K_a^I, \tag{4.2}$$

where using (3.2) and (3.10),

$$K_a^I = U^T K^E G^+(GKG^+)^{-1}GK^E U = U^T K^E G^T(GKG^T)^{-1}GK^E U. \tag{4.3}$$

Therefore, the slow subsystem can be rewritten in a form similar to that of the original system (2.9), namely,

$$M_a dy_s/dt = (K_a + K_a^I)y_s. \tag{4.4}$$

First, we note that $K_a$ and $K_a^I$ are symmetric. Second, the row sums of each of these two matrices are zero. This can be verified by post-multiplying $K_a$ and $K_a^I$ by the r-dimensional vector

$$\underline{1} = [1 \;\; 1 \;\; \dots \;\; 1]^T.$$

Since $U\underline{1}$ yields an n-dimensional vector $\underline{1}$ and the row sums of $K^E$ are zero, we have

$$K_a\underline{1} = 0, \qquad K_a^I\underline{1} = 0. \tag{4.5}$$

Thus $(K_a + K_a^I)$ has a zero eigenvalue with an invariant subspace spanned by $\underline{1}$. The symmetry and zero row sum properties imply that the slow subsystem (4.4) is also a dynamic network model of the same nature as the original system (2.9). This model is referred to as the complete aggregate of (2.9).

We observe from (4.1) that $K_a = U^TK^EU$, that is the $(\alpha,\beta)$-entry of $K_a$ is the sum of all the connections between the areas $\alpha$ and $\beta$. Hence, $K_a$ is formed by the external connections only. An important conclusion from (4.1) and (4.2) is that

$$\|M^{-1}K_\alpha^I\|/\|M^{-1}K^I\| = O(\underline{c}^I\delta d)/O(\underline{c}^I\delta) = O(d). \tag{4.6}$$

If we assume d=0, the complete aggregate (4.4) reduces to the so-called rigid aggregate

$$M_a d\bar{y}_s/dt = K_a\bar{y}_s, \tag{4.7}$$

which represents a network with infinitely stiff internal connections. Its name makes explicit its role of describing, in the slow time-scale, the motion of the areas as rigid bodies. The rigid aggregate (4.7) is appealing because of its conceptual and computational simplicity. However, pursuing this mass-spring analog a little further, we may recall a classical result [13, p. 331] that every increase of spring stiffness is accompanied by an increase of some or all mode frequencies. The modes of (4.7) are indeed faster than the slow modes of the original system (2.9). That the presence of $K_a^I$ in the complete aggregate (4.4) accounts for the fact that the internal connections are not infinitely stiff is clear from the dependence of $K_a^I$ in (4.3) on K and, hence, on the internal connections $K^I$. Moreover, when d is small, $G(K^I+K^E)G^+$ can be approximated by $GK^IG^+$, yielding

$$K_a^I \cong U^TK^EG^T(GK^IG^T)^{-1}GK^EU, \tag{4.8}$$

which makes the effect of the internal connections more explicit. The computation of even this simplified form of $K_a^I$ is time-consuming for large networks. To avoid it, and also to improve the accuracy of the aggregate (4.7), we propose a <u>compensated aggregate</u>

$$M_a d\tilde{y}_s/dt = cK_a\tilde{y} \tag{4.9}$$

where c is a compensation factor chosen to minimize the discrepancies between the eigenvalues of (4.9) and the actual slow eigenvalues. The actual slow eigenvalues are known from the grouping algorithm [6], while $M_a$, $K_a$ and the eigenvalues of $cM_a^{-1}K_a$ are straightforward to calculate. Therefore, the choice of c does not require much additional computation and, as the examples will show, leads to a major improvement of the accuracy of the compensated aggregate over the rigid aggregate (4.4).

4.2 <u>Local Models</u>

The fast subsystem (3.28) represents the relative motions of the individual variables in the areas with respect to their local reference states. The local motions can be simplified by recognizing that

$$A_{22} = A_{220} + dA_{22d} \tag{4.10}$$

where

$$A_{220} = (GK^I G^+)/\underline{c}^I, \tag{4.11}$$

$$A_{22d} = (GK^E G^+)/\bar{c}^E, \tag{4.12}$$

are O(1). Since $A_{220}$ is a block-diagonal matrix, then for d small, the fast subsystem (3.28) decouples into r <u>local</u> models

$$dz_f^\alpha/dt_f = A_{220}^\alpha\, z_f^\alpha, \quad \alpha = 1,2,\ldots,r, \tag{4.13}$$

where $z_f^\alpha$ are the area $\alpha$ states of z and $A_{220} = \mathrm{diag}\,(A_{220}^1, \ldots, A_{220}^r)$. These models are obtained by discarding the connections between the areas, and the frequencies of the modes of (4.12) will be somewhat lower than those of the fast modes of the full system [13, p. 331].

A correction for the mode shift is to include the diagonal blocks, $A_{22d}^\alpha$, of $A_{22d}$ in the <u>compensated local</u> models

$$dz_f^\alpha/dt_f = (A_{220}^\alpha + dA_{22d}^\alpha)z_f^\alpha \,. \tag{4.14}$$

Note that the off-diagonal blocks of $A_{22d}$ have only $O(d^2)$ contributions since each $A_{220}^\alpha$ is a block-diagonally dominant matrix. As $G$ and $G^+$ are block-diagonal, $A_{22d}^\alpha$ depends only on the diagonal entries of $K^E$. Thus, the terms $A_{22d}^\alpha$ represent only the effects of the external connections from area $\alpha$ to the other areas. In power system models, including the $A_{22d}^\alpha$ terms is equivalent to modeling the external connections with other areas as shunts to ground. For networks with $d$ not sufficiently small, it is important to retain the $dA_{22d}^\alpha$ terms.

## 4.3 Examples

Let us illustrate the time-scale separation properties and the decomposition-aggregation procedure with the 10-node network in Figure 2.1 and the 8-node longitudinal network in Figure 2.2.

For the 10-node network with the given 2-area partition, we have

$$K_1^I = \begin{bmatrix} -3 & 1 & 0 & 1 & 1 \\ 1 & -4 & 1 & 1 & 1 \\ 0 & 1 & -3 & 1 & 1 \\ 1 & 1 & 1 & -4 & 1 \\ 1 & 1 & 1 & 1 & -4 \end{bmatrix}$$

$$K_2^I = \begin{bmatrix} -4 & 1 & 1 & 1 & 1 \\ 1 & -3 & 0 & 1 & 1 \\ 1 & 0 & -3 & 1 & 1 \\ 1 & 1 & 1 & -4 & 1 \\ 1 & 1 & 1 & 1 & -4 \end{bmatrix}$$

$$K_{12}^E = \begin{bmatrix} 0 & 0 & 0 & 0 & 0 \\ 0 & 0 & 0 & 0 & 0 \\ 1 & 0 & 0 & 0 & 0 \\ 0 & 0 & 0 & 0 & 0 \\ 0 & 0 & 0 & 0 & 1 \end{bmatrix} = (K_{21}^E)^T$$

$K_{11}^E = \text{diag}\,(0,\ 0,\ -1,\ 0,\ -1)$, $\quad K_{22}^E = \text{diag}\,(-1,\ 0,\ 0,\ 0,\ -1)$.

The slow and fast eigenvalues of the full system and their approximations by various simplified models are given in Tables 4.1

Table 4.1

Eigenvalues of Slow Models (10-Node Network)

| Slow Model | Matrix | Eigenvalues |
|---|---|---|
| Exact | $K$ | 0.00, -0.61 |
| Complete Aggregate | $A_0$ | 0.00, -0.63 |
| Rigid Aggregate | $A_{11}$ | 0.00, -0.80 |

Table 4.2

Eigenvalues of Fast Models (10-Node Network)

| Fast Model | Matrix | Eigenvalues | |
|---|---|---|---|
| | | Area 1 | Area 2 |
| Exact | $K$ | -3.34,-5.00,-5.00,-6.16 | -3.00,-5.00,-5.00,-6.90 |
| Fast Subsystem | $A_{22}$ | -3.31,-5.00,-5.00,-6.00 | -3.00,-5.00,-5.00,-6.89 |
| Compensated Local Models | $A^{\alpha}_{220}+dA^{\alpha}_{22d}$ | -3.42,-5.00,-5.28,-5.89 | -3.00,-5.00,-5.60,-6.00 |
| Local Models | $A^{\alpha}_{220}$ | -3.00,-5.00,-5.00,-5.00 | -3.00,-5.00,-5.00,-5.00 |

and 4.2. Note that the fast eigenvalues of $K$ and $A_{22}$, while assigned to the areas, in general, depend on the other areas. The time scale separation is

$$|-3.00|/|-0.61| = 4.9 = O(1/\delta).$$

The relative errors in the slow eigenvalue approximations by the rigid aggregate and the fast eigenvalue approximations by the local models are of $O(d)$, and the relative errors of the approximations by the slow and fast subsystems are of $O(\delta d)$. By choosing $c = 0.61/0.80 = 0.76$, the eigenvalues of the compensated aggregate are the exact slow eigenvalues, which is achievable for all 2-area aggregates. The eigenvalue approximations of the fast subsystem by the compensated local models are of $O(d^2)$. Thus, the time-scale results are verified for this example.

The 8-node longitudinal network in Figure 2.2 has two slow eigenvalues and six fast eigenvalues (Tables 4.3 and 4.4). For the 2-area partition given in Figure 2.2, we have previously determined that $d=1$, $\delta=0.25$. Although $d$ is not small, the time-scale separation still holds:

$$|-0.586|/|-0.152| = 3.86 = O(1/\delta).$$

The eigenvalue approximations of various aggregate and local models are given in Tables 4.3 and 4.4. The eigenvalues of the slow and fast subsystems are good approximations to those of the full model, while the approximations of the rigid aggregate and the local models are poor. With $c = 0.152/0.500 = 0.304$, the eigenvalues of the compensated aggregate are identical to the slow eigenvalues.

Table 4.3
Eigenvalues of the Slow Models (Longitudinal Network)

| Slow Model | Eigenvalues |
|---|---|
| Exact | 0, -0.152 |
| Complete Aggregate | 0, -0.182 |
| Rigid Aggregate | 0, -0.500 |

Table 4.4
Eigenvalues of Fast Models (Longitudinal Network)

| Fast Model | Eigenvalues |
|---|---|
| Exact | -0.586, -1.235, -2.000, -2.765, -3.414, -3.848 |
| Fast Subsystem | -0.586, -1.111, -2.000, -2.645, -3.414, -3.745 |
| Local Models | Area 1: -0.586, -2.000, -3.414<br>Area 2: -0.586, -2.000, -3.414 |
| Compensated Local Models | Area 1: -0.924, -2.306, -3.520<br>Area 2: -0.924, -2.306, -3.520 |

## 5. A LARGE SCALE SYSTEM APPLICATION

We now describe the application of the decomposition-aggregation methodology to a large, realistic power network model. This experience motivates the modification of the theoretical results for practical applications to large networks. The power network considered is a 411-machine, 1750-bus, 2800-line model of the Western Systems Coordinating Council (WSCC) system (Figure 5.1). Most of the machines are shown as dots on the map. The total number of nodes is 411 + 1750 = 2161. The bus nodes are not reduced in order to preserve the overall sparsity of the network structure, and thus allow the use of the sparsity-based grouping algorithm, which is summarized in Appendix C for completeness.

The WSCC system consists of several major load centers with dispersed generation sites. Connections are dense about the load centers. The centers are interconnected with a few long transmission lines along the Pacific Coast and around the eastern portion of the system, forming the so-called "donut" pattern [14]. These long transmission lines are sparse, but each individually strong, and cannot be modeled as $\varepsilon$-connections.

The 2161x2161 connection matrix K was constructed using sparse storage. Then all the non-zero off-diagonal entries of K were set to 1, while the diagonal entries were set to be the negative of the row sums of the off-diagonal entries. The inertias at the nodes were all set to 1. The computation time required to calculate the 15 slowest eigenvalues and their eigenvectors using the sparse algorithm was about 6 minutes on a VAX 11/780 computer. The eigenvectors were used to partition WSCC into several different numbers of areas. We will use the 9-area partition shown in Figure 5.1 to illustrate the procedure. The same technique can be applied to other partitions.

The algorithm partitioned the 9 areas of WSCC along boundaries of well-defined geographical regions approximately corresponding to utility systems. Since the connections within the utility systems are denser than the connections between different utility systems, the 9 areas are partitioned along boundaries of sparse connections. The 9-area partition will be compared to the 11-area partition based on actual connection strength in Section 7 to identify the weak connections.

Let us now examine the sparsity pattern using the area and node parameter $\delta$ and d. The numbers of nodes, and internal and external connections for the areas are shown in Table 5.1, and the numbers of external connections between the areas are shown by the $K_a$ matrix in Table 5.2. In this system, some internal nodes have single internal connections, that is, $\underline{c}^I = 1$. Since $\underline{m} = 104$ is much larger than $\bar{\gamma}^E$, we still have small $\delta = 20/104 = 0.192$. However, with $\bar{c}^E = 2$, $d = 2$ violates the requirement $d \ll 1$, the assumption $\underline{c}^I \gg \bar{c}^E$ is no longer

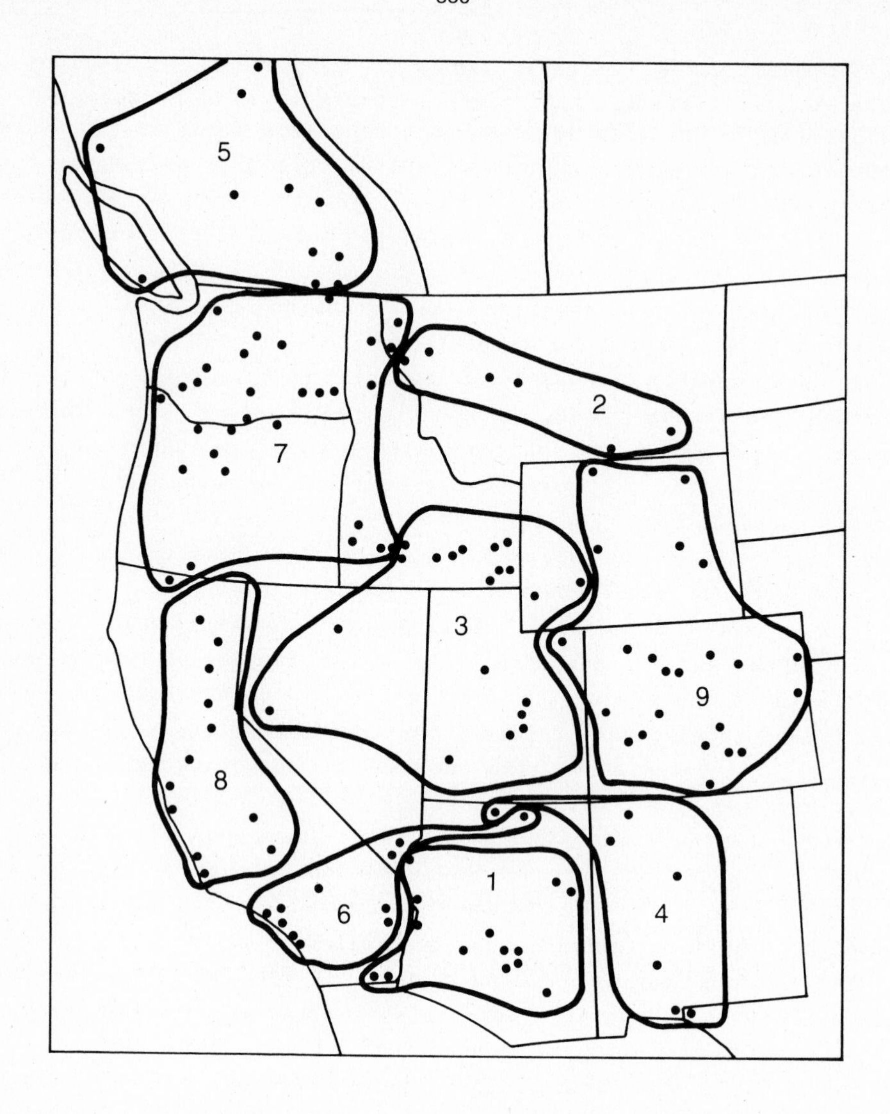


Figure 5.1 A 9-Area Partition of WSCC

satisfied, and the results of Theorem 3.1 do not apply. The situation $\underline{c}^I = 1$ is not uncommon in large networks, which motivates the following modification of the definition of d.

Instead of using $\underline{c}^I$ and $\bar{c}^E$, we introduce the averages of internal and external connections:

Table 5.1

Numbers of Nodes, and Internal and External Connections, for 9-Area Partition of WSCC

| Area $\alpha$ | Number of Nodes, $m_\alpha$ | Number of Internal Connections, $\gamma_\alpha^I$ | Average Number of Internal Connections, $(c_\alpha^I)_{ave}$ | Number of External Connections, $\gamma_\alpha^E$ | Average Number of External Connections, $(c_\alpha^E)_{ave}$ |
|---|---|---|---|---|---|
| 1 | 274 | 352 | 2.57 | 20 | 0.073 |
| 2 | 104 | 124 | 2.38 | 13 | 0.125 |
| 3 | 176 | 221 | 2.51 | 18 | 0.102 |
| 4 | 234 | 290 | 2.48 | 17 | 0.073 |
| 5 | 166 | 208 | 2.51 | 3 | 0.018 |
| 6 | 283 | 357 | 2.52 | 11 | 0.039 |
| 7 | 405 | 542 | 2.68 | 15 | 0.037 |
| 8 | 232 | 306 | 2.64 | 5 | 0.022 |
| 9 | 287 | 371 | 2.59 | 12 | 0.042 |

Table 5.2

Aggregate Connection Matrix $K_a$

| Row | Column | | | | | | | | |
|---|---|---|---|---|---|---|---|---|---|
| | 1 | 2 | 3 | 4 | 5 | 6 | 7 | 8 | 9 |
| 1 | -20.0 | 0.0 | 1.0 | 10.0 | 0.0 | 9.0 | 0.0 | 0.0 | 0.0 |
| 2 | 0.0 | -13.0 | 3.0 | 0.0 | 0.0 | 0.0 | 6.0 | 0.0 | 4.0 |
| 3 | 1.0 | 3.0 | -18.0 | 2.0 | 0.0 | 1.0 | 5.0 | 3.0 | 3.0 |
| 4 | 10.0 | 0.0 | 2.0 | -17.0 | 0.0 | 0.0 | 0.0 | 0.0 | 5.0 |
| 5 | 0.0 | 0.0 | 0.0 | 0.0 | -3.0 | 0.0 | 3.0 | 0.0 | 0.0 |
| 6 | 9.0 | 0.0 | 1.0 | 0.0 | 0.0 | -11.0 | 0.0 | 1.0 | 0.0 |
| 7 | 0.0 | 6.0 | 5.0 | 0.0 | 3.0 | 0.0 | -15.0 | 1.0 | 0.0 |
| 8 | 0.0 | 0.0 | 3.0 | 0.0 | 0.0 | 1.0 | 1.0 | -5.0 | 0.0 |
| 9 | 0.0 | 4.0 | 3.0 | 5.0 | 0.0 | 0.0 | 0.0 | 0.0 | -12.0 |

$$(c_\alpha^I)_{ave} = 2\gamma_\alpha^I/m_\alpha, \qquad (c_\alpha^E)_{ave} = 2\gamma_\alpha^E/m_\alpha, \qquad \alpha = 1,2,\ldots,r,$$

and define the average node parameter as

$$d_{ave} = \max_\alpha \{(c_\alpha^E)_{ave}\}/\min_\alpha \{(c_\alpha^I)_{ave}\}.$$

The averages $(c_\alpha^I)_{ave}$ and $(c_\alpha^E)_{ave}$ for WSCC are given in Table 5.1. While $(c_\alpha^I)_{ave}$ is about 2.5, $(c_\alpha^E)_{ave}$ is very small, usually less than 5% of $(c_\alpha^I)_{ave}$. Thus

$$d_{ave} = 0.125/2.38 = 0.053,$$

is small, indicating the overall sparsity of the external connections.

The slow modes and the modes of the rigid aggregate (4.7) are listed in Table 5.3. The power system model is in the second order form

$$\ddot{x} = K\,x$$

so that the modes given in Table 5.3 are the square roots of the eigenvalues of K and $K_a$. The average error in the approximations of the aggregate modes to the system slow modes is 174%. When the $K_a$ matrix is multiplied by $c = 1/(1 + 1.74)^2 = 0.133$, the modes of the compensated aggregate are much better approximations (Table 5.3). The value c is small since the areas cover large geographical regions and the center of area nodes are not close to the area boundaries.

Table 5.3

Comparison of System Slow Modes and Aggregate Modes (in rad/sec)

| System Slow Modes | Rigid Aggregate (4.7) | | Compensated Aggregate (4.9) | |
|---|---|---|---|---|
| | Modes | Percent Error | Modes | Percent Error |
| 0.0000 | 0.0000 | – | 0.0000 | – |
| 0.0376 | 0.1026 | 172.8 | 0.0375 | -0.3 |
| 0.0583 | 0.1413 | 142.3 | 0.0516 | -12.9 |
| 0.0640 | 0.1571 | 145.6 | 0.0574 | -10.3 |
| 0.0823 | 0.1967 | 138.9 | 0.0719 | -12.7 |
| 0.0904 | 0.2475 | 174.0 | 0.0905 | 0.1 |
| 0.1111 | 0.3312 | 198.0 | 0.1210 | 9.0 |
| 0.1149 | 0.3473 | 202.3 | 0.1269 | 10.5 |
| 0.1207 | 0.3805 | 215.3 | 0.1391 | 15.2 |

## 6. NETWORKS WITH SPARSE AND WEAK CONNECTIONS

In our presentation of the area decomposition results for sparse connections, we have simplified the analysis by assuming that each connection $k_{ij}^{\alpha\beta}$ between node i in area $\alpha$ and node j in area $\beta$ normalized with respect to the inertias is unity. To be more precise in determining time-scale decomposition of areas, the actual connection strength and inertia values have to be used. Of particular importance are the weak and strong connections. For many practical power systems, while the "cores" of the areas are largely determined by the sparse and dense connection patterns, the boundaries of the areas are determined by the connection strength and inertias. In some cases, weak and strong connections may cause large perturbations in the area partition. For example, an area determined from sparse connections may split into several areas when actual connection strength is considered, because it possesses weak internal connections. As another example, areas determined from sparse connections may combine into a single area because the external connection strength between the areas is strong. The purpose of this section is to present time-scale decomposition of areas taking into account simultaneously sparse and dense connections, and weak and strong connections. We will summarize the weak connection results, analyze a special class of systems with sparse and weak connections, and use the WSCC example to discuss a general approach to analyze practical systems using the grouping algorithm in Appendix C.

### 6.1 Weak Connections

The area decomposition results for networks with weak connections in [1] can be readily derived following the analysis in Sections 2, 3 and 4 for sparse connections.

In the weak connection analysis, we use the actual inertia values and connection strength. The network dynamic equations for an r-area system are

$$m_i^\alpha \dot{x}_i^\alpha = \sum_{\beta=1}^{r} \sum_{j=1}^{m_\beta} k_{ij}^{\alpha\beta}(x_j^\beta - x_i^\alpha), \text{ for } i = 1,\ldots,m_\alpha,$$

$$\alpha = 1,2,\ldots,r, \text{ and } \beta \neq \alpha \text{ when } j = i, \tag{6.1}$$

where $m_i^\alpha$ is the inertia or capacitance of the storage element of $x_i^\alpha$, and $k_{ij}^{\alpha\beta}$ is the actual connection strength between node i of area $\alpha$ and node j of area $\beta$. Following the notation of (2.9) and (2.10), we write (6.1) in the compact form

$$M\dot{x} = Kx = (K^I + K^E)x \tag{6.2}$$

where the inertia matrix is

$$M = \text{diag}\,(M_1, M_2, \ldots, M_r), \quad M_\alpha = \text{diag}(m_1^\alpha, \ldots, m_{m_\alpha}^\alpha), \quad \alpha = 1, 2, \ldots, r. \tag{6.3}$$

When the external connection $K^E$ is weak, we represent each entry of $K^E$ as a multiple of a small parameter $\varepsilon$,

$$K^E = \varepsilon \tilde{K}^E \tag{6.4}$$

where $\tilde{K}^E$ is O(1). The O(1) internal connection $K^I$ is strong compared to $K^E$, and to be consistent, we assume that every node in an area is strongly connected directly or indirectly via other nodes to every other node in the same area. Although it is possible to allow weak connections in $K^I$, for simplicity they will be neglected.

In contrast to (3.1), when the actual parameter values are used, the aggregate variable becomes

$$y_\alpha = \sum_{i=1}^{m_\alpha} m_i^\alpha x_i^\alpha / m^\alpha, \quad m^\alpha = \sum_{i=1}^{m_\alpha} m_i^\alpha, \quad \alpha = 1, 2, \ldots, r. \tag{6.5}$$

The difference variable $z^\alpha$ remains unchanged. In matrix form, the aggregate and local variable transformation is

$$\begin{bmatrix} y \\ z \end{bmatrix} = \begin{bmatrix} C \\ G \end{bmatrix} x \tag{6.6}$$

where

$$C = (U^T M U)^{-1} U M \tag{6.7}$$

and G is as in (3.6).

In the aggregate and local variables (6.6), system (6.2) becomes, because of the null space properties (3.14),

$$\begin{bmatrix} \dot{y} \\ \dot{z} \end{bmatrix} = \begin{bmatrix} \varepsilon A_{11} & \varepsilon A_{12} \\ \varepsilon A_{21} & A_{22} \end{bmatrix} \begin{bmatrix} y \\ z \end{bmatrix} \tag{6.8}$$

where

$$A_{11} = C\tilde{K}^E U, \quad A_{12} = C\tilde{K}^E G^+,$$
$$A_{21} = G\tilde{K}^E U, \quad A_{22} = G(K^I + \varepsilon\tilde{K}^E)G^+. \tag{6.9}$$

System (6.8) is a singularly perturbed system in the fast time scale $t_f = t$, and is similar to system (3.22) for sparse connections, except that the weak connection parameter $\varepsilon$ functions as both the area parameter $\delta$ and the node parameter d.

In the slow time-scale $t_s = \varepsilon t_f$, system (6.8) has the explicit singular perturbation form

$$dy/dt_s = A_{11}y + A_{12}z,$$
$$\varepsilon dz/dt_s = \varepsilon A_{21}y + A_{22}z. \tag{6.10}$$

The time-scale results are now in terms of $\varepsilon$ instead of d and $\delta$ as in Theorem 3.1.

<u>Theorem 6.1</u>. There exists $\varepsilon^*$ such that for all $0 < \varepsilon \leq \varepsilon^*$, the system (6.8) has r slow eigenvalues

$$\lambda_s = \lambda[(A_{11} - \varepsilon A_{12}A_{22}^{-1}A_{21}) + O(\varepsilon^2)]$$
$$= \lambda\,[A_{11} + O(\varepsilon)] \tag{6.11}$$

and n-r fast eigenvalues

$$\lambda_f = \lambda[A_{22} + O(\varepsilon^2)]/\varepsilon$$
$$= \lambda\,[GK^I G^+ + O(\varepsilon)]/\varepsilon \tag{6.12}$$

and, hence, $|\lambda_s|/|\lambda_f| = O(\varepsilon)$. Furthermore, the slow and fast subsystems of (6.8) are

$$dy_s/dt_s = (A_{11} - \varepsilon A_{12}A_{22}^{-1}A_{21})\, y_s = A_0 y_s, \quad y_s(0) = y(0), \tag{6.13}$$

$$dz_f/dt_f = A_{22}z_f, \quad z_f(0) = z(0) + \varepsilon A_{22}^{-1}A_{21}y(0). \tag{6.14}$$

respectively, and their solutions approximate the solution y,z of (6.8) as

$$y = y_s(t_s) + O(\varepsilon^2), \tag{6.15}$$

$$z = -\varepsilon A_{22}^{-1}A_{21}y_s(t_s) + z_f(t_f) + O(\varepsilon^2). \tag{6.16}$$

Theorem 6.1 immediately follows from Theorem 3.1. Aggregates and local models can be derived from the slow subsystem (6.13) and the fast subsystem (6.14) following the analysis of Section 4, and will not be repeated. However, it is important to note that the small parameter $\varepsilon$ plays the role of the small node parameter d in obtaining the rigid aggregate and decoupled local models. This is in contrast to its predominant role of time-scale separation in Theorem 6.1.

6.2 <u>Networks with Sparse and Weak Connections</u>

Theorems 3.1 and 6.1 indicate that areas can be decomposed along either sparse connections or weak connections. Both types of connections can be expected to appear in applications. Thus sparse and weak connections should be accounted for simultaneously in determining the areas. There are many different ways in which sparse and weak connections can occur in a network. For example, some sparse connections can be stronger than the dense connections, while some dense connections are weak such that eliminating the weak connections would separate an area into several smaller areas. To consider all the different combinations of sparse and weak connections would not be practical. Instead we will use the WSCC example in Section 6.3 to illustrate some of the possibilities. Only the simplest combination of sparse and weak connections will be analyzed here.

We consider the case when the sparse external connections are small in magnitude. We will again assume that internal connections normalized with respect to inertias are unity, but the normalized external connections are now $\varepsilon$ small. This is a case in which we can readily combine our previous results on sparse and weak connections. From the inequalities (3.15)-(3.19), the null space properties (3.14) and the relation (6.4), the network dynamic equations (6.8) expressed in terms of the aggregate and local variables are

$$dy/dt_f = \varepsilon\delta A_{11}y + \varepsilon\delta A_{12}z$$
$$dz/dt_f = \varepsilon d A_{21}y + A_{22}z \qquad (6.17)$$

where $t_f = \underline{c}^I t$ and

$$A_{11} = C\tilde{K}^E U/(\underline{c}^I\delta), \quad A_{12} = C\tilde{K}^E G^+/(\underline{c}^I\delta),$$
$$A_{21} = G\tilde{K}^E U/(\underline{c}^I d), \quad A_{22} = G(K^I + \varepsilon\tilde{K}^E)G^+/\underline{c}^I. \qquad (6.18)$$

System (6.17) is a singularly perturbed system expressed in the fast time scale $t_f$ where the product $\varepsilon\delta$ denotes the time-scale separation and the product $\varepsilon d$ denotes weak coupling. In the slow time-scale $t_s = (\varepsilon\delta)t_f$, system (6.17) becomes the explicit singular perturbation form

$$dy/dt_s = A_{11}y + A_{12}z,$$
$$\varepsilon\delta dz/dt_s = \varepsilon d A_{21}y + A_{22}z. \qquad (6.19)$$

Theorems 3.1 and 6.1 can now be combined to obtain the following result.

Theorem 6.2. There exist $\delta^*$, $d^*$ and $\varepsilon^*$ such that for all $0 < \delta \leq \delta^*$, $0 < d \leq d^*$ and $0 < \varepsilon \leq \varepsilon^*$, the system (6.19) has r slow eigenvalues

$$\lambda_s = \lambda[(A_{11} - \varepsilon d A_{12}A_{22}^{-1}A_{21}) + O(\varepsilon^2\delta d)]$$
$$= \delta[A_{11} + O(\varepsilon d)] \qquad (6.20)$$

and n-r fast eigenvalues

$$\lambda_f = \lambda[A_{22} + O(\varepsilon^2\delta d)]/\varepsilon\delta$$
$$= \lambda[GK^I G^+/\underline{c}^I + O(\varepsilon d)]/\varepsilon\delta, \qquad (6.21)$$

and, hence, $|\lambda_s|/|\lambda_f| = O(\varepsilon\delta)$. Furthermore, the slow and fast subsystems of (6.19) are

$$dy_s/t_s = (A_{11} - \varepsilon d A_{12} A_{22}^{-1} A_{21})\, y_s = A_0 y_s, \quad y_s(0) = y(0), \tag{6.22}$$

$$dz_f/dt_f = A_{22} z_f, \quad z_f(0) = z(0) + \varepsilon d A_{22}^{-1} A_{21} y(0), \tag{6.23}$$

respectively, and their solutions approximate the solution y,z of (6.19) as

$$y = y_s(t_s) + O(\varepsilon^2 \delta d), \tag{6.24}$$

$$z = -\varepsilon d A_{22}^{-1} A_{21} y_s(t_s) + z_f(t_f) + O(\varepsilon^2 \delta d). \tag{6.25}$$

Various aggregate and local models can be obtained for (6.19) following the steps in Section 4 and will not be repeated here. We will only use the 10-node, 2-area network to illustrate the results. The external connections are assumed to be $\varepsilon$ with $\varepsilon = 0.1$. The internal connection matrix $K^I$ remains unchanged, as given in Section 4.3, but the external connection matrix $K^E$ becomes

$$K_{12}^E = \begin{bmatrix} 0 & 0 & 0 & 0 & 0 \\ 0 & 0 & 0 & 0 & 0 \\ 0.1 & 0 & 0 & 0 & 0 \\ 0 & 0 & 0 & 0 & 0 \\ 0 & 0 & 0 & 0 & 0.1 \end{bmatrix} = (K_{21}^E)^T$$

$$K_{11}^E = \mathrm{diag}(0,\ 0,\ -0.1,\ 0,\ -0.1), \quad K_{22}^E = \mathrm{diag}(-0.1,\ 0,\ 0,\ 0,\ -0.1).$$

The slow and fast eigenvalues of the full system and their approximations by various simplified models are given in Tables 6.1 and 6.2. Because of the composite effect of $\varepsilon$, $\delta$ and d, the eigenvalue approximations by all models are excellent.

Table 6.1

Eigenvalues of Slow Models (10-Node Network with Weak Connections)

| Slow Model | Matrix | Eigenvalues |
|---|---|---|
| Exact | $K$ | 0.00, -0.0778 |
| Complete Aggregate | $A_0$ | 0.00, -0.0778 |
| Rigid Aggregate | $A_{11}$ | 0.00, -0.0800 |

Table 6.2
Eigenvalues of Fast Models (10-Node Network with Weak Connections)

| Fast Model | Matrix | Eigenvalues | |
|---|---|---|---|
| | | Area 1 | Area 2 |
| Exact | $K$ | -3.05,-5.00,-5.00,-5.09 | -3.00,-5.00,-5.00,-5.18 |
| Fast Subsystem | $A_{22}$ | 3.05,-5.00,-5.00,-5.09 | -3.00,-5.00,-5.00,-5.18 |
| Compensated Local Models | $A^{\alpha}_{220}+\varepsilon dA^{\alpha}_{22d}$ | -3.05,-5.00,-5.02,-5.09 | -3.00,-5.00,-5.07,-5.07 |
| Local Models | $A^{\alpha}_{220}$ | -3.00,-5.00,-5.00,-5.00 | -3.00,-5.00,-5.00,-5.00 |

From the results of Theorem 6.2 in which the product $\varepsilon\delta$ denotes the time-scale separation, we can readily infer the possible existence of multiple time scales. Consider the case when some external connections are sparse, some weak and others both sparse and weak. Assuming that $\delta << \varepsilon$, the slow time-scale has a three-level hierarchy

$$t^1_s = \varepsilon\delta t_f, \; t^2_s = \delta t_f, \; t^3_s = \varepsilon t_f \tag{6.26}$$

in which $t^1_s$ is the slowest time-scale. In this four time-scale system, a nested aggregation starting from $t^3_s$ can be used to obtain subsystems for the different time-scales.

## 6.3 A Procedure to Identify Sparse and Weak Connections in Large Scale Networks

Large scale power networks often contain both sparse and weak connections which together determine the time scales and area decomposition. Information on sparse and weak connections is useful for power system planning and operation, such as the siting and setting of power system stabilizers and protective relays. The results in Theorem 6.2 only deal with the simplest situation where the sparse connections are also weak. For more general situations we propose to use the grouping algorithm to identify the weak and sparse connections. We will use the WSCC example as an illustration.

The identification procedure consists of three steps. In the first step, the grouping algorithm is applied to the system with connections normalized to unity to find the areas based on sparse connections. In the second step, the grouping algorithm is applied to the system with actual connection strength and inertia values to find the areas based on both sparse and weak connections. For computational efficiency in case of a large system, the bus nodes would also be retained in this step and a small inertia is added to each bus node [6]. In the last step, the area partitions from the first two steps are compared and contrasted to identify the weak and sparse connections.

For the WSCC system, step 1 has been performed in Section 5 and we have found the sparse connections in the 9-area decomposition (Figure 5.1). Step 2 requires about the same amount of computation time as step 1 and an 11-area partition (Figure 6.1) is found. In step 3 we examine the two different area partitions.

First, note that the 11 areas found based on the actual parameters are very close to the 9 areas found based on sparse connections. With the 9 areas found from sparse connections regarded as the "cores," the transition from the 9 areas to the 11 areas involve mostly some adjustment of the boundaries. Second, two areas in the 9-area partition separate into smaller areas in the 11-area partition, namely, area 5 in Figure 5.1 separates into two areas, areas 5 and 11, in Figure 6.1, and area 6 separates into three areas, areas 4, 6 and 10. Areas 4, 10 and 11 are small and are connected to the bulk of the system with weak links. Thus they are grouped separately when actual connection strength is used in the grouping algorithm. Third, two areas in the 9-area partition, namely, areas 1 and 4, merge to form a single area, area 1, in the 11-area partition. This situation illustrates that although the connections may be sparse, if they are strong, they will cause two areas to combine into a single area.

The slow modes and the models of the rigid aggregate with actual parameters are listed in Table 6.3. The average error in the approximations of the aggregate modes to the system slow modes is 120%. When the $K_a$ matrix is multiplied by $c = 1/(1 + 1.20)^2 = 0.207$, the modes of the compensated aggregate are much better approximations (Table 6.3). This value of c is comparable to c = 0.133 used for sparse connections in Section 5, indicating that sparse connections are the dominating factor in the area decomposition of the WSCC system.

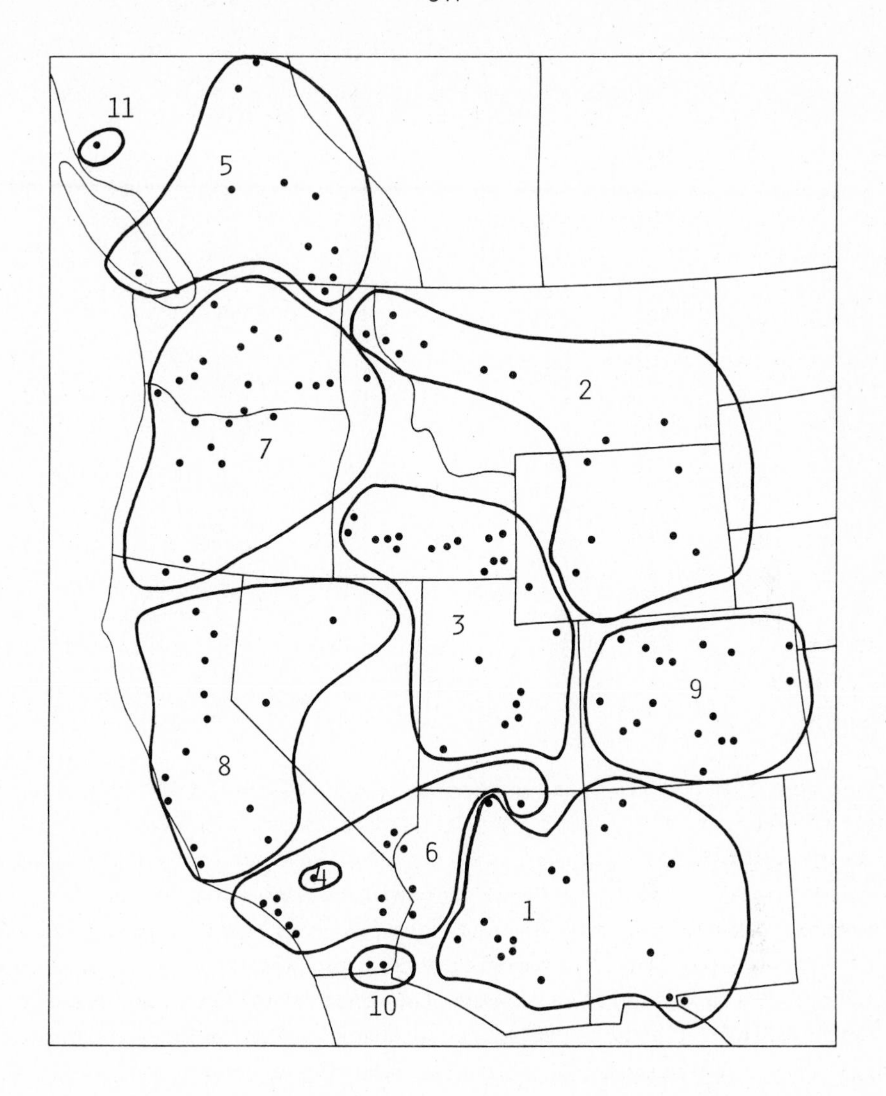


Figure 6.1 An 11-Area Partition of WSCC

Figures 5.1 and 6.1 of the WSCC system illustrate that our 3-step identification approach provides valuable information on sparse and weak connections. The results here would have been useful in the design of the WSCC Intertie Generator Dropping and Controlled Separation scheme for large disturbances described in [22]. Knowing the sparse and weak connections will also be helpful for future system planning, expansion and control design.

Table 6.3
Comparison of System Slow Modes and Aggregate Modes (in Hz)

| System Slow Modes | Rigid Aggregate | | Compensated Aggregate | |
|---|---|---|---|---|
| | Modes | Percent Error | Modes | Percent Error |
| 0.0000 | 0.0000 | – | 0.0000 | – |
| 0.3424 | 0.6994 | 104.2 | 0.3179 | -7.2 |
| 0.4405 | 0.9971 | 126.4 | 0.4532 | 2.9 |
| 0.6430 | 1.1443 | 78.0 | 0.5202 | -19.1 |
| 0.6562 | 1.4538 | 121.5 | 0.6608 | 0.7 |
| 0.7575 | 1.5392 | 103.2 | 0.6997 | -7.6 |
| 0.7783 | 1.5877 | 104.0 | 0.7217 | -7.3 |
| 0.8488 | 1.8911 | 122.8 | 0.8596 | 1.3 |
| 0.8598 | 1.9844 | 130.8 | 0.9020 | 4.9 |
| 0.9178 | 2.3419 | 155.2 | 1.0645 | 16.0 |
| 0.9503 | 2.4121 | 153.8 | 1.0965 | 15.4 |

## 7. NONLINEAR DYNAMIC NETWORKS

The application of time-scale decomposition and aggregation is not limited only to linear dynamic network models (2.8) or (6.1) which are suitable for small disturbance analysis of more accurate nonlinear models. For nonlinear dynamic networks with weak external connection, the decomposition and aggregation results are contained in [5.1]. In this section, we will use power system models to present the underlying properties allowing the extension of these results to nonlinear networks with sparse connections. The detailed derivation of the results is beyond the scope of this paper.

For an n-machine power system with r areas, the nonlinear electromechanical model without conductance terms is

$$m_i^\alpha \ddot{x}_i^\alpha = -d_i^\alpha \dot{x}_i^\alpha + P_i^\alpha - \sum_{\beta=1}^{r} \sum_{j=1}^{m_\beta} V_i^\alpha V_j^\beta B_{ij}^{\alpha\beta} \sin(x_j^\beta - x_i^\alpha), \text{ for } i=1,2,\ldots,m_\alpha, \tag{7.1}$$

$\alpha = 1,2,\ldots,r$, and $j \neq i$ when $\beta = \alpha$, where $x_i^\alpha$ is the machine angle, $m_i^\alpha$ the inertia constant, $d_i^\alpha$ the damping constant, $P_i^\alpha$ the mechanical input power and $v_i^\alpha$ the voltage at node i in area $\alpha$. As in the linear case, we assume that the admittance $B_{ij}^{\alpha\beta}$ between node i in area $\alpha$ and node j in area $\beta$ is dense within an area and sparse between the areas such that the node parameter d (2.3) and the area parameter $\delta$ (2.7) defined according to the number of connections are small.

Without the external connections, each of the r areas is isolated. The dynamic equation of area $\alpha$ is

$$m_i^\alpha \ddot{x}_i^\alpha = -d_i^\alpha \dot{x}_i^\alpha + P_i^\alpha - \sum_{j=1, j\neq i}^{m_\alpha} v_i^\alpha v_j^\alpha B_{ij}^{\alpha\alpha} \sin(x_j^\alpha - x_i^\alpha) \quad , \quad i=1,2,\dots,m_\alpha. \tag{7.2}$$

Area $\alpha$ has a continuum of equilibrium points

$$\dot{x}^\alpha = 0, \; x^\alpha = x_0^\alpha + c_\alpha u_\alpha \tag{7.3}$$

where $x_0^\alpha$ is an equilibrium point, $c_\alpha$ a scalar, and $u_\alpha$ as given in (2.11)

$$u_\alpha = [1 \; 1 \dots 1]^T. \tag{7.4}$$

Equation (7.3) implies that if (7.2) is at equilibrium, then (7.2) is still at equilibrium if all the components of $x^\alpha$ are increased or decreased by an equal amount. This is the <u>equilibrium property</u> and (7.3) defines the <u>equilibrium manifold</u> for area $\alpha$. On this manifold, the potentials within an area are equalized.

A second property is the <u>conservation property</u> that the state

$$y_\alpha = \sum_{i=1}^{m_\alpha} m_i^\alpha x_i^\alpha / m^\alpha, \; m^\alpha = \sum_{i=1}^{m_\alpha} m_i^\alpha \tag{7.5}$$

remains unchanged when

$$\sum_{i=1}^{m_\alpha} P_i^\alpha = 0, \tag{7.6}$$

$$d_i^\alpha = 0, \quad i = 1,2,\dots,m_\alpha. \tag{7.7}$$

Because of the symmetry $B_{ij}^{\alpha\beta} = B_{ji}^{\beta\alpha}$ and given (7.6), (7.7), we obtain from (7.2)

$$\ddot{y}_\alpha = \sum_{i=1}^{m^\alpha} m_i^\alpha \ddot{x}_i^\alpha / m^\alpha = 0, \tag{7.8}$$

that is, given any initial condition $x^\alpha(0)$, the motions of (7.2) satisfy

$$y_\alpha(t) - y_\alpha(0) = 0. \tag{7.9}$$

Equation (7.9) is the <u>dynamic manifold</u> on which the equalization of potentials takes place.

Using these two properties, we propose the aggregate and local variable transformations

$$\begin{bmatrix} y \\ \\ z \end{bmatrix} = \begin{bmatrix} C \\ \\ G \end{bmatrix} x - \begin{bmatrix} 0 \\ \\ G \end{bmatrix} x_0^\alpha \tag{7.10}$$

for the interconnected system (7.1), where the matrices C and G are as given in (6.7) and (3.6). Although (7.1) is nonlinear, (7.10) is the same linear transformation used for linear networks, except for the shift of the equilibrium.

The two-time-scale properties can then be shown for (7.1) expressed in the aggregate and local variables

$$\begin{aligned} \ddot{y} &= f(y, z, x_0^\alpha), \\ \ddot{z} &= g(y, z, x_0^\alpha). \end{aligned} \tag{7.11}$$

Assuming that the internal connections are dense and the external connections are sparse, we can show that

$$\|f\| / \|g\| = O(\delta) \tag{7.12}$$

for y,z in a domain D containing the equilibrium. Weak dynamic coupling can also be shown as

$$\|\partial g/\partial y\| / \|\partial g/\partial z\| = O(d). \tag{7.13}$$

From these results, we can derive the nonlinear slow and fast subsystems. It is also possible to derive various aggregate and local models following the analysis in Section 4.

The time-scale decomposition and aggregation analysis for nonlinear dynamic networks with both sparse and weak connections will also make use of the conservation and equilibrium properties. The derivation can follow the steps in Section 6.

## 8. CONCLUSIONS

The time-scale approach to the decomposition-aggregation of sparse dynamic networks developed here extends and complements the method [1] for dynamic networks with $\varepsilon$-connections. These new results are more broadly applicable, since most practical large scale networks have the sparsity pattern assumed in this paper and are characterized in terms of an area parameter $\delta$ and a node parameter d. Although the results are for a separation into two time-scales, they can be readily extended to networks with multiple time-scales using a nested application of the decomposition-aggregation transformation. Furthermore, the assumption of linear connections can be removed in a manner similar to the nonlinear networks with $\varepsilon$-connections [5,1].

Using the node parameter d, new aggregate and local models describing the slow and fast dynamics are proposed. These models reveal the roles played by the internal and external connections. Some of the results also improve the models proposed earlier for networks with $\varepsilon$-connections. The application of the results to large networks is illustrated with a 2000-node power system.

The sparse connection results are then combined with the weak connection results to provide a more general treatment of networks whose time-scales are due to either weak or sparse connections or both. Only the special case where sparse connections are also weak is analyzed. More general situations are illustrated with the 2000-node power system. A three-step procedure using the grouping algorithm for identifying the sparse connections and the weak connections has been proposed and shown to provide useful information about the 2000-node system.

# APPENDIX A
## DERIVATION OF BOUNDS (3.15)-(3.19)

The norm used to derive the bounds (3.15)-(3.19) is the row norm or $\|\bullet\|_\infty$.

The expression $A_{11} = CK^E U$ can be written as

$$A_{11} = M_a^{-1} K_a. \tag{A.1}$$

In each row of $A_{11}$, the diagonal entry is $-\gamma_\alpha^E/m_\alpha$, and the off-diagonal entries are all positive and sum to $\gamma_\alpha^E/m_\alpha$. Therefore

$$\|A_{11}\| = \max_\alpha \{2\gamma_\alpha^E/m_\alpha\} \leq 2\bar{\gamma}^E/\underline{m}. \tag{A.2}$$

In the $\alpha$-th row of $CK^E$, the entry corresponding to node i in area $\alpha$ is $c_{\alpha i}^E$, the number of connections from node i to the nodes in all the other areas, and the entry corresponding to node j in area $\beta$ is $c_{\alpha\beta j}^E$, the number of connections from node j to the nodes in area $\alpha$. Thus, in the $\alpha$-th row of $A_{12} = CK^E G^+$, the entry corresponding to $z_i^\alpha$ is

$$(c_{\alpha\beta(j+1)}^E - \gamma_\alpha^E/m_\alpha)/m_\alpha \tag{A.3}$$

while the entry corresponding to $z_j^\beta$ in area $\beta$ is

$$(c_{\alpha\beta(j+1)}^E - \gamma_{\alpha\beta}^E/m_\beta)/m_\alpha \tag{A.4}$$

where $\gamma_{\alpha\beta}^E$ is the number of connections between areas $\alpha$ and $\beta$. Hence

$$\|A_{12}\| = \max_\alpha \{(1/m_\alpha)[\sum_{\substack{\beta=1 \\ \beta\neq\alpha}}^{r} (m_\beta - 1)\gamma_{\alpha\beta}^E/m_\beta + \sum_{j=2}^{m_\beta} c_{\alpha\beta j}^E)$$

$$+ (m_\alpha - 1)\gamma_\alpha^E/m_\alpha + \sum_{i=2}^{m_\alpha} c_{\alpha i}^E\}$$

$$\leq \max_\alpha \{4\gamma_\alpha^E/m_\alpha\} \leq 4\bar{\gamma}^E/\underline{m}\ . \tag{A.5}$$

In the row of $GK^E$ corresponding to the variable $z_i^\alpha$, the entry corresponding to node i+1 of area $\alpha$ is $c_\alpha^E(i+1)$, the entry corresponding to the first node of area $\alpha$ is $-c_{\alpha 1}^E$, and the entry corresponding to node j in area $\beta$ is $c_{\alpha(i+1)\beta j}^E - c_{\alpha 1\beta j}^E$, where $c_{\alpha i\beta j}^E$ is the number of connections between node i of area $\alpha$ and node j of area $\beta$. Then in the row of $A_{21} = GK^E U$ corresponding to the variable $z_i^\alpha$, the entry corresponding to $y_\alpha$ is

$$c_{\alpha(i+1)}^E - c_{\alpha 1}^E \tag{A.6}$$

and the entry corresponding to $y_\beta$ is

$$c_{\alpha(i+1)\beta}^E - c_{\alpha 1\beta}^E. \tag{A.7}$$

Hence

$$\|A_{21}\| = \max_{\alpha,i} \{ \Sigma \; (c_{\alpha(i+1)\beta}^E + c_{\alpha 1\beta}^E) + c_{\alpha(i+1)}^E + c_{\alpha 1}^E \}$$

$$= \max_{\alpha,i} \{2c_{\alpha(i+1)}^E + 2c_{\alpha 1}^E\} \leq 4\bar{c}^E. \tag{A.8}$$

In the row of $GK^E G^+$ corresponding to the variable $z_i^\alpha$, the entry corresponding to $z_i^\alpha$ is

$$c_{\alpha(i+1)}^E + (c_{\alpha 1}^E - c_{\alpha(i+1)}^E)/m_\alpha \; , \tag{A.9}$$

the entry corresponding to $z_k^\alpha$, $k \neq i$, is

$$(c_{\alpha 1}^E - c_{\alpha(k+1)}^E)/m_\alpha \; , \tag{A.10}$$

and the entry corresponding to $z_j^\beta$ is

$$c_{\alpha(i+1)\beta(j+1)}^E - c_{\alpha 1\beta(j+1)}^E + (c_{\alpha 1\beta}^E - c_{\alpha(i+1)\beta}^E)/m_\beta. \tag{A.11}$$

Hence

$$\|GK^E G^+\| = \max_{\alpha,i} \{ \sum_{\beta=1}^{r} (c^E_{\alpha 1\beta} - c^E_{\alpha(i+1)\beta})$$

$$+ \sum_{\beta=1}^{r} \sum_{j=1}^{m_\beta -1} (c^E_{\alpha(i+1)\beta(j+1)} - c^E_{\alpha 1\beta(j+1)})$$

$$+ (m_\alpha - 1)(c^E_{\alpha 1} + c^E_{\alpha(i+1)})/m_\alpha + c^E_{\alpha(i+1)}\}$$

$$\leq \max_{\alpha,i} \{3c^E_{\alpha 1} + 4c^E_{\alpha(i+1)}\}$$

$$\leq 7\bar{c}^E. \tag{A.12}$$

Since $GK^I G^+$ is block-diagonal, we will establish its norm using the diagonal blocks $G_\alpha K^I_\alpha G^+_\alpha$. The product $K^I_\alpha G^+_\alpha$ is an $m_\alpha x(m_\alpha -1)$ matrix which is equal to $K^I_\alpha$ with the first column deleted. Then $G_\alpha(K^I_\alpha G^+_\alpha)$ is an $(m_\alpha -1)x(m_\alpha -1)$ matrix obtained by deleting the first row of $K^I_\alpha G^+_\alpha$ by subtracting it from the other rows. Assuming that each node is connected to at least 3/4 of the other nodes in the same area, the lower bound of $\|GK^I G^+\|$ is

$$\|GK^I G^+\| \geq \min_{\alpha,i} \{c^I_{\alpha(i+1)}/3\} \geq \underline{c}^I/3. \tag{A.13}$$

Different lower bounds will be obtained if other densities of internal connections are assumed.

# APPENDIX B
## PROOF OF THEOREM 3.1

We show the $O(\delta d)$ approximations of $\lambda_s$ and $\lambda_f$ in (3.25) and (3.26) by an iterative upper block triangularization process in [1]. Introducing the new fast variables

$$\eta_1 = z + dA_{22}^{-1}A_{21}y \tag{B.1}$$

we transform (3.24) into

$$\begin{aligned} dy/dt_s &= A_{11}^1 y + A_{12}\eta_1, \\ \delta\, d\eta_1/dt_s &= \delta d\, A_{21}^1 y + A_{22}^1\eta_1, \end{aligned} \tag{B.2}$$

where

$$\begin{aligned} A_{11}^1 &= A_{11} - dA_{12}A_{22}^{-1}A_{21}, \\ A_{21}^1 &= A_{22}^{-1}A_{21}A_{11}^1, \\ A_{22}^1 &= A_{22} + \delta d\, A_{22}^{-1}A_{21}A_{12}. \end{aligned} \tag{B.3}$$

A second transformation using the revised fast variables

$$\eta_2 = \eta_1 + \delta d(A_{22}^1)^{-1}A_{21}^1\, y \tag{B.4}$$

yields

$$\begin{aligned} dy/dt_s &= A_{11}^2 y + A_{12}\eta_2, \\ \delta d\eta_2/dt_s &= \delta^2 d\, A_{21}^2 y + A_{22}^2\eta_2, \end{aligned} \tag{B.5}$$

where

$$\begin{aligned} A_{11}^2 &= A_{11}^1 - \delta d\, A_{12}(A_{22}^1)^{-1}A_{21}^1, \\ A_{21}^2 &= (A_{22}^1)^{-1}A_{21}^1A_{11}^2, \\ A_{22}^2 &= A_{22}^1 + \delta^2 d(A_{22}^1)^{-1}A_{21}^1A_{12}. \end{aligned} \tag{B.6}$$

For $\delta$ and d sufficiently small, this process converges to

$$dy/dt_s = A_{11}^{\infty}y + A_{12}\eta_{\infty},$$
$$\delta d\eta_{\infty}/dt_s = A_{22}^{\infty}\eta_{\infty}, \quad (B.7)$$

where

$$A_{11}^{\infty} = A_{11} - dA_{12}A_{22}^{-1}A_{21} + O(\delta d) ,$$
$$A_{22}^{\infty} = A_{22} + O(\delta d) , \quad (B.8)$$

thus proving (3.25) and (3.26). A suitable set of bounds, ($\delta^*$, $d^*$), can be computed using the results in [20]. The approximations (3.27)-(3.30) follow by performing the complete block diagonalization [1,20] of (B.7).

# APPENDIX C
## GROUPING ALGORITHM

For a large scale dynamic network, an automated tool for identifying sparsely connected areas is the grouping algorithm in [1,6], which is summarized as follows:

Step 1: Choose the number of areas, $r$.

Step 2: Compute a basis matrix $V$ of the eigenspace of the $r$ slowest eigenvalues, using either EISPACK [15] for dense matrices or the Lanczos algorithm in [16] for sparse matrices.

Step 3: Apply Gaussian elimination with complete pivoting to $V$ and obtain the states used for the pivots as the reference states.

Step 4: Assign a state $i$ to reference state $\alpha$ if the row in $V$ corresponding to state $\alpha$ is, among all the reference states, closest to that corresponding to state $i$.

When the number of areas $r$ is not known, we choose a sufficiently large $r$ for Step 2. Then Step 3 can be repeated for various values of $r$ to find an $r$ that yields a partition with small node parameter $d$ and area parameter $\delta$. Typical computation times on a VAX 11/780 computer for a 2000-node network with an average of 4 non-zero entries per row of $K$ are about 4 minutes of CPU per 10 eigenvalues/eigenvectors.

REFERENCES


1. Time-Scale Modeling of Dynamic Networks with Applications to Power Systems, J.H. Chow, Editor, Lecture Notes in Control and Information Sciences, Vol. 46, Springer-Verlag, Berlin, 1982.

2. P.V. Kokotovic, "Subsystems, Time Scales, and Multi-Modeling," Automatica, Vol. 17, pp. 789-795, 1981.

3. F. Delebecque and J.P. Quadrat, "Optimal Control of Markov Chains Admitting Strong and Weak Interactions," Automatica, Vol. 17, pp. 281-296, 1981.

4. R.G. Phillips and P.V. Kokotovic, "A Singular Perturbation Approach to Modeling and Control of Markov Chains," IEEE Transactions on Automatic Control, Vol. AC-26, pp. 1087-1094, 1982.

5. G. Peponides, P.V. Kokotovic and J.H. Chow, "Singular Perturbations and Time Scales in Nonlinear Models of Power Systems," IEEE Transactions on Circuits and Systems, Vol. CAS-29, pp. 758-767, 1982.

6. J.H. Chow, J. Cullum and R.A. Willoughby, "A Sparsity-Based Technique for Identifying Slow-Coherent Areas in Large Power Systems," IEEE Transactions on Power Apparatus and Systems, Vol. PAS-103, pp. 463-473, 1984.

7. D.M. Cvetkovic, M. Doob and H. Sachs, Spectra of Graphs - Theory and Application, VEB Deutscher Verlag der Wissenschaften, Berlin, 1982.

8. J.H. Chow and P.V. Kokotovic, "Sparsity and Time Scales," Proceedings of 1983 ACC, San Francisco, pp. 656-661.

9. P.V. Kokotovic, R.E. O'Malley, Jr. and P. Sannuti, "Singular Perturbations and Order Reduction in Control Theory - An Overview," Automatica, Vol. 12, pp. 123-132, 1976.

10. G.M. Peponides, "Nonexplicit Singular Perturbations and Interconnected Systems," Ph.D. Dissertation, Report R-960, Coordinated Science Laboratory, University of Illinois, Urbana, Illinois, 1982.

11. F. Delebecque, "A Reduction Process for Perturbed Markov Chains," SIAM J. Applied Mathematics, Vol. 43, pp. 325-350, 1983.

12. M. Coderch, A.S. Willsky, S.S. Sastry, D.A. Castanon, "Hierarchical Aggregation of Linear Systems with Multiple Time Scales," IEEE Transactions on Automatic Control, Vol. AC-28, pp. 1017-1030, 1983.

13. F.R. Gantmacher, The Theory of Matrices, Chelsea Publishing Company, New York, 1959.

14. R.L. Cresap and J.F. Hauer, "Emergence of a New Swing Mode in the Western Power System," IEEE Transactions on Power Apparatus and Systems, Vol. PAS-100, pp. 2037-2045, 1981.

15. EISPACK Guide - Matrix Eigensystem Routines, edited by B.T. Smith, et al., Springer-Verlag, New York, 1976.

16. J. Cullum and R.A. Willoughby, "Computing Eigenvalues of Very Large Symmetric Matrices," J. Computational Physics, Vol. 44, pp. 329-358, 1981.

17. H.A. Simons, "The Architecture of Complexity," Proceedings of the American Philosophical Society, Vol. 104, pp. 467-482, 1962.

18. R.M. May, Stability and Complexity in Model Ecosystems, Princeton University Press, 1973.

19. D.G. Feingold and R.S. Varga, "Block Diagonally Dominant Matrices and Generalizations of the Gerschgorin Circle Theorem," Pacific Journal of Mathematics, Vol. 12, pp. 1241-1250, 1962.

20. P.V. Kokotovic, "A Riccati Equation for Block-Diagonalization of Ill-Conditioned Systems," IEEE Transactions on Automatic Control, Vol. AC-20, pp. 812-814, 1975.

21. J.H. Chow, J.J. Allemong and P.V. Kokotovic, "Singular Perturbation Analysis of Systems with Sustained High Frequency Oscillations," Automatica, Vol. 14, pp. 271-279, 1978.

22. L.H. Fink, "Emergency Control Practices," Task Force on Emergency Control Report, Paper 85WM034-4, IEEE Winter Power Meeting, 1985.

# Part III: STABILITY AND AVERAGING

# STABILITY ANALYSIS OF SINGULARLY PERTURBED SYSTEMS

*H.K. Khalil*[†]

## INTRODUCTION

The time scale decomposition of a singularly perturbed system into reduced and boundary layer systems provides a strong tool for stability analysis. KLIMUSHCHEV and KRASOVSKII (1961) employed Lyapunov functions to show that asymptotic stability of the equilibrium of a singularly perturbed system can be established, for sufficently small perturbation parameter $\varepsilon$, by investigating the equilibria of the reduced and boundary-layer systems. Similar results for linear systems were given in DESOER and SHENSA (1970) and WILDE and KOKOTOVIC (1972). HOPPENSTEADT (1966) gave what are probably the weakest "conceptual" conditions under which uniform asymptotic stability of the equilibrium of a singularly perturbed system is confirmed for sufficiently small $\varepsilon$. The conditions comprise uniform asymptotic stability of the equilibrium of the reduced system, uniform asymptotic stability of the equilibrium of the boundary-layer system, uniformly in the frozen slow parameters, and growth conditions on right-hand side functions. Stability investigations in which conditions, guaranteeing those of HOPPENSTEADT (1966), are imposed on Lyapunov functions for the reduced and boundary-layer systems have been pursued by a few researchers. Examples can be found in HABETS (1974), CHOW (1978), GRUJIC (1981) and SABERI and KHALIL (1984). Stability results for linear multiparameter singularly perturbed systems are reported in KHALIL and KOKOTOVIC (1979), LADDE and SILJAK (1983) and ABED (1985), and for a class of nonlinear systems which are linear in

[†]Electrical Engineering Department, Michigan State University, East Lansing, MI.

the fast variables in KHALIL (1981). Other stability investigations are reported in the surveys KOKOTOVIC, O'MALLEY and SANNUTI (1976) and SAKSENA, O'REILLY and KOKOTOVIC (1984). In Section 1 we present the stability analysis of nonlinear nonautonomous singularly perturbed systems. Our presentation is based on SABERI and KHALIL (1984). In Section 2 we extend the analysis to multiparameter perturbations generalizing the results of KHALIL (1981) by allowing nonlinearities in the fast variables. Bounds on perturbation parameters are given but are not illustrated by numerical examples. Numerical examples can be found in KHALIL (1981), SABERI and KHALIL (1984) and KOKOTOVIC, KHALIL and O'REILLY (1985).

## 1. Stability Analysis

We consider a nonlinear nonautonomous singularly perturbed system

$$\dot{x} = f(t,x,z,\varepsilon), \qquad x \in R^n$$
$$\varepsilon\dot{z} = g(t,x,z,\varepsilon), \qquad \varepsilon > o, \qquad z \in R^m \tag{1.1}$$

where the functions f and g are smooth enough to ensure that, for specified initial conditions, (1.1) has a unique solution. Suppose that (1.1) has an isolated equilibrium point at the origin. Stability of the origin is investigated by examining the reduced system

$$\dot{x} = f(t,x,h(t,x),o) \tag{1.2}$$

where $z = h(t,x)$ is an isolated root of $o = g(t,x,z,o)$ and the boundary-layer system

$$\frac{dz}{d\tau} = g(t,x,z(\tau),o), \qquad \tau = t/\varepsilon \tag{1.3}$$

where t and x are treated as fixed parameters. Theorem 1 states, essentially, that if $x=o$ is a uniformly asymptotically stable equilibrium of the reduced system (1.2), $z = h(t,x)$ is an asymptotically stable equilibrium of the boundary-layer system (1.3), uniformly in t and x, and f and g satisfy certain growth conditions, then the origin is a uniformly asymptotically stable equilibrium of the singularly perturbed system (1.1), for sufficiently small $\varepsilon$. In Theorem 1, asymptotic stability requirements on the reduced and boundary-layer systems are expressed by requiring the existence of Lyapunov functions for each system. The growth requirements on f and g take the form of inequalities satisfied by the Lyapunov functions, which we call interconnection conditions.

We assume that the following conditions hold for all

$(t,x,z,\varepsilon) \in [t_o,\infty) \times B_x \times B_z \times [o,\varepsilon_1]$; $B_x \subseteq R^n$ and $B_z \subseteq R^m$.

Assumption 1.1: The origin ($x = 0$, $z = 0$) is the unique equilibrium of (1.1), i.e.,

$$0 = f(t,0,0,\varepsilon) \tag{1.4}$$

$$0 = g(t,0,0,\varepsilon) \tag{1.4}$$

Moreover, the equation

$$0 = g(t,x,z,0) \tag{1.5}$$

has a unique root $z = h(t,x)$ and there exists a class $\kappa$ function $p(\cdot)$ such that

$$||h(t,x)|| \leq p(||x||) \tag{1.6}$$

Assumption 1.2: Reduced System

There exists a Lyapunov function $V(t,x)$ satisfying

(i) $V(t,x)$ is positive-definite and decrescent, that is,

$$0 < q_1(||x||) \leq V(t,x) \leq q_2(||x||),$$

for some class $\kappa$ function $q_1(\cdot)$ and $q_2(\cdot)$

(ii) $$\frac{\partial V}{\partial t} + \frac{\partial V}{\partial x} f(t,x,h(t,x),0) \leq -\alpha_1 \psi^2(x), \qquad \alpha_1 > 0 \tag{1.7}$$

where $\psi(\cdot)$ is a continuous scalar function of $x$ which vanishes only at $x = 0$.

Assumption 1.3: Boundary-Layer System

There exists a Lyapunov function $W(t,x,z)$ satisfying

(i) $0 < q_3(||z - h(t,x)||) \leq W(t,x,z) \leq q_4(||z - h(t,x)||)$

for some class $\kappa$ functions $q_3(\cdot)$ and $q_4(\cdot)$

(ii) $$\frac{\partial W}{\partial z} g(t,x,z,0) \leq -\alpha_2 \phi^2(z - h(t,x)), \quad \alpha_2 > 0 \tag{1.8}$$

where $\phi(\cdot)$ is a continuous function of an $R^m$-vector $z_f$ which vanishes only at $z_f = 0$.

Assumption 1.4: Interconnection Conditions

V and W satisfy the following inequalities:

(i) $\frac{\partial V}{\partial x}[f(t,x,z,\varepsilon) - f(t,x,h(t,x),0] \leq \beta_1\psi(x)\phi(y - h(t,x)) + \varepsilon\gamma_1\psi^2(x)$ (1.9)

(ii) $\frac{\partial W}{\partial z}[g(t,x,z,\varepsilon) - g(t,x,z,0)] \leq \varepsilon\gamma_2'\phi^2(z - h(t,x)) + \varepsilon\beta_2'\psi(x).$
$\phi(z - h(t,x))$ (1.10)

(iii) $\frac{\partial W}{\partial t} + \frac{\partial W}{\partial x}f(t,x,z,\varepsilon) \leq \gamma_2''\phi^2(z - h(t,x)) + \beta_2''\psi(x)\phi(z - h(t,x))$ (1.11)

For simplicity the constants $\beta_1$, $\beta_2'$, $\beta_2''$, $\gamma_1$, $\gamma_2'$ and $\gamma_2''$ are assumed to be nonnegative. The term $\varepsilon\gamma_1\psi^2$ in (1.9) allows for more general dependence of f on $\varepsilon$. It drops out when f is independent of $\varepsilon$. Similarly, (1.10) drops out when g is independent of $\varepsilon$. In arriving at the above inequalities, one might have to obtain uniform bounds on $\varepsilon$-dependent terms like $K_1 + \varepsilon K_2$; that is why $\varepsilon$ may be restricted to an interval $[0, \varepsilon_1]$. With the Lyapunov functions of the reduced and boundary-layer systems, V and W, in hand, we consider a Lyapunov function candidate $\nu(t,x,z)$ defined by a weighted sum of V and W

$$\nu(t,x,z) = (1 - d)V(t,x) + dW(t,x,z), \qquad 0<d<1 \tag{1.12}$$

From the properties of V and W and inequality (1.6), it follows that $\nu(t,x,z)$ is positive-definite and decrescent. Computing $\dot{\nu}$ with respect to (1.1) and using (1.7) - (1.11) we obtain

$$\dot{\nu} \leq - \begin{bmatrix} \psi \\ \phi \end{bmatrix}^T \begin{bmatrix} (1-d)(\alpha_1-\varepsilon\gamma_1) & -\frac{1}{2}(1-d)\beta_1 - \frac{1}{2}d\beta_2 \\ -\frac{1}{2}(1-d)\beta_1 - \frac{1}{2}d\beta_2 & \frac{d}{\varepsilon}(\alpha_2-\varepsilon\gamma_2) \end{bmatrix} \begin{bmatrix} \psi \\ \phi \end{bmatrix} \tag{1.13}$$

where $\beta_2 = \beta_2' + \beta_2''$ and $\gamma_2 = \gamma_2' + \gamma_2''$.

It can be easily seen that for all

$$\varepsilon < \varepsilon_d = \frac{\alpha_1\alpha_2}{\alpha_1\gamma_2+\alpha_2\gamma_1+\frac{1}{4(1-d)d}[(1-d)\beta_1+d\beta_2]^2} \tag{1.14}$$

the right-hand side of (1.13) is a negative-definite quadratic form in $\psi$ and $\phi$. Hence there exists a positive constant K (possibly dependent on $\varepsilon$ and d) such that

$$\dot{\nu} \leq - K[\psi^2(x) + \phi^2(z - h(t,x))]$$

Since $\phi^2(z_f)$ is a continuous positive-definite function of $z_f$, there exists a class $\kappa$ function $q(||z_f||)$ such that

$$\phi^2(z_f) \geq q(||z_f||)$$

Therefore

$$\dot{\nu} \leq - K[\psi^2(x) + q(||z - h(t,x)||)]$$

Using (1.6), it can be shown that $\psi^2(x) + q(||z - h(t,x)||)$ is positive-definite. Therefore, $\dot{\nu}$ is negative-definite. Moreover,

$$\varepsilon_d \Big|_{d=\beta_1/(\beta_1+\beta_2)} = \varepsilon^* = \frac{\alpha_1\alpha_2}{\alpha_1\gamma_2+\alpha_2\gamma_1+\beta_1\beta_2} \tag{1.15}$$

<u>Theorem 1</u>: Suppose that Assumptions 1.1 - 1.4 hold. Then the origin ($x = 0$, $z = 0$) is a uniformly asymptotically stable equilibrium of the singularly perturbed system (1.1) for all $\varepsilon \in (0,\min(\varepsilon^*,\varepsilon_1))$, where $\varepsilon^*$ is given by (1.15). Moreover, for every $d \in (0,1)$, $\nu(t,x,z) = (1 - d)V(t,x,) + dW(t,x,z)$ is a Lyapunov function for all $\varepsilon \in (0,\min(\varepsilon_d,\varepsilon_1))$, where $\varepsilon_d$ is given by (1.14).

Various special cases can be drawn from Theorem 1. For the autonomous system

$$\begin{aligned} \dot{x} &= f(x,z) \\ \varepsilon\dot{z} &= g(x,z) \end{aligned} \tag{1.16}$$

we have $\gamma_1 = \gamma_2' = \beta_2' = 0$, $\beta_2 = \beta_2''$, $\gamma_2 = \gamma_2''$, and the bounds $\varepsilon_d$ and $\varepsilon^*$ reduce to

$$\varepsilon_d = \frac{\alpha_1 \alpha_2}{\alpha_1 \gamma_2 + \frac{1}{4(1-d)d} [(1-d)\beta_1 + d \beta_2]^2} \tag{1.17}$$

$$\varepsilon^* = \frac{\alpha_1 \alpha_2}{\alpha_1 \gamma_2 + \beta_1 \beta_2} \tag{1.18}$$

Several examples of antonomous systems were studied in SABERI and KHALIL (1984), including estimating the region of attraction of a synchronous generator connected to an infinite bus.

For the linear time-varying system

$$\begin{aligned} \dot{x} &= A_{11}(t)\, x + A_{12}(t) z \\ \varepsilon \dot{z} &= A_{21}(t)\, x + A_{22}(t) z \end{aligned} \tag{1.19}$$

where $A_{ij}(t)$ are continuously differentiable, $\dot{A}_{ij}(t)$ are bounded, $\text{Re}\,\lambda(A_{22}(t)) \leq - C < o$ and the reduced system

$$\dot{x} = [A_{11}(t) - A_{12}(t)\, A_{22}^{-1}(t)\, A_{21}(t)]x \overset{\Delta}{=} A_o(t)\, x \tag{1.20}$$

is uniformly asymptotically stable, the functions V and W can be taken as

$$V(t,x) = x^T P_s(t) x \tag{1.21}$$

$$W(t,x,z) = (Z + L_o(t)\, x)^T\, P_f(t)\, (z + L_o(t)\, x) \tag{1.22}$$

where $P_s(t) \geq cI_n > o$ and $P_f(t) \geq cI_m > o$ satisfy the Lyapunov equations

$$-\dot{P}_s(t) = P_s(t) A_o(t) + A_o^T(t)\, P_s(t) + I_n \tag{1.23}$$

$$0 = P_f(t) A_{22}(t) + A_{22}^T(t)\, P_f(t) + I_m \tag{1.24}$$

Then Assumptions 5.2 - 5.4 are satisfied with $\psi(x) = ||x||$ (Euclidean norm), $\phi\,(z-h(t\,,x)) = ||z+L_o(t)x||$, where $L_o = A_{22}^{-1} A_{21}$, $\alpha_1$, $\alpha_2 = 1$, $\gamma_1 = o$, and $\beta_1$, $\beta_2$ and $\gamma_2$ are bounds on the time functions

$2||P_s(t)\ A_{12}(t)||$ , $2||\ P_f(t)(\dot{L}_o(t) + L_o(t)A_o\ (t))||$ and $||\dot{P}_f(t) + 2\ P_f(t)L_o(t)A_{12}(t)||$ , respectively. Hence (1.19) is uniformly asymptotically stable for all

$$\varepsilon < \varepsilon^* = \frac{1}{\gamma_2 + \beta_1\beta_2} \tag{1.25}$$

This bound is in most cases less conservative than the one given by JAVID (1978), as illustrated by examples in KOKOTOVIC, KHALIL and O'REILLY (1985).

For the linear time invariant system

$$\dot{x} = A_{11}x + A_{12}z$$

$$\varepsilon\dot{z} = A_{21}x + A_{22}z \tag{1.26}$$

Where Re $\lambda(A_o) < o$ and Re $\lambda(A_{22}) < o$, the functions V and W can be taken as in (1.21) and (1.22) with $P_s$ and $P_f$ as the solutions of the algebraic Lyapunov equations

$$P_s\ A_o + A_o^T\ P_s = -\ I_n \tag{1.27}$$

$$P_f\ A_{22} + A_{22}^T P_f = -\ I_m \tag{1.28}$$

The system (1.26) is asymptotically stable for all $\varepsilon$ satisfying (1.25) where $\beta_1 = 2||P_sA_{12}||$ , $\beta_2 = 2||L_oA_o||$ and $\gamma = 2||P_fL_oA_{12}||$ .

This bound is compared in SABERI (1983) with the bound given by ZIEN (1973) and is shown to be less conservative in most cases, especially in high-dimensional problems.

## 2. Multiparameter Perturbations

The analysis of Section 1 can be extended to multiparameter singularly perturbed systems. For convenience, we consider an autonomous system

$$\dot{x} = f(x, z_1, \ldots, z_n) , \qquad x \in R^n$$
$$\varepsilon_i \dot{z} = g_i(x, z_1, \ldots, z_n), \qquad \varepsilon_i > o, \; z_i \in R^{m_i}, i=1,\ldots N \tag{2.1}$$

If $\varepsilon_i$'s are known, they can be represented as known multiples of a single parameter, i.e., $\varepsilon_i = \alpha_i \varepsilon$, and the problem reduces to the single parameter case treated in section 1. In some applications it is important to perform stability analysis without assuming the knowledge of the mutual ratios of the perturbation parameters. In this section we extend the stability analysis of section 1 to the multiparameter system (2.1) when the parameters $\varepsilon_i$'s are unknown.

We want to study the stability of an isolated equilibrium at the origin ($x=o$, $z_i=o \; \forall i$). The following assumptions are required to hold for all $(x, z_1, \ldots, z_N) \in B_x x B_{z_1} x \ldots x B_{z_N}$.

Assumption 2.1: The origin is the unique equilibrium of (2.1) and the N simultaneous algebraic equations

$$0 = g_i(x, z_1, \ldots, z_N) \tag{2.2}$$

have a unique root $z_i = h_i(x)$ such that $h_i(o) = o$.

The reduced system is given by

$$\dot{x} = f(x, h_1(x), \ldots, h_N(x)) \triangleq f_o(x) \tag{2.3}$$

and has equilibrium at $x=o$.

Assumption 2.2: There exists a Lyapunov function V(x) for (2.3) such that

$$\frac{\partial V}{\partial x} f_o(x) \leq - \alpha_o \quad \psi^2 \tag{2.4}$$

where $\psi = \psi(x)$ is a scalar continuous function of x which vanishes only at x = o.

The boundary-layer system is defined in a stretched time scale

$$\tau = t/\varepsilon, \qquad \varepsilon = \max_i \varepsilon_i \tag{2.5}$$

to be

$$\frac{dz_i}{d\tau} = \mu_i g_i(x, z_1, \ldots, z_N), \qquad \mu_i = \frac{\varepsilon}{\varepsilon_i} \geq 1 \tag{2.6}$$

where x is treated as a fixed parameter. The main difference between the single parameter and multiparameter cases is in studying the stability of the boundary-layer system. In the current case we need the equilibrium $z_i = h_i(x)$ of (2.6) to be asymptotically stable, uniformly in x, for all values of the parameters $\mu_i$'s. This amounts to the generalization of the concept of block D-stability of a matrix that was employed in KHALIL and KOKOTOVIC (1979) and KHALIL (1981). To achieve this we view (2.6) as an interconnection of N subsystem and employ composite stability methods as in MICHEL and MILLER (1977), SILJAK (1978) and ARAKI (1978). Let

$$p_i(x, z_i) = g_i(x, z_1, \ldots, z_N) \Big|_{z_j = h_j(x),\ j \neq i} \tag{2.7}$$

and

$$g_i(x, z_1, \ldots, z_N) = g_i(x, z_1, \ldots, z_N) - p_i(x, z_i) \tag{2.8}$$

Then, (2.6) can be rewritten as

$$\frac{dz_i}{d\tau} = \mu_i [p_i(x, z_i) + g_i(x, z_1, \ldots, z_N) \tag{2.9}$$

Viewing $g_i$ as the interconnection term, the ith isolated subsystem is given by

$$\frac{dz_i}{d\tau} = \mu_i p_i\ (x,\ z_i) \qquad (2.10)$$

which has an equilibrium at $z_i = h_i(x)$. Notice that the positive constant $\mu_i$ does not affect the stability of the equilibrium of (2.10). Suppose now that we can find a Lyapunov function $W_i\ (x, z_i)$ satisfying

$$\frac{\partial W_i}{\partial z_i} p_i\ (x, z_i) \leq -\ a_i\ \phi_i^2, \qquad a_i > o \qquad (2.11)$$

where $\phi_i = \phi_i(z_i - h_i(x))$ is a continusous scalar function of $z_i - h_i(x)$ which vanishes only at $z_i = h_i(x)$. Suppose further that $W_i$'s satisfy the "spatial" interconnection condition

$$\frac{\partial W_i}{\partial z_i} q_i(x,\ z_1, \ldots,\ z_N) \leq \sum_{j=1}^{N} b_{ij}\phi_i\phi_j, \qquad b_{ij} \geq o \qquad (2.12)$$

Taking

$$W\ (x,\ z_1, \ldots, z_N) = \sum_{i=1}^{N} e_i W_i(x, z_i), \qquad e_i > o \qquad (2.13)$$

with unspecified $e_i$'s as a Lyapunov function candidate for (2.9), it can be shown that the derivative of W along the trajectory of (2.9) satisfies

$$\frac{dW}{d\tau} \leq -\ \phi^T(E\ N\ R + R^T\ E\ N\ )\ \phi \qquad (2.14)$$

where $\phi^T = (\ \phi_1, \ldots,\ \phi_N)$, $E = \text{diag}\ (e_1, \ldots, e_n)$, $N = \text{diag}\ (\mu_1, \ldots, \mu_N)$ and

$$R = (r_{ij}); \qquad r_{ij} = \begin{cases} \tfrac{1}{2}a_i\ , & j=i \\ -\tfrac{1}{2}b_{ij}, & j \neq i \end{cases}$$

Assume that R is an M-matrix, i.e.,

$$\det \begin{bmatrix} r_{11} & \cdots & r_{1j} \\ \vdots & & \vdots \\ r_{j1} & \cdots & r_{jj} \end{bmatrix} > o, \quad j=1, \ldots, N \qquad (2.15)$$

Then, we can choose $d_i > o$ and $D = \mathrm{diag}\,(d_1, \ldots, d_N)$ such that

$$DR + R^T D > 0 \qquad (2.16)$$

Thus taking $e_i = d_i / \mu_i$, we see that the right-hand side of (2.14) is negative definite and

$$W(x, z_1, \ldots, z_N) = \sum_{i=1}^{N} \frac{d_i}{\mu_i} W_i(x, z_i) \qquad (2.17)$$

is a Lyapunov function for the boundary-layer system (2.6) (or (2.9)), showing that the equilibrium of (2.6) is asymptotically stable uniformly in x and $\mu_i$. In summary we have made the following assumption on the boundary-layer system.

Assumption 2.3: There exist Lyapunov functions $W(x, z_i)$ for the isolated subsystems (2.11) such that inequalities (2.10), (2.12) and (2.15) are satisfied.

We now assume that V and $W_i$ satisfy "temporal" interconnection conditions.

Assumption 2.4: V and $W_i$, i=1, ..., N, satisfy the inequalities

$$\frac{\partial V}{\partial x} [f(x, z_1, \ldots, z_N) - f(x, h_1(x), \ldots, h_N(x))] \leq \sum_{i=1}^{N} \beta_i \psi \phi_i \qquad (2.18)$$

$$\frac{\partial W_i}{\partial x} f(x, z_1, \ldots, z_N) \leq c_i \psi \phi_i + \sum_{j=1}^{N} \gamma_{ij} \phi_i \phi_j \qquad (2.19)$$

for some nonnegative numbers $\beta_i$, $c_i$ and $\gamma_{ij}$.

Inequalities (2.18) and (2.19) are stright-forward extensions of (1.9) and (1.11) when restricted to the autonomous case. A Lyapunov function candidate for the singularly perturbed system (2.1) is taken as

$$\nu(x,z_1,\ldots,z_N) = d_o\, V(x) + \sum_{i=1}^{N} \frac{d_i \varepsilon_i}{\varepsilon} W_i\,(x,\, z_i)\,, \tag{2.20}$$

for some $d_o > 0$. The derivative of $\nu$ along the trajectory of (2.1) satisfies

$$\dot{\nu} \leq - [\psi \,|\, \phi] \begin{bmatrix} d_o\alpha_o & -u^T \\ -u & \frac{1}{\varepsilon} S \end{bmatrix} \begin{bmatrix} \psi \\ \phi \end{bmatrix} \tag{2.21}$$

where $u^T = (u_1,\ldots,u_N)$, $u_i = \frac{1}{2}(d_o\beta_i + \frac{d_i \varepsilon_i c_i}{\varepsilon})$,

$$S = DR + R^T D - DE\,\Gamma - \Gamma^T E\, D,$$

$E = \text{diag}\ (\varepsilon_1,\ldots,\ \varepsilon_N)$, and $\Gamma = (\frac{1}{2}\gamma_{ij})$.

Since $D\,R + R^T\,D$ is positive-definite, $S$ is positive-definite for sufficiently small $\varepsilon_i$. Moreover, since $\frac{\varepsilon_i}{\varepsilon} \leq 1$, the elements of $u$ are bounded and for any $d_o > 0$, $d_o\gamma_o - \varepsilon\, u^T S^{-1} u > o$ as $\varepsilon_i \to o$. Thus there exists $\varepsilon_i^* > o$ such that whenever $\varepsilon_i < \varepsilon_i^*\ \forall_i$, $\dot{\nu} < o$ and the equilibrium of (2.1) is asymptotically stable. The conclusion is summarized in Theorem 2.

Theorem 2: Suppose that Assumptions 2.1 - 2.4 hold. Then the origin ($x = o$, $z_i = o\ \forall i$) is an asymptotically stable equilibrium of the singularly perturbed system (2.1) for sufficiently small $\varepsilon_i$.

Calculating the bounds $\varepsilon_i^*$ can be done by determining the largest numbers $\varepsilon_i$ for which the matrix on the right-hand side of (2.21) is positive-definite. An easier to calculate, yet more conservative, bound can be obtained as follows. Using that $\varepsilon_i \leq \varepsilon$ we obtain

$$\dot{v} \leq - [\,|\psi| : |\phi_1|, \ldots, |\phi_N|\,] \left[\begin{array}{c|c} d_o\alpha_o & -\tilde{u}^T \\ \hline -\tilde{u} & \frac{1}{\varepsilon} S \end{array}\right] \left[\begin{array}{c} |\psi| \\ \hline |\phi_1| \\ \vdots \\ |\phi_N| \end{array}\right] \qquad (2.22)$$

where $\tilde{u}_i = \frac{1}{2}(d_o\beta_i + d_i c_i)$. Now

$$\lambda_{min}(S) \geq \lambda_{min}(DR + R^T D) - 2\varepsilon \, ||D\Gamma||$$

and

$$d_o\alpha_o - \varepsilon\tilde{u}^T S^{-1} \tilde{u} \geq d_o\alpha_o - \frac{\varepsilon||\tilde{u}||^2}{\lambda_{min}(DR + R^T D) - 2\varepsilon||D\Gamma||}$$

Thus, the right-hand side of (2.22) is negative-definite if

$$\varepsilon < \varepsilon^* = \frac{\alpha_o\lambda_{min}(DR + R^T D)}{2\alpha_o||D\Gamma|| + \frac{1}{4d_o}\sum_{i=1}^{N}(d_o\beta_i + d_i C_i)^2} \qquad (2.23)$$

since $\varepsilon = \max_i \varepsilon_i$, the bound (2.23) is a uniform bound on all $\varepsilon_i$'s.

Remarks:

1. Extension of Theorem 2 to nonautonomous systems is straightforward.

2. For N=1, the bound (2.23) reduces to the bound (1.17) derived in the single parameter case.

3. The analysis is valid for all $\varepsilon_i > o$ including cases like $\varepsilon_i << \varepsilon_j$. However, when the perturbation parameters are of different orders of magnituate, less conservative results can be obtained by treating the problem as nested single parameter perturbations.

4. The analysis does not involve bounds on the ratio between parameters, i.e.,

$$o < m_{ij} \leq \frac{\varepsilon_i}{\varepsilon_j} \leq M_{ij} < \infty \tag{2.24}$$

as in the earlier work of KHALIL and KOKOTOVIC (1979) and KHALIL (1981). This should come as no surprise. The use of the (2.24) bounds in KHALIL and KOKOTOVIC (1979) and KHALIL (1981) was a matter of convenience. It was a way of excluding cases when perturbation parameters are of different orders of magnitude since, as we pointed out in Remark 3, such cases are better treated as nested single parameter perturbations. Actually in KHALIL and KOKOTOVIC (1979) and KHALIL (1981) the numbers $m_{ij}$ and $M_{ij}$ were allowed to take any finite positive values. Recently, ABED (1985) emphasized that bounds like (2.24) are not needed by performing the analysis for the linear case without such bounds.

References

Abed, E.H. (1985), Multiparameter singular perturbation problems: iterative expansions and asymptotic stability, Systems and Control Letters.

Araki, M. (1978), stability of large-scale nonlinear systems-quadratic-order theory of composite-system method using M-matrices, IEEE Trans. on Auto. Control, AC - 23, pp. 129-142.

Chow, J.H. (1978), Asymptotic stability of a class of nonlinear singularly perturbed systems, J. Franklin Inst., 306, pp.275-278.

Desoer, C.A. and M.J. Shensa (1970), Network with very small and very large parasitics: natural frequencies and stability, Proc. IEEE 58, pp. 1933-1938.

Grujic, L.T. (1981), Uniform asymptotic stability of nonlinear singularly perturbed large-scale systems, Int. J. Control, 33, pp 481-504.

Habets, P. (1974), Stabilite asympototique pour des problemes de perturbations singulieres, In C.I.M.E. Stability Problems, Bressanone, Edizioni Cremonese, Rome, Italy, pp. 3-18.

Hoppensteadt, F. (1966), Singular perturbation on the infinite interval, Trans. Amer. Math. Soc., 123, pp. 521-535

Javid, H. (1978), Uniform asymptotic stability of linear time varying singularly perturbed systems, J. Franklin Inst., 305, pp 27-37.

Khalil, H.K. and P.V. Kokotovic (1979), D-stability and multi-parameter singular perturbations, SIAM J. Control and OPT., 17, pp. 56-65.

Khalil, H.K. (1981), Asymptotic stability of a class of non-linear multiparameter singularly perturbed systems, Automatica, 17, pp. 797-804.

Klimushchev, A.I. and N.N. Krasovskii (1961), Uniform asympto-tic stability of systems of differential equations with a small parameter in the derivative terms, PMM 25, pp. 680-690.

Kokotovic, P.V., R.E. O'Malley, Jr., and P. Sannuti (1976), singular perturbations and order reduction in control theory-an overview, Automatica, 12, pp. 123-132.

Kokotovic, P.V., H.K. Khalil and J. O'Reilly (1985), Singular perturbation methods in control, Academic Press

Ladde, G.S. and D.D. Siljak (1983), Multiparameter singular perturbations of linear systems with multiple time scales, Automatica, 19, pp. 385-394.

Michel, A.N. and R. K. Miller (1977), Qualitative Analysis of Large Scale Dynamical Systems, Academic Press.

Saberi, A. (1983), Stability and Control of Nonlinear Singularly Perturbed Systems, with Application to High-Gain Feedback, Ph.D. Dissertation, Michigan State University.

Saberi, A. and H. Khalil (1984), Quadratic-type Lyapunov functions for singularly perturbed systems, IEEE Trans. on Auto. Control, AC-29, pp. 542-550.

Saksena, V.R., J.O'Reilly and P.V. Kokotovic (1984), Singular perturbations and time-scale methods in control theory: survey 1976-1983, Automatica, 20, pp 273-293.

Siljak, D.D. (1978), Large-Scale Dynamic Systems: Stability and Structure, New York: North-Holland.

Wilde, R.R. and P.V. Kokotovic (1972), Stability of singularly perturbed systems and networks with parasitics, IEEE Trans. on Auto. Control, AC-17, pp. 245-246.

Zien, L. (1973), An upper bound for the singular parameter in a stable singularly perturbed system, J. Franklin Inst., 295, pp. 373-381.

# NEW STABILITY THEOREMS FOR AVERAGING AND THEIR APPLICATION TO THE CONVERGENCE ANALYSIS OF ADAPTIVE IDENTIFICATION AND CONTROL SCHEMES

*L.-C. Fu*[†] *M. Bodson*[†] *S. Sastry*[††]

## 1. Introduction

The method of averaging is concerned with differential equations of the form

$$\dot{x} = \epsilon f(t, x) \tag{1.1}$$

and relates the properties of solutions of system (1.1) with solutions of the autonomous "averaged" system

$$\dot{x}_{av} = \epsilon f_{av}(x_{av}) \tag{1.2}$$

$$f_{av}(x) = \lim_{T \to \infty} \frac{1}{T} \int_s^{s+T} f(t, x)\, dt \tag{1.3}$$

for sufficiently small values of the parameter $\epsilon$. The method was proposed originally by Kryloff, Bogoliuboff, and Mitropolskii [1], reformulated by Hale [2,3], developed subsequently by Sethna [4,5] and stated in a geometric form in Arnold [6], Guckenheimer and Holmes [7]. These results constitute a generalization of classical singular perturbation techniques such as those in Hoppensteadt [8] and are an extremely important tool for the state space analysis of systems with multiple time scales. These results have been used extensively in mathematical physics. From our viewpoint, as control theorists, we feel that the technique bears the promise of evolving into a "frequency domain" technique for the *state space* trajectory analysis of some classes of nonlinear systems--- to be distinguished from the Volterra approach for *input/output* functional expansions for nonlinear systems.

The current paper has two sets of contributions:

(A) We develop new theorems for averaging. With the exception of [4,5], all the aforementioned references make the assumption of almost periodicity for the right-hand

[†]Electronics Research Laboratory, Department of Electrical Engineering & Computer Science, University of California, Berkeley CA 94720.

[††]Electronics Research Laboratory, Department of Electrical Engineering & Computer Science, University of California, Berkeley CA 94720. This research was supported in part by NASA under grant NAG 2-243, the IBM Corporation under a Faculty Development Award and the Semiconductor Research Corporation. We would like to thank B. Riedle, P. Kokotovic, K. Astrom, S. Boyd, E. Bai, H. Ping, B. Anderson, I. Mareels, and B. Bitmead for several valuable discussions.

side of (1.2). We relax this assumption in Section 2 of this paper. Our theorems are rather different in hypothesis and rather simpler than those of [4,5] in this regard. Another important contribution of Section 2 is to relate the exponential stability of the averaged system (1.2) to the exponential stability of the unaveraged system (1.1), using a converse theorem of Lyapunov. As such, these theorems are a considerable extension of the local stability theorems of Hale.

In Section 4, we extend all of these results to two-time scale state space systems and the results are generalizations in the sense mentioned above of those of Hale and Sethna.

(B) Our development of these theorems on averaging was heavily motivated by recent literature on the application of averaging techniques to adaptive control--notably the work of Krause et al [9], Astrom [10], Riedle and Kokotovic [11]. Averaging methods have been more prevalent in the stochastic adaptive control literature, eg. Ljung [12] and the first attempts to apply averaging were made heuristically in [9], and increasingly rigorously in [10] and [11].* The primary focus of the efforts in [9-11] is to use averaging to explain instability mechanisms in adaptive control arising from unmodelled dynamics, a phenomenon popularized by Rohrs et al [13]. In this paper, we content ourselves with applying our results on averaging theory along with techniques of generalized harmonic analysis introduced in Boyd and Sastry [14]. We study convergence rates of adaptive identification schemes and linearized adaptive control schemes without unmodelled dynamics and in the presence of persistent excitation. Estimates of convergence rates are of interest in the determination of optimal input signals for identification. In earlier work (Bodson and Sastry [15]), we also showed how persistent excitation guarantees a margin of robustness to unmodelled dynamics and established connections between the rate of convergence of the adaptive schemes and their robustness margins. A more detailed study of instability theorems for averaging and their application to understanding the mechanism of slow drift instability pointed out by Riedle and Kokotovic [16] is an interesting avenue of future work.

In adaptive systems, averaging has usually been associated with *slow adaptation*. Since the parameter $\epsilon$ appears in the right-hand side of the differential equation governing the adaptive parameters, and since averaging is considered as a perturbation technique, it is frequently understood that the results are valid only for $\epsilon$ small (if not infinitesimal), i.e. for slow adaptation. However, simulations of adaptive systems show that averaging often provides a good approximation for relatively large (of order 1) values of the parameter $\epsilon$.

*After this manuscript was written, new and related work of Kosut and Anderson [17] was communicated to us for system (1.1) with $f(t,x)$ linear in $x$, but with weaker conditions in the limit in (1.3).

In practice, the requirement of slow adaptation appears to be only a requirement that parameter adaptation be slower than the time varying inputs and than the time constants of the underlying identified or controlled system. Simulations also show that convergence rates estimates obtained through averaging are tight estimates of the actual convergence rates. Although guaranteed rates of convergence of some adaptive systems can be obtained by techniques other than averaging, such bounds are usually conservative. Averaging also allows to replace the *nonautonomous* ordinary differential equation (ODE) describing the system by an *autonomous* ODE describing the evolution of the parameters. Far more being known for autonomous ODE, the understanding of the convergence and dynamical behavior of the whole parameter vector trajectory in the state space is improved significantly by this method.

The results of Section 3 on the application of averaging theory to obtaining estimates of the convergence rates for adaptive identifiers are to our knowledge new , while those of Section 5 on convergence rates for adaptive control schemes in the relative degree 1 case are a small generalization of the results of [11] with a somewhat different focus.

## 2. Basic Averaging Theory

In this section, we consider differential equations of the form:

$$\dot{x} = \epsilon f(t, x, \epsilon) \qquad x(0)=x_0 \tag{2.1}$$

where $x \in R^n$, $t \geq 0$, $0 < \epsilon \leq \epsilon_0$, and $f$ is piecewise continuous with respect to time. We will concentrate our attention on the behavior of the solutions in some closed ball $B_h$ of radius $h$, centered at the origin.

For small $\epsilon$, the variation of $x$ with time is slow, as compared to the rate of time variation of $f$. Such systems can be conveniently studied using the *method of averaging* (see e.g. [1], [3], [6], [7]). The theory relies on the assumption of the existence of the mean value of $f(t, x, 0)$ defined by the limit:

$$f_{av}(x) = \lim_{T \to \infty} \frac{1}{T} \int_t^{t+T} f(\tau, x, 0)\, d\tau \tag{2.2}$$

assuming that the limit exists uniformly in $t$ and $x$. This is formulated more precisely in the following definition:

**Definition 2.1 Mean Value of a Function, Convergence Function**

The function $f(t, x, 0)$ is said to have mean value $f_{av}(x)$ if there exists a continuous function $\gamma(T)$: $R_+ \to R_+$, strictly decreasing, such that $\gamma(T) \to 0$ as $T \to \infty$, and:

$$\| \frac{1}{T} \int_t^{t+T} f(\tau, x, 0)\, d\tau - f_{av}(x) \| \leq \gamma(T) \tag{2.3}$$

for all $t, T \geq 0$, $x \in B_h$.

The function $\gamma(T)$ will be called the *convergence function*.

Note that the function $f(t, x, 0)$ has mean value $f_{av}(x)$ if and only if the function:

$$d(t, x) = f(t, x, 0) - f_{av}(x) \tag{2.4}$$

has zero mean value.

The following definition ([20], p 7) will also be useful:

**Definition 2.2 Class K Function**

A function $\alpha(\epsilon)$: $R_+ \to R_+$ belongs to class $K$ ($\alpha(\epsilon) \in K$), if it is continuous, strictly increasing, and $\alpha(0)=0$.

It is common, in the literature on averaging, to assume that the function $f(t, x, \epsilon)$ is periodic in $t$, or almost periodic in $t$. Then, the existence of the mean value is guaranteed, without further assumption ([3], theorem 6, p 344). We do not make the assumption of (almost) periodicity, but consider instead the assumption of the existence of the mean value as the starting point of our analysis.

Note that if the function $d(t,x)$ is periodic in $t$, and is bounded, then the integral of the function $d(t,x)$ is also a bounded function of time. This is equivalent to saying that there exists a convergence function $\gamma(T)=a/T$ (i.e. of the order of $1/T$) such that (2.3) is satisfied. On the other hand, if the function $d(t,x)$ is bounded, but is only required to be almost periodic, then the integral of the function $d(t,x)$ need not be a bounded function of time, even if its mean value is zero ([3], p 346). The function $\gamma(T)$ is bounded (by the same bound as $d(t,x)$), and converges to zero as $T\to\infty$, but the convergence function need not be bounded by $a/T$ as $T\to\infty$ (it may be of order $1/\sqrt{T}$ for example). In general, a zero mean function need not have a bounded integral, although the converse is true. In this paper, we do not make the distinction between the periodic, and the almost periodic case, but we do distinguish the bounded integral case from the general case, and indicate the importance of the function $\gamma(T)$ in the subsequent development.

Assuming the existence of the mean value for the original system (2.1), the *averaged system* is defined to be:

$$\dot{x}_{av}=\epsilon f_{av}(x_{av}) \qquad x_{av}(0)=x_0 \tag{2.5}$$

Note that the averaged system is autonomous and, for $T$ fixed and $\epsilon$ varying, the solutions over intervals $[0,T/\epsilon]$ are identical, modulo a simple time scaling by $\epsilon$.

We address the following two questions:

(i) the closeness of the response of the original and averaged systems,

(ii) the relationships between the stability properties of the two systems.

To compare the solutions of the original and of the averaged system, it is convenient to transform the original system in such a way that it becomes a *perturbed* version of the averaged system. An important lemma that leads to this result is attributed to Bogoliuboff and Mitropolskii ([1], p 450, and [3], lemma 4, p 346). We state a generalized version of this lemma.

**Lemma 2.1 Approximate Integral of a Zero Mean Function**

**If:** $d(t,x): R_+\times B_h\to R^n$ is a bounded function, piecewise continuous with respect to $t$, and has zero mean value with convergence function $\gamma(T)$.

**Then:** There exists $\xi(\epsilon)\in K$, and a function $w_\epsilon(t,x): R_+\times B_h\to R^n$ such that:

$$\|\epsilon w_\epsilon(t,x)\|\leqslant\xi(\epsilon) \tag{2.6}$$

$$\left\| \frac{\partial w_\epsilon(t,x)}{\partial t} - d(t,x) \right\| \leqslant \xi(\epsilon) \tag{2.7}$$

for all $t \geqslant 0$, $x \in B_h$. Moreover, $w_\epsilon(0,x)=0$, for all $x \in B_h$.

**If, moreover:** $\gamma(T)=a/T^r$ for some $a \geqslant 0$, $r \in (0,1]$,

**Then:** The function $\xi(\epsilon)$ can be chosen to be $2a\epsilon^r$.

The proof of *Lemma 2.1* is provided in the appendix. The construction of the function $w_\epsilon(t,x)$ is identical to that in [2.1], but the proof of (2.6), (2.7) is different, and leads to the relationship between the convergence function $\gamma(T)$ and the function $\xi(\epsilon)$.

The main point of *Lemma 2.1* is that, although the exact integral of $d(t,x)$ may be an unbounded function of time, there exists a bounded function $w_\epsilon(t,x)$, whose first partial derivative with respect to $t$ is arbitrarily close to $d(t,x)$. Although the bound on $w_\epsilon(t,x)$ may increase as $\epsilon \to 0$, it increases slower than $1/\epsilon$, as indicated by (2.6).

It is necessary to obtain a function $w_\epsilon(t,x)$, as in *Lemma 2.1*, that has some additional smoothness properties. A useful lemma is given by Hale in [3] (lemma 5, p 349). For the price of additional assumptions on the function $d(t,x)$, the following lemma leads to stronger conclusions that are useful in the sequel.

**Lemma 2.2 Smooth Approximate Integral of a Zero Mean Function**

**If:** $d(t,x)$: $R_+ \times B_h \to R^n$ is piecewise continuous with respect to t, has bounded and continuous first partial derivatives with respect to x, and $d(t,0)=0$ for all $t \geqslant 0$. Moreover, $d(t,x)$ has zero mean value, with convergence function $\gamma(T)\|x\|$, and $\frac{\partial d(t,x)}{\partial x}$ has zero mean value, with convergence function $\gamma(T)$.

**Then:** There exists $\xi(\epsilon) \in K$, and a function $w_\epsilon(t,x)$: $R_+ \times B_h \to R^n$, such that:

$$\| \epsilon w_\epsilon(t,x) \| \leqslant \xi(\epsilon) \|x\| \tag{2.8}$$

$$\left\| \frac{\partial w_\epsilon(t,x)}{\partial t} - d(t,x) \right\| \leqslant \xi(\epsilon) \|x\| \tag{2.9}$$

$$\left\| \epsilon \frac{\partial w_\epsilon(t,x)}{\partial x} \right\| \leqslant \xi(\epsilon) \tag{2.10}$$

for all $t \geqslant 0$, $x \in B_h$. Moreover, $w_\epsilon(0,x)=0$, for all $x \in B_h$.

**If, moreover:** $\gamma(T)=a/T^r$ for some $a \geqslant 0$, $r \in (0,1]$,

**Then:** the function $\xi(\epsilon)$ can be chosen to be $2a\epsilon^r$.

The proof of *Lemma 2.2* is provided in the appendix. The difference from *Lemma 2.1* is in the condition on the partial derivative of $w_\epsilon(t,x)$ with respect to $x$ in (2.10), and the dependence on $\| x \|$ in (2.8), (2.9). These results will be necessary to derive the following theorems.

Note that if the original system is linear, i.e.:

$$\dot{x} = A(t)x \qquad x(0)=x_0 \tag{2.11}$$

for some $A(t): R_+ \to R^{n \times n}$, then the main assumption of *Lemma 2.2* is that there exists $A_{av}$ such that $A(t)-A_{av}$ has zero mean value.

The following assumptions will hence forth be in effect.

(A1) $x=0$ is an equilibrium point of system (2.1), i.e. $f(t,0,0)=0$ for all $t \geqslant 0$. $f(t,x,\epsilon)$ is Lipschitz in $x$, i.e.:

$$\| f(t,x_1,\epsilon)-f(t,x_2,\epsilon)\| \leqslant l_1 \| x_1-x_2\| \tag{2.12}$$

for all $t \geqslant 0$, $x_1,x_2 \in B_h$, $\epsilon \leqslant \epsilon_0$.

(A2) $f(t,x,\epsilon)$ is Lipschitz in $\epsilon$, linearly in $x$, i.e.:

$$\| f(t,x,\epsilon_1)-f(t,x,\epsilon_2)\| \leqslant l_2 \| x\| \, |\epsilon_1-\epsilon_2| \tag{2.13}$$

for all $t \geqslant 0$, $x \in B_h$, $\epsilon_1,\epsilon_2 \leqslant \epsilon_0$.

(A3) $f_{av}(0)=0$, and $f_{av}(x)$ is Lipschitz in $x$, i.e.:

$$\| f_{av}(x_1)-f_{av}(x_2)\| \leqslant l_{av} \| x_1-x_2\| \tag{2.14}$$

for all $x_1$, $x_2 \in B_h$.

(A4) the function $d(t,x)=f(t,x,0)-f_{av}(x)$ satisfies the conditions of *Lemma 2.2*.

**Lemma 2.3 Perturbation Formulation of Averaging**

**If:** The systems (2.1), and (2.5) satisfy assumptions (A1)-(A4).

**Then:** There exist functions $w_\epsilon(t,x)$ and $\xi(\epsilon)$, as in *Lemma 2.2*, and a transformation of the form:

$$x = z + \epsilon w_\epsilon(t,z) \tag{2.15}$$

under which system (2.1) becomes:

$$\dot{z} = \epsilon f_{av}(z) + \epsilon p(t,z,\epsilon) \qquad z(0)=x_0 \tag{2.16}$$

where $p(t,z,\epsilon)$ satisfies:

$$\| p(t,z,\epsilon)\| \leqslant \psi(\epsilon) \| z\| \tag{2.17}$$

for some $\psi(\epsilon) \in K$, $\epsilon_1>0$, and for all $\epsilon \leqslant \epsilon_1$. Further, $\psi(\epsilon)$ is of the order of $\epsilon+\xi(\epsilon)$.

**Comments**

The proof of *Lemma 2.3* is provided in the appendix. A similar lemma can be found in [3] (lemma 3.2, p 192). Inequality (2.17) is a Lipschitz type of condition on $p(t,z,\epsilon)$, which is not found in [3], and results from the stronger conclusions of *Lemma 2.2*.

*Lemma 2.3* is fundamental to the theory of averaging presented hereafter. It separates the error in the approximation of the original system by the averaged system $(x-x_{av})$ into two components: $x-z$ and $z-x_{av}$. The first component results from a pointwise (in time) transformation of variable. This component is guaranteed to be small by inequality (2.8). For $\epsilon$ sufficiently small $(\epsilon \leqslant \epsilon_1)$, the transformation $z \rightarrow x$ is invertible, and as $\epsilon \rightarrow 0$, it tends to the identity transformation. The second component is due to the perturbation term $p(t,z,\epsilon)$. Inequality (2.17) guarantees that this perturbation is small as $\epsilon \rightarrow 0$.

At this point, we can relate the convergence of the function $\gamma(T)$ to the order of the two components of the error $x-x_{av}$ in the approximation of the original system by the averaged system. The relationship between the functions $\gamma(T)$ and $\xi(\epsilon)$ was indicated in *Lemma 2.1*. *Lemma 2.3* relates the function $\xi(\epsilon)$ to the error due to the averaging. If $d(t,x)$ has a bounded integral (i.e. $\gamma(T) \sim 1/T$), then both $x-z$ and $p(t,z,\epsilon)$ are of the order of $\epsilon$ with respect to the main term $f_{av}(z)$. In general, these terms go to zero as $\epsilon \rightarrow 0$, but possibly more slowly than linearly ( as $\sqrt{\epsilon}$ for example). The proof of *Lemma 2.1* provides a direct relationship between the order of the convergence to the mean value, and the order of the error terms.

We now focus attention on the approximation of the original system by the averaged system. Consider first the following assumption:

(A5) $x_0$ is sufficiently small that, for fixed $T$, and some $h'<h$, $x_{av}(t) \in B_{h'}$ for all $t \in [0, T/\epsilon]$ (this is possible, from (A3)).

**Theorem 2.4 Basic Averaging Theorem**

**If:** The original system (2.1), and the averaged system (2.5) satisfy assumptions (A1)-(A5),

**Then:** There exists $\psi(\epsilon)$ as in *Lemma 2.3* such that, given $T \geqslant 0$:

$$\| x(t) - x_{av}(t) \| \leqslant \psi(\epsilon)\, b_T \tag{2.18}$$

for some $b_T$, $\epsilon_T > 0$, and for all $t \in [0, T/\epsilon]$, $\epsilon \leqslant \epsilon_T$.

**Proof:** From *Lemma 2.2* and *Lemma 2.3*, we have that:

$$\| x - z \| \leqslant \xi(\epsilon) \| z \| \leqslant \psi(\epsilon) \| z \| \tag{2.19}$$

for $\epsilon \leqslant \epsilon_1$. On the other hand, we have that:

$$\frac{d}{dt}(z - x_{av}) = \epsilon\,(f_{av}(z) - f_{av}(x_{av})) + \epsilon\, p(t, z, \epsilon) \qquad z(0) - x_{av}(0) = 0 \tag{2.20}$$

for all $t \in [0, T/\epsilon]$, $x_{av} \in B_{h'}$, $h' < h$. We will now show that, on this time interval, and for as long as $x, z \in B_h$, the errors $(z - x_{av})$ and $(x - x_{av})$ can be made arbitrarily small by reducing $\epsilon$.

Integrating (2.20):

$$\| z(t) - x_{av}(t) \| \leqslant \epsilon l_{av} \int_0^t \| z(\tau) - x_{av}(\tau) \| \, d\tau + \epsilon\, \psi(\epsilon) \int_0^t \| z(\tau) \| \, d\tau \tag{2.21}$$

Using the *Generalized Bellman-Gronwall Lemma* (see appendix):

$$\| z(t) - x_{av}(t) \| \leqslant \epsilon \psi(\epsilon) \int_0^t \| z(\tau) \| \, e^{\epsilon l_{av}(t-\tau)} d\tau \leqslant \psi(\epsilon)\, h \left( \frac{e^{\epsilon l_{av} T} - 1}{l_{av}} \right)$$

$$:= \psi(\epsilon)\, a_T \tag{2.22}$$

Combining these results:

$$\begin{aligned} \| x(t) - x_{av}(t) \| &\leqslant \| x(t) - z(t) \| + \| z(t) - x_{av}(t) \| \\ &\leqslant \psi(\epsilon) \| x_{av}(t) \| + (1 + \psi(\epsilon)) \| z(t) - x_{av}(t) \| \\ &\leqslant \psi(\epsilon)\,(h + (1 + \psi(\epsilon_1))\, a_T) \\ &:= \psi(\epsilon) b_T \end{aligned} \tag{2.23}$$

By assumption, $\| x_{av}(t) \| \leqslant h' < h$. Let $\epsilon_T$ (with $0 < \epsilon_T \leqslant \epsilon_1$) such that $\psi(\epsilon_T)\, b_T < h - h'$. It follows, from a simple contradiction argument, that the estimate in (2.23) is valid for all $t \in [0, T/\epsilon]$, whenever $\epsilon \leqslant \epsilon_T$.

### Comments

*Theorem 2.4* establishes that the trajectories of the original and the averaged system are arbitrarily close on intervals $[0, T/\epsilon]$, as $\epsilon$ is reduced. The error is of the order of $\psi(\epsilon)$, and the order is related to the order of convergence of $\gamma(T)$. If $d(t, x)$ has a bounded integral (i.e. $\gamma(T) \sim 1/T$), then the error is of the order of $\epsilon$.

It is important to remember that, although the intervals $[0, T/\epsilon]$ are unbounded, *Theorem 2.4* does not state that:

$$\| x(t) - x_{av}(t) \| \leqslant \psi(\epsilon) b \tag{2.24}$$

for all $t \geqslant 0$, and some $b$. Consequently, *Theorem 2.4* does not allow us to relate the stability of the original and of the averaged system. This relationship is investigated in *Theorem 2.5*, after a preliminary definition.

**Definition 2.3 Exponential Stability, Rate of Convergence**

The equilibrium point $x = 0$ of a differential equation is said to be *exponentially stable*, with *rate of convergence* $\alpha$ ($\alpha > 0$), if:

$$\| x(t) \| \leqslant m \| x(t_0) \| e^{-\alpha(t - t_0)} \tag{2.25}$$

for all $t \geqslant t_0 \geqslant 0$, $x(t_0) \in B_{h_0}$, and some $m \geqslant 1$.

We assume that $h_0 \leqslant h / m$, so that all trajectories are guaranteed to remain in $B_h$.

**Theorem 2.5 Exponential Stability Theorem**

**If:** The original and averaged systems satisfy assumptions (A1)-(A5), the function $f_{av}(x)$ has continuous and bounded first order partial derivatives in $x$, and $x = 0$ is an exponentially stable equilibrium point of the averaged system,

**Then:** There exists $\epsilon_2 > 0$ such that the equilibrium point $x = 0$ of the original system is exponentially stable for all $\epsilon \leqslant \epsilon_2$.

**Proof:** The proof relies on a converse theorem of Lyapunov for exponentially stable systems (see for example [20], p 273). Under the hypotheses, there exists a function $v(x_{av}): R^n \to R_+$, and strictly positive constants $\alpha_1, \alpha_2, \alpha_3, \alpha_4$, such that, for all $x_{av} \in B_{h_0}$:

$$\alpha_1 \| x_{av} \|^2 \leqslant v(x_{av}) \leqslant \alpha_2 \| x_{av} \|^2 \tag{2.26}$$

$$\dot{v}(x_{av})|_{(2.5)} \leqslant -\epsilon\, \alpha_3 \| x_{av} \|^2 \tag{2.27}$$

$$\left\| \frac{\partial v}{\partial x_{av}} \right\| \leqslant \alpha_4 \| x_{av} \| \tag{2.28}$$

The derivative in (2.27) is to be taken along the trajectories of the averaged system (2.5). The function $v$ is now used to study the stability of the perturbed system (2.16). Considering $v(z)$, inequalities (2.26) and (2.28) are still verified, with $z$ replacing $x_{av}$. The derivative of $v(z)$ along the trajectories of (2.16) is given by:

$$\dot{v}(z)|_{(2.16)} = \dot{v}(z)|_{(2.5)} + \left( \frac{\partial v}{\partial z} \right) (\epsilon\, p(t, z, \epsilon)) \tag{2.29}$$

and, using previous inequalities (including those from *Lemma 2.3* ):

$$\dot{v}(z)|_{(2.16)} \leqslant -\epsilon\alpha_3\|z\|^2 + \epsilon\alpha_4\psi(\epsilon)\|z\|^2$$

$$\leqslant -\epsilon\left(\frac{\alpha_3-\psi(\epsilon)\alpha_4}{\alpha_2}\right)v(z) \tag{2.30}$$

for all $\epsilon \leqslant \epsilon_1$. Let $\epsilon'_2$ be such that $\alpha_3-\psi(\epsilon'_2)\alpha_4>0$, and define $\epsilon_2=\min(\epsilon_1,\epsilon'_2)$. Denote:

$$\alpha(\epsilon) := \frac{\alpha_3 - \psi(\epsilon)\alpha_4}{2\alpha_2} \tag{2.31}$$

Consequently, (2.30) implies that:

$$v(z) \leqslant v(z(t_0))\, e^{-2\epsilon\alpha(\epsilon)(t-t_0)} \tag{2.32}$$

and:

$$\|z(t)\| \leqslant \sqrt{\frac{\alpha_2}{\alpha_1}}\, \|z(t_0)\|\, e^{-\epsilon\alpha(\epsilon)(t-t_0)} \tag{2.33}$$

Since $\alpha(\epsilon)>0$ for all $\epsilon \leqslant \epsilon_2$, system (2.15) is exponentially stable. Using (L16), it follows that:

$$\|x(t)\| \leqslant \frac{1+\xi(\epsilon)}{1-\xi(\epsilon)} \sqrt{\frac{\alpha_2}{\alpha_1}}\, \|x(t_0)\|\, e^{-\epsilon\alpha(\epsilon)(t-t_0)} \tag{2.34}$$

for all $t \geqslant t_0 \geqslant 0$, $\epsilon \leqslant \epsilon_2$, and $x(t_0)$ sufficiently small that all signals remain in $B_h$. In conclusion, the original system is exponentially stable, with rate of convergence (at least) $\epsilon\alpha(\epsilon)$.

**Comments**

1) *Theorem 2.5* is a *local* exponential stability result. The original system will be *globally* exponentially stable, if the averaged system is globally exponentially stable, and provided that *all* assumptions are valid globally.

2) The proof of *Theorem 2.5* gives a useful bound on the rate of convergence of the original system. As $\epsilon$ tends to zero, $\epsilon\alpha(\epsilon)$ tends to $\frac{\epsilon}{2}\frac{\alpha_3}{\alpha_2}$, which is the bound on the rate of convergence of the averaged system that one would obtain using (2.26)-(2.27). In other words, the proof provides a bound on the rate of convergence, and this bound gets arbitrarily close to the corresponding bound for the averaged system, provided that $\epsilon$ is sufficiently small. This is a useful conclusion because it is in general very difficult to obtain a guaranteed rate of convergence for the original, nonautonomous system. The proof assumes the existence of a Lyapunov function satisfying (2.26)-(2.28), but does not depend on the specific function chosen. Since the averaged system is autonomous, such a function is usually easier to find than for the original system,

and any such function will provide a bound on the rate of convergence of the original system for $\epsilon$ sufficiently small.

3) The conclusion of *Theorem 2.5* is quite different from the conclusion of *Theorem 2.4*. Since both $x$ and $x_{av}$ go to zero exponentially with $t$, the error $x - x_{av}$ also goes to zero exponentially with $t$. Yet, *Theorem 2.5* does not relate the bound on the error to $\epsilon$. It is possible, however, to combine *Theorem 2.4* and *Theorem 2.5* to obtain a uniform approximation result, with an estimate similar to (2.24).

## 3. Averaging Theory Applied to Adaptive Identifiers

### *Brief Review of a Simple Adaptive Identifier*

We consider an unknown plant, described by a single-input single-output, exponentially stable transfer function:

$$\hat{p}(s)=\frac{\hat{n}_p(s)}{\hat{d}_p(s)} \tag{3.1}$$

where $\hat{d}_p(s)$ is a monic polynomial of degree n (n is assumed to be known), and $\hat{n}_p(s)$ is a polynomial of degree less than or equal to $n$. The coefficients of the polynomials are unknown, and are to be obtained from the identifier.

The identifier considered here is an adaptive observer/identifier (see e.g. [21], [22]), and its structure is shown in Fig. 3.1. The filter blocks $F_1$ and $F_2$ generate smoothed derivatives of the input $r$, and of the output $y_p$ of the plant. Each of these blocks has a transfer function:

$$(sI-\Lambda)^{-1}b=\frac{1}{\det(sI-\Lambda)}\begin{bmatrix}1\\ s\\ \cdot\\ \cdot\\ s^{n-1}\end{bmatrix}\in R^n(s) \tag{3.2}$$

where $\Lambda \in R^{n \times n}$, $b \in R^n$, and $\det(sI-\Lambda)$ is a Hurwitz polynomial. The outputs of the filters are respectively $v^{(1)}, v^{(2)} \in R^n$. The signal $y_0$ is obtained through the adaptive gains $c(t), d(t) \in R^n$, and $c_{n+1}(t) \in R$:

$$y_0=c^T v^{(1)}+d^T v^{(2)}+c_{n+1}r \tag{3.3}$$

and it may be verified that there exists a unique choice of the adaptive gains, denoted $c^*$, $d^*$, and $c_{n+1}^*$ such that the transfer function from the input $r$ to the output $y_0$ is identical to the plant transfer function $\hat{p}(s)$. We define the parameter vector $\theta \in R^{2n+1}$:

$$\theta^T=(c^T, d^T, c_{n+1}) \tag{3.4}$$

and the signal vector $w \in R^{2n+1}$:

$$w^T=(v^{(1)T}, v^{(2)T}, r) \tag{3.5}$$

so that:

$$y_0=\theta^T w \tag{3.6}$$

In the sequel, we will neglect the effect of the initial conditions of the plant, and of the filter blocks $F_1$, $F_2$. The results can be modified to take them into account, without any fundamental differences in the conclusions. We simply assume that the dynamics of

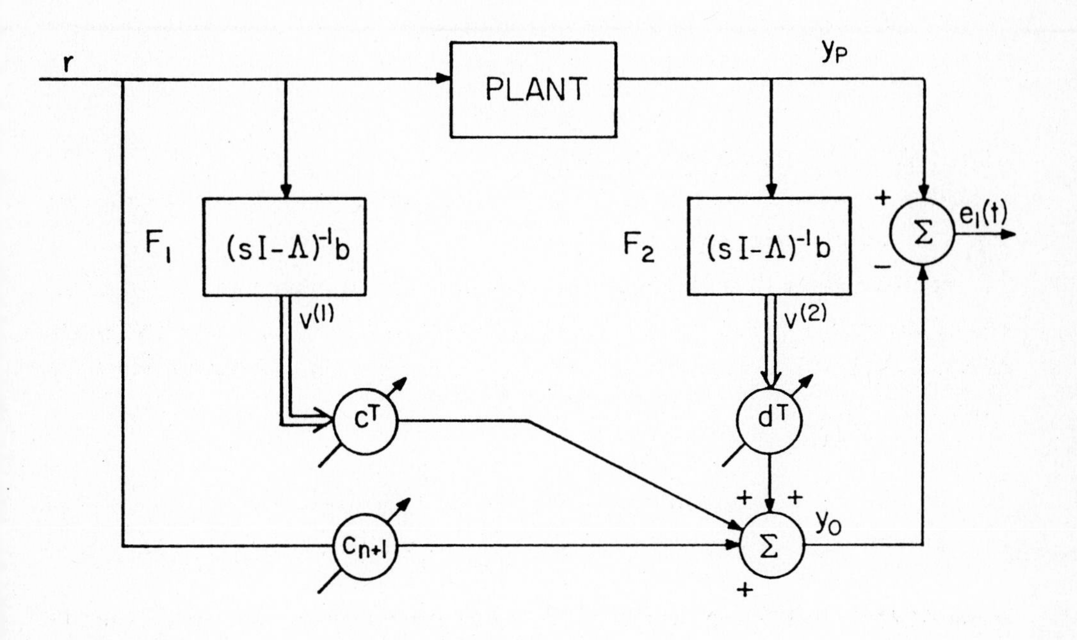


Fig 3.1 Block diagram of adaptive identifier.

the observers (determined by the eigenvalues of $\Lambda$) are faster than those of the identifier. The output of the plant is then given by an equation similar to that of the identifier:

$$y_p = \theta^{*T} w \tag{3.7}$$

where $\theta^*$ is the vector of "true" parameters corresponding to $\hat{p}(s)$. Defining the parameter error:

$$\phi = \theta - \theta^* \tag{3.8}$$

the output error $e_1 = y_p - y_0$ is given by:

$$e_1 = \phi^T w \tag{3.9}$$

It can be shown ([21], [22]) that, with the following adaptation law:

$$\dot{\phi} = -\Gamma e_1 w \tag{3.10}$$

where $\Gamma \in R^{n \times n} > 0$, the following propositions are true:

(i) if $r$ is bounded, then $\lim_{t \to \infty} e_1(t) = 0$

(ii) if, moreover, $w$ is *persistently exciting*, that is, if there exist constants $\alpha_1, \alpha_2, \delta > 0$, such that:

$$\alpha_1 I \leqslant \int_s^{s+\delta} w\, w^T\, dt \leqslant \alpha_2 I \quad \text{for all } s > 0 \tag{3.11}$$

then the parameter error also tends to zero, i.e.:

$$\lim_{t \to \infty} \phi(t) = 0 \tag{3.12}$$

and the convergence is exponential.

***Application of the Averaging Theory***

To apply the averaging theory developed in section 2, we will study the case when $\Gamma = \epsilon I$, i.e. when the update law (3.10) is given by:

$$\dot{\phi} = -\epsilon e_1 w \tag{3.13}$$

or, using (3.9):

$$\dot{\phi} = -\epsilon\, w\, w^T \phi \tag{3.14}$$

Eq. (3.14) leads us to the following definition:

**Definition 3.1 (Stationarity, Autocovariance):** A signal $u : R_+ \to R^n$ is said to be *stationary* if the following limit exists uniformly in $s$:

$$R_u(\tau) := \lim_{T \to \infty} \frac{1}{T} \int_s^{s+T} u(t) u^T(t+\tau)\, d\tau \quad \in R^{n \times n} \tag{3.15}$$

in which instance, the limit $R_u(\tau)$ is called the *autocovariance* of $u$.

It may be verified that the autocovariance matrix of a stationary signal $w$ is a positive semidefinite function $R_w(\tau)$, and that $w$ is persistently exciting if and only if the autocovariance at 0 is positive definite [14]. Also, $R_w(\tau)$ can be written as the inverse Fourier transform of a positive *spectral measure* $S_w(d\nu)$:

$$R_w(\tau) = \int_{-\infty}^{\infty} e^{i\nu\tau} S_w(d\nu) \tag{3.16}$$

Further, if the input $r$ is also stationary, $S_w(d\nu)$ can be computed, using the fact that the transfer function from $r$ to $w$ is given by:

$$q(s) := \begin{pmatrix} (sI-\Lambda)^{-1}b \\ (sI-\Lambda)^{-1}b\ p(s) \\ 1 \end{pmatrix} \in R^{2n+1}(s) \tag{3.17}$$

so that:

$$S_w(d\nu) = \hat{q}(j\nu)\hat{q}^*(j\nu)\, s_r(d\nu) \tag{3.18}$$

Using eqns. (3.16) and (3.17), we can conclude that:

$$R_w(0) = \int_{-\infty}^{\infty} \hat{q}(j\nu)\hat{q}^*(j\nu)\, s_r(d\nu) > 0 \tag{3.19}$$

This in turn is assured [14] if the support of $s_r(d\nu)$ is greater than or equal to $2n+1$ points (the dimension of $w$ = the number of unknown parameters = $2n+1$).

With these definitions, the averaged system corresponding to (3.14) is simply:

$$\dot{\phi}_{av} = -\epsilon R_w(0)\phi_{av} \tag{3.20}$$

This system is particularly easy to study, since it is linear, and when $w$ is persistently exciting, $R_w(0)$ is a positive definite matrix.

A natural Lyapunov function for (3.14) is:

$$V(\phi_{av}) = \frac{1}{2}\|\phi_{av}\|^2 \tag{3.21}$$

and:

$$-\epsilon\lambda_{min}(R_w(0))\|\phi_{av}\|^2 \leqslant -\dot{V}(\phi_{av}) \leqslant -\epsilon\lambda_{max}(R_w(0))\|\phi_{av}\|^2 \tag{3.22}$$

where $\lambda_{min}$ and $\lambda_{max}$ are respectively the minimum and maximum eigenvalues of $R_w(0)$. Thus, the rate of exponential convergence of the averaged system is at least $\epsilon\lambda_{min}(R_w(0))$, and at most $\epsilon\lambda_{max}(R_w(0))$. By the comments after theorem 2.5, we can conclude that the rate of convergence of the unaveraged system for $\epsilon$ small enough is close to the interval $[\epsilon\lambda_{min}(R_w(0)), \epsilon\lambda_{max}(R_w(0))]$.

Eq. (3.19) gives an interpretation of $R_w(0)$ in the frequency domain, and also a mean of computing an estimate of the rate of convergence of the adaptive algorithm, given the spectral content of the reference input. If the input $r$ is periodic or almost periodic, the integral in (3.19) may be replaced by a summation. Since the transfer function $q(s)$ depends on the unknown plant being identified, the use of the averaged equation to determine the rate of convergence is more conceptual than practical. It would be interesting to determine the spectral content of the reference input that will optimize the rate of convergence of the identifier, given the physical constraints on $r$. Such a problem is very reminiscent of the procedure indicated in [23] (chapter 6) for the design of input signals in identification. The autocovariance matrix defined here is similar to the *average information matrix* defined in [23] (p 134). Our interpretation is, however, in terms of rates of parameter convergence of the averaged system rather than in terms of parameter covariance.

To illustrate the conclusions of this section, we consider the following example:

$$\hat{p}(s) = \frac{2s+2}{s+3} \tag{3.23}$$

The filter is chosen to be $\det(sI-\Lambda) = (s+5)$. The "true" values of the parameters $c_1$, $d_1$, $c_2$ are -1.6, 0.4, and 2. Denote the parameter error as

$$\phi_1 = c_1 - c_1^* \ , \quad \phi_2 = d_1 - d_1^* \ , \quad \phi_3 = c_2 - c_2^*$$

Since the number of unknown parameters is 3, parameter convergence will occur when the support of $s_r(d\nu)$ is greater than or equal to 3 points. For the simulations, we considered an input of the form $a_0 + a_1 \sin(\omega t)$. By virtue of (3.18) and (3.19), (3.20) now becomes

$$\begin{bmatrix} \dot{\phi}_{av\,1} \\ \dot{\phi}_{av\,2} \\ \dot{\phi}_{av\,3} \end{bmatrix} = -\epsilon \begin{bmatrix} a_0^2 + \frac{25a_1^2}{2(25+\omega^2)} & \frac{2}{3}a_0^2 + \frac{25(3+\omega^2)a_1^2}{(9+\omega^2)(25+\omega^2)} & a_0^2 + \frac{25a_1^2}{2(25+\omega^2)} \\ \frac{2}{3}a_0^2 + \frac{25(3+\omega^2)a_1^2}{(9+\omega^2)(25+\omega^2)} & \frac{4}{9}a_0^2 + \frac{50(1+\omega^2)a_1^2}{(9+\omega^2)(25+\omega^2)} & \frac{2}{3}a_0^2 + \frac{5(15+7\omega^2)a_1^2}{(9+\omega^2)(25+\omega^2)} \\ a_0^2 + \frac{25a_1^2}{2(25+\omega^2)} & \frac{2}{3}a_0^2 + \frac{5(15+7\omega^2)a_1^2}{(9+\omega^2)(25+\omega^2)} & a_0^2 + \frac{a_1^2}{2} \end{bmatrix} \cdot \begin{bmatrix} \phi_{av\,1} \\ \phi_{av\,2} \\ \phi_{av\,3} \end{bmatrix} \tag{3.24}$$

With $a_0 = 2$, $a_1 = 2$ and $\omega = 4$, the eigenvalues of the averaged system (3.24) are computed to be $-0.28\epsilon$, $-0.64\epsilon$ and $-15.39\epsilon$. Figs 3.2 and 3.3 show the plots of the parameter errors of $c_1$ and $d_1$ for both the original and averaged systems with three different adaptation gains $\epsilon$=0.1, 0.5, 1. Fig 3.4 is a plot of the Lyapunov function of (3.21) for both systems using a log scale. It illustrates the closeness of the rate of convergence of the two systems.

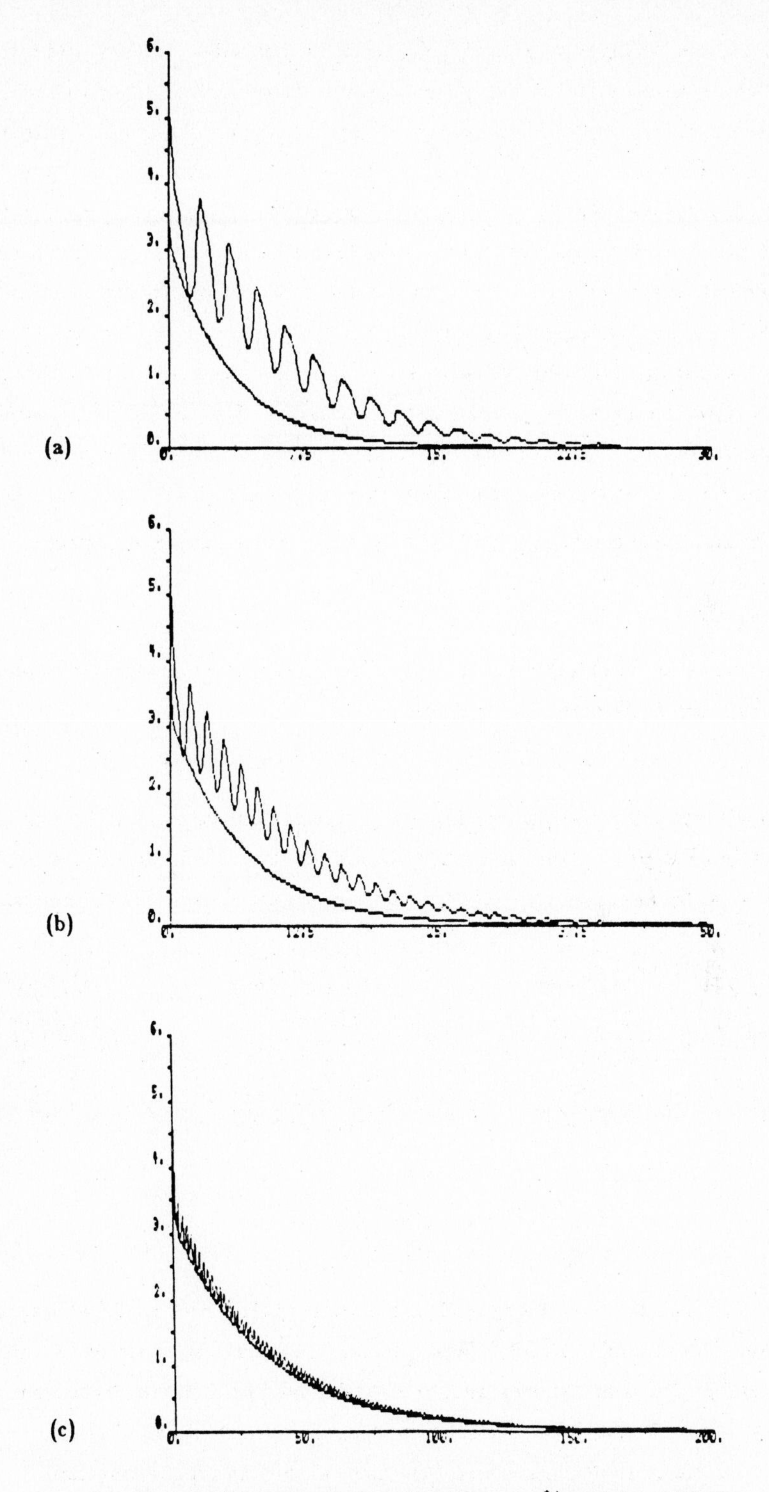


Fig 3.2 Trajectories of parameter error $\phi_1 (= c_1 - c_1^*)$ and $\phi_{av1}$ with three different adaptation gains (a) $\epsilon=1$ (b) $\epsilon=0.5$ (c) $\epsilon=0.1$

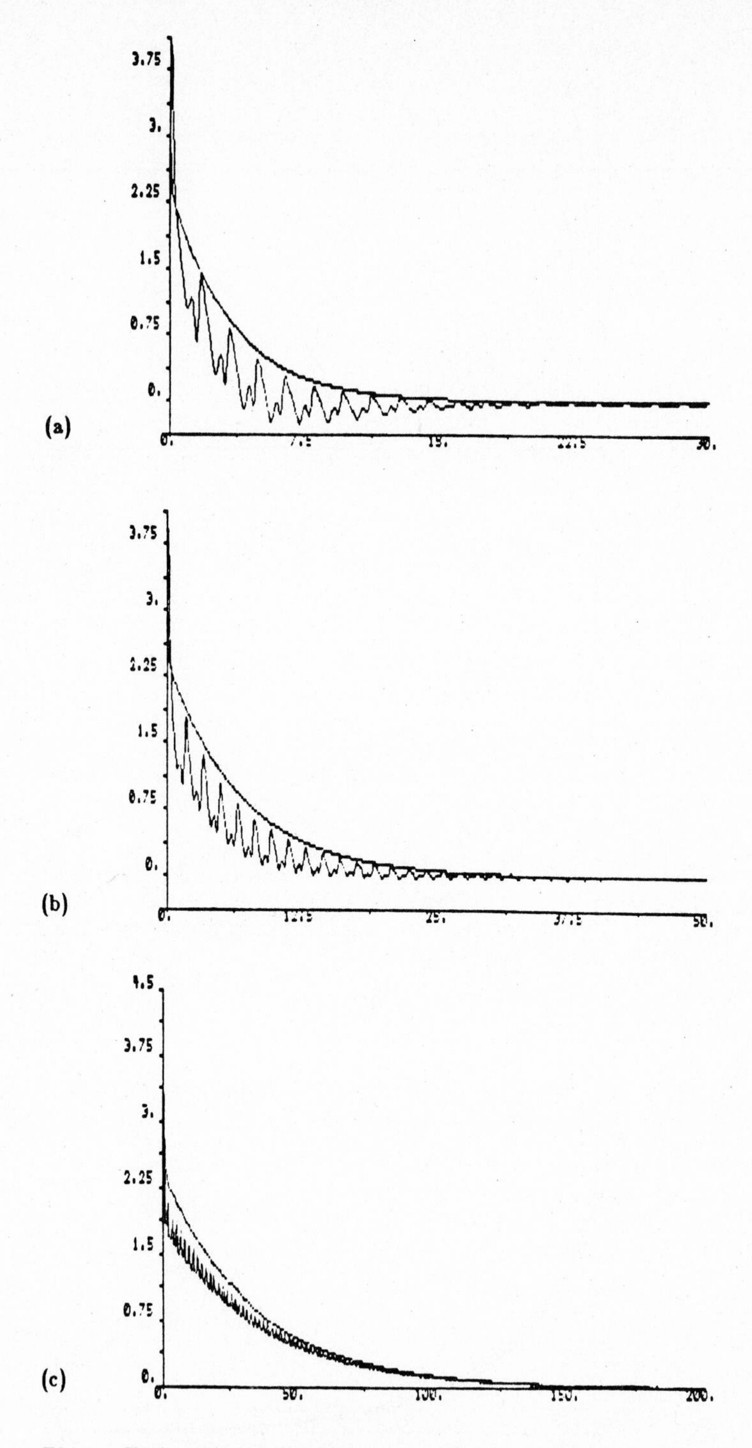


Fig 3.3 Trajectories of parameter error $\phi_2(= d_1 - d_1^*)$ and $\phi_{av2}$ with three different adaptation gains (a) $\epsilon=1$ (b) $\epsilon=0.5$ (c) $\epsilon=0.1$

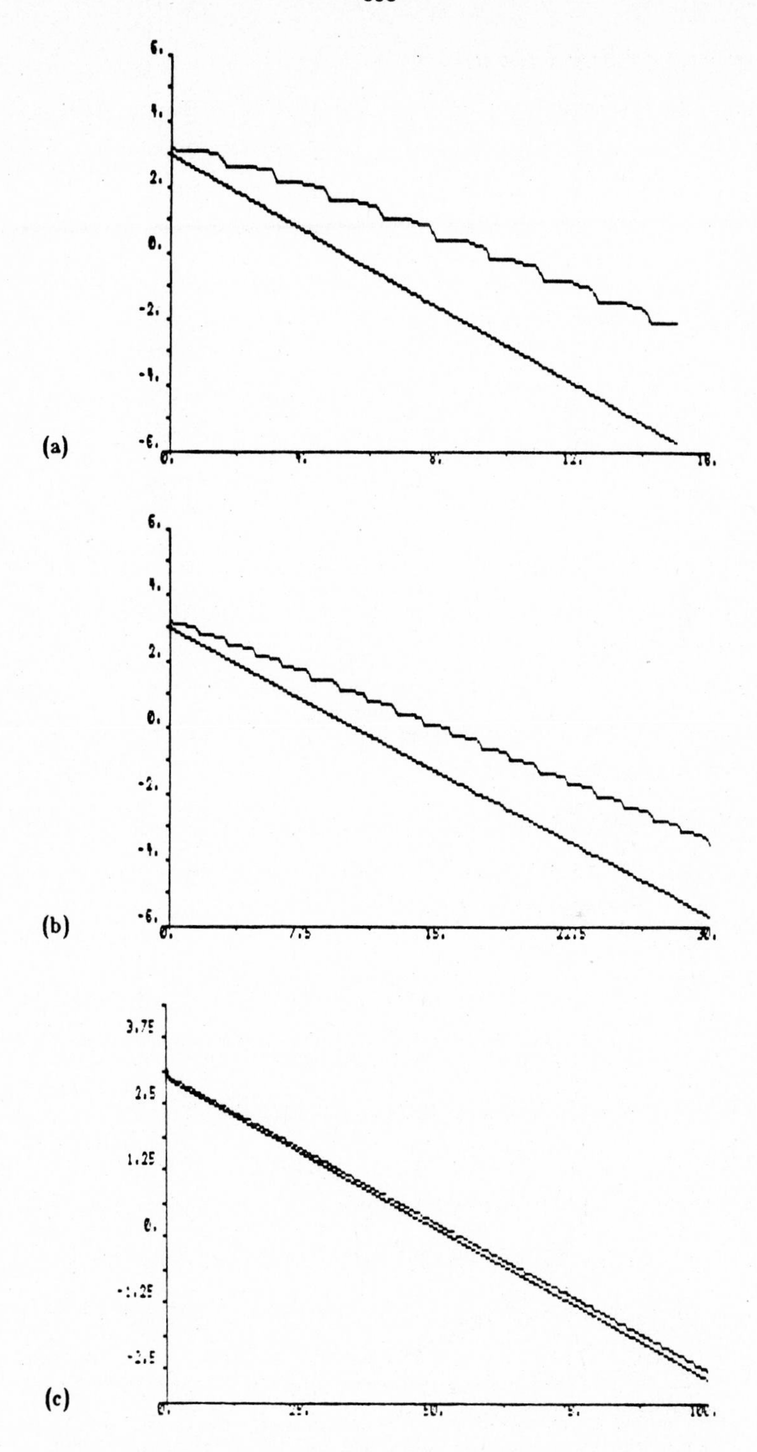


**Fig 3.4 Trajectories of Lyapunov function $V(\phi)$ and $V(\phi_{av})$ with three adaptation gains (a) $\epsilon=1$ (b) $\epsilon=0.5$ (c) $\epsilon=0.1$ using log scale.**

## 4. Averaging of Two-Time Scale Systems

Systems of the form (2.1) studied in section 2 are to be thought of as one time scale systems in that the entire state variable $x$ is varying slowly in comparison with the rate of time variation of the right hand side of the differential equation. In this section, we will study averaging for the case when only some of the state variables are slowly varying.

Consider, for example, the system:

$$\dot{x} = \epsilon f(t,x,y) \qquad x(0)=x_0 \tag{4.1}$$

$$\dot{y} = Ay + \epsilon g(t,x,y) \qquad y(0)=y_0 \tag{4.2}$$

where $x \in R^n$ is called the *slow* state, $y \in R^m$ is called the *fast* state, and $f, g$ are piecewise continuous functions of time.

The goal of averaging will be to approximate the evolution of the slow state. The system (4.1), (4.2) is not the most general two-time scale system. In fact, it is easily seen to be decoupled and linear at $\epsilon = 0$. The study of this special form is motivated by several applications. We will also study another special form later in this section. It is easy to see, from the proofs of this section and those of section 2, that $f$ and $g$ may be allowed to depend smoothly on $\epsilon$ as in (A2).

The averaged system for the slow state is:

$$\dot{x}_{av} = \epsilon f_{av}(x_{av}) \qquad x_{av}(0) = x_0 \tag{4.3}$$

where $f_{av}$ is defined by the limit:

$$f_{av}(x) = \lim_{T\to\infty} \frac{1}{T}\int_t^{t+T} f(\tau,x,0)\, d\tau \tag{4.4}$$

assuming that the limit exists uniformly in $t$ and $x$.

The following assumptions will be in effect for (4.1), (4.2):

(B1) $x=0$, $y=0$ is an equilibrium point of system (4.1), (4.2), i.e. $f(t,0,0)=0$ and $g(t,0,0)=0$ for all $t \geqslant 0$. Both $f$ and $g$ are Lipschitz in $x$ and $y$, i.e.:

$$\| f(t,x_1,y_1) - f(t,x_2,y_2) \| \leqslant l_1 \| x_1 - x_2 \| + l_2 \| y_1 - y_2 \| \tag{4.5}$$

$$\| g(t,x_1,y_1) - g(t,x_2,y_2) \| \leqslant l_3 \| x_1 - x_2 \| + l_4 \| y_1 - y_2 \| \tag{4.6}$$

for all $t \geqslant 0$, $x_1, x_2 \in B_h$, $y_1, y_2 \in B_h$.

(B2) $f_{av}(0) = 0$, and $f_{av}$ is Lipschitz in $x$, i.e.:

$$\| f_{av}(x_1) - f_{av}(x_2) \| \leqslant l_{av} \| x_1 - x_2 \| \tag{4.7}$$

for all $x_1, x_2 \in B_h$.

(B3) the function $d(t,x) = f(t,x,0) - f_{av}(x)$ satisfies the conditions of *Lemma* 2.2

(B4) $A \in R^{m \times m}$ is Hurwitz.

(B5) $x_0$ is sufficiently small that, for $T$ fixed, and some $h' < h$, $x_{av}(t) \in B_{h'}$ for all $t \in [0, T/\epsilon]$ (this is possible, from (B2)). We will also assume that $y_0 \in B_{h'}$, the corresponding closed ball in $R^m$.

**Theorem 4.1 Basic Averaging Theorem for Two-Time Scale System**

**If:** The original system (4.1), (4.2), and the averaged system (4.3), satisfy assumptions (B1)-(B5),

**Then:** There exists $\psi(\epsilon) \in K$ such that, given $T \geq 0$:

$$\| x(t) - x_{av}(t) \| \leq \psi(\epsilon)\, b_T \tag{4.8}$$

for some $b_T$, $\epsilon_T > 0$, and for all $t \in [0, T/\epsilon]$, $\epsilon \leq \epsilon_T$, and $y_0$ sufficiently small. Further, $\psi(\epsilon)$ is of the order of $\epsilon + \xi(\epsilon)$ (as defined in *Lemma* 2.2).

**Proof:** We first apply *Lemma* 2.2, and obtain a result similar to *Lemma* 2.3. Consider the transformation of variable:

$$x = z + \epsilon w_\epsilon(t,z) \tag{4.9}$$

with $\epsilon \leq \epsilon_1$. This transformation leads to:

$$\dot{z} = (I + \epsilon \frac{\partial w_\epsilon}{\partial z})^{-1} \epsilon \left\{ f_{av}(z) + ( f(t,z,0) - f_{av}(z) - \frac{\partial w_\epsilon}{\partial t} ) \right.$$

$$+ ( f(t,z+\epsilon w_\epsilon,0) - f(t,z,0) )$$

$$\left. + ( f(t,z+\epsilon w_\epsilon,y) - f(t,z+\epsilon w_\epsilon,0) ) \right\} \tag{4.10}$$

or:

$$\dot{z} = \epsilon f_{av}(z) + \epsilon p_1(t,z,\epsilon) + \epsilon p_2(t,z,y,\epsilon) \qquad z(0) = x_0 \tag{4.11}$$

where:

$$\| p_1(t,z,\epsilon) \| \leq \frac{1}{1-\xi(\epsilon_1)} (\xi(\epsilon) l_{av} + \xi(\epsilon) + \xi(\epsilon) l_1) \| z \| := \xi(\epsilon)\, k_1 \| z \| \tag{4.12}$$

and:

$$\| p_2(t,z,y,\epsilon) \| \leq \frac{1}{1-\xi(\epsilon_1)} l_2 \| y \| := k_2 \| y \| \tag{4.13}$$

We now estimate the error $x - x_{av}$, following a proof similar to the proof of *Theorem* 2.4. First, we have that:

$$\| x - z \| \leqslant \xi(\epsilon) \| z \| \tag{4.14}$$

Then, the error $z - x_{av}$ can be estimated from:

$$\frac{d}{dt}(z - x_{av}) = \epsilon(f_{av}(z) - f_{av}(x_{av})) + \epsilon p_1(t, z, \epsilon) + \epsilon p_2(t, z, y, \epsilon)$$

$$z(0) - x_{av}(0) = 0 \tag{4.15}$$

for all $t \in [0, T/\epsilon]$, $x_{av}(t) \in B_{h'}$, $h' < h$. As in the proof of *Theorem 2.4*, we will show that, on this interval, and for as long as $x, z \in B_h$, the errors $z - x_{av}$, and $x - x_{av}$ can be made arbitrarily small by reducing $\epsilon$.

Integrating (4.15):

$$\| z(t) - x_{av}(t) \| \leqslant \epsilon l_{av} \int_0^t \| z(\tau) - x_{av}(\tau) \| d\tau + \epsilon \xi(\epsilon) k_1 \int_0^t \| z(\tau) \| d\tau$$

$$+ \epsilon k_2 \int_0^t \| y(\tau) \| d\tau \tag{4.16}$$

Further, $y(t)$ can be calculated from (4.2):

$$y(t) = e^{At} y_0 + \epsilon \int_0^t e^{A(t-\tau)} g(\tau, x, y) d\tau \tag{4.17}$$

Since A is Hurwitz, we have that:

$$\| e^{At} \| \leqslant m\, e^{-\lambda t} \tag{4.18}$$

for some $m, \lambda > 0$, and:

$$\| y(t) \| \leqslant m \| y_0 \| e^{-\lambda t} + \epsilon m \int_0^t e^{-\lambda(t-\tau)} (l_3 \| x(\tau) \| + l_4 \| y(\tau) \|) d\tau \tag{4.19}$$

or:

$$\| e^{\lambda t} y(t) \| \leqslant m \| y_0 \| + \epsilon m l_3 \int_0^t e^{\lambda \tau} \| x(\tau) \| d\tau + \epsilon m l_4 \int_0^t \| e^{\lambda \tau} y(\tau) \| d\tau \tag{4.20}$$

Applying the *Generalized Bellman-Gronwall Lemma*:

$$\| e^{\lambda t} y(t) \| \leqslant m \| y_0 \| e^{\epsilon m l_4 t} + \int_0^t \epsilon m l_3 e^{\lambda \tau} \| x(\tau) \| e^{\epsilon m l_4 (t-\tau)} d\tau \tag{4.21}$$

Define $\lambda(\epsilon) = \lambda - \epsilon m l_4$, and $\epsilon_3$ $(0 < \epsilon_3 \leqslant \epsilon_1)$ so that $\lambda(\epsilon) > 0$ for $\epsilon \leqslant \epsilon_3$. It follows that:

$$\| y(t) \| \leqslant m\, h\, e^{-\lambda(\epsilon) t} + \epsilon m l_3 h / \lambda(\epsilon) \tag{4.22}$$

Using this estimate in (4.16), and using the *Generalized Bellman-Gronwall Lemma* again:

$$\| z(t)-x_{av}(t)\| \leqslant \int_0^t \left[ \xi(\epsilon)k_1 h + mk_2 h e^{-\lambda(\epsilon)\tau} + \frac{\epsilon m k_2 l_3 h}{\lambda(\epsilon)} \right] \epsilon e^{\epsilon l_{av}(t-\tau)} d\tau$$

$$\leqslant (\epsilon+\xi(\epsilon)) \left[ k_1 h + \frac{mk_2 h\, l_{av}}{\lambda(\epsilon)+\epsilon l_{av}} + \frac{mk_2 l_3 h}{\lambda(\epsilon)} \right] \left[ \frac{e^{l_{av}T}}{l_{av}} \right]$$

$$:= \psi(\epsilon)\, a_T \tag{4.23}$$

As in *Theorem 2.4*, it follows that, for some $b_T$:

$$\| x(t)-x_{av}(t)\| \leqslant \psi(\epsilon)\, b_T \tag{4.24}$$

By assumption, $\| x_{av}(t)\| \leqslant h' < h$. Let $\epsilon_T$ $(0<\epsilon_T \leqslant \epsilon_3)$ such that $\psi(\epsilon_T) b_T < h-h'$. Further, let $y_0$, and $\epsilon_T$ sufficiently small that, by (4.22), $y(t) \in B_h$, for all $t \in [0, T/\epsilon]$. It follows, from a simple contradiction argument, that the estimate in (4.24) is valid for all $t \in [0, T/\epsilon]$, whenever $\epsilon \leqslant \epsilon_T$.

**Theorem 4.2 Exponential Stability Theorem for Two-Time Scale Systems**

**If:** The original system (4.1), (4.2), and the averaged system (4.3) satisfy assumptions (B1)-(B4), the function $f_{av}(x)$ has continuous and bounded first partial derivatives in $x$, and $x=0$ is an exponentially stable equilibrium point of the averaged system,

**Then:** There exists $\epsilon_4>0$ such that the equilibrium point $x=0$ of the original system is exponentially stable for all $\epsilon \leqslant \epsilon_4$.

**Proof:** Since $x_{av}=0$ is an exponentially stable equilibrium point of the averaged system, there exists a function $v(x_{av})$ satisfying (2.26)-(2.28). On the other hand, since $A$ is Hurwitz, there exist matrices $P, Q>0$, such that $A^T P + PA = -Q$. Denote by $p_1, p_2, q_1, q_2$ the minimum and maximum eigenvalues of the $P$ and $Q$ matrices. We now study the stability of the system (4.11), (4.2), and consider the following Lyapunov function:

$$v_1(z,y) = v(z) + \frac{\alpha_2}{p_2} y^T P\, y \tag{4.25}$$

so that:

$$\alpha'_1 (\| z\|^2 + \| y\|^2) \leqslant v_1(z,y) \leqslant \alpha_2 (\| z\|^2 + \| y\|^2) \tag{4.26}$$

where $\alpha'_1 = \min(\alpha_1, \frac{\alpha_2}{p_2} p_1)$. The derivative of $v_1$ along the trajectories of (4.11), (4.2) can be estimated, using the previous results:

$$\dot v_1(z,y) \leqslant -\epsilon\alpha_3 \| z\|^2 + \epsilon k_1 \xi(\epsilon)\alpha_4 \| z\|^2$$

$$+ \epsilon k_2 \alpha_4 \| z \| \, \| y \| - \frac{\alpha_2}{p_2} q_1 \| y \|^2$$

$$+ 4\epsilon l_3 \alpha_2 \| z \| \, \| y \| + 2\epsilon l_4 \alpha_2 \| y \|^2 \tag{4.27}$$

for $\epsilon \leq \epsilon_1$ (so that, in particular, $\| x \| \leq 2 \| z \|$). Note that since $ab \leq (a^2+b^2)/2$ for all $a, b \in R$, we have:

$$\epsilon \| z \| \, \| y \| \leq \frac{1}{2} \left( \epsilon^{4/3} \| z \|^2 + \epsilon^{2/3} \| y \|^2 \right) \tag{4.28}$$

so that:

$$\dot{v}_1(z, y) \leq -\epsilon \left( \alpha_3 - \xi(\epsilon) k_1 \alpha_4 - \epsilon^{1/3} \frac{k_2 \alpha_4}{2} - 2\epsilon^{1/3} l_3 \alpha_2 \right) \| z \|^2$$

$$- \left( \frac{\alpha_2}{p_2} q_1 - 2\epsilon l_4 \alpha_2 - \epsilon^{2/3} \frac{k_2 \alpha_4}{2} - 2\epsilon^{2/3} l_3 \alpha_2 \right) \| y \|^2$$

$$:= -2\epsilon \alpha_2 \, \alpha(\epsilon) \| z \|^2 - q(\epsilon) \| y \|^2 \tag{4.29}$$

Note that, with this definition, $\alpha(\epsilon) \rightarrow \frac{1}{2} \frac{\alpha_3}{\alpha_2}$ as $\epsilon \rightarrow 0$.

Let $\epsilon_4 (0 < \epsilon_4 \leq \epsilon_1)$ be sufficiently small that $\alpha(\epsilon) > 0$, $q(\epsilon) > 0$, and $2\epsilon \alpha_2 \, \alpha(\epsilon) \leq q(\epsilon)$ whenever $\epsilon \leq \epsilon_4$. Consequently:

$$\dot{v}_1(z, y) \leq -2\epsilon \alpha(\epsilon) v_1(z, y) \tag{4.30}$$

and:

$$v_1(z, y) \leq v_1(z(t_0), y(t_0)) \, e^{-2\epsilon \alpha(\epsilon)(t - t_0)} \tag{4.31}$$

As in *Theorem 2.5*, this implies the exponential convergence of the original system, with rate of convergence $\epsilon \alpha(\epsilon)$. Also, for $x(t_0)$, $y(t_0)$ sufficiently small, all signals are guaranteed to remain in $B_h$, so that all assumptions are applicable.

### Comments

The comments of *Theorem 2.5* apply similarly to *Theorem 4.2*. In particular, the proof gives a useful bound on the rate of convergence of the original system, and this bound again tends to the bound on the rate of convergence of the averaged system.

**Mixed Time Scales**

We now discuss a more general class of two-time scale systems, arising in adaptive control:

$$\dot{x} = \epsilon f'(t, x, y') \tag{4.32}$$

$$\dot{y}' = Ay' + h(t, x) + \epsilon g'(t, x, y') \tag{4.33}$$

We will show that system (4.32)-(4.33) can be transformed into the system described in the previous section. In this case, $x$ is a slow variable, but $y'$ has both a fast, and a slow component.

The averaged system corresponding to (4.32), (4.33) is obtained as follows. Define the function:

$$v(t, x) = \int_0^t e^{A(t-\tau)} h(\tau, x)\, d\tau \tag{4.34}$$

and assume that the following limit exists uniformly in $t$ and $x$:

$$f_{av}(x) = \lim_{T \to \infty} \frac{1}{T} \int_t^{t+T} f'(\tau, x, v(\tau, x))\, d\tau \tag{4.35}$$

Intuitively, $v(t, x)$ represents the steady-state value of the variable $y$ with $x$ frozen and $\epsilon=0$ in (4.33).*

To show that the averaged system of (4.35) is the right one, we transform the system (4.32), (4.33) to the form (4.1), (4.2), using the transformation:

$$y = y' - v(t, x) \tag{4.36}$$

From (4.34), $v(t, x)$ satisfies:

$$\frac{\partial}{\partial t} v(t, x) = Av(t, x) + h(t, x) \qquad v(t, 0) = 0 \tag{4.37}$$

Differentiating (4.36), we have that:

$$\dot{y} = Ay + \epsilon \left[ -\frac{\partial v(t, x)}{\partial x} f'(t, x, y + v(t, x)) + g'(t, x, y + v(t, x)) \right] \tag{4.38}$$

so that system (4.32), (4.33), is of the form (4.1), (4.2), with:

$$f(t, x, y) = f'(t, x, y + v(t, x)) \tag{4.39}$$

$$g(t, x, y) = -\frac{\partial v(t, x)}{\partial x} f'(t, x, y + v(t, x)) + g'(t, x, y + v(t, x)) \tag{4.40}$$

*This choice of transformation was pointed out to us by B. Riedle & P. Kokotovic.

The averaged system is obtained by averaging the right-hand side of (4.39) with $y=0$, so that the definitions (4.4), and (4.35) agree.

To apply *Theorem 4.1*, and *Theorem 4.2*, we require that assumptions (B1)-(B5) be satisfied. In particular, we assume similar Lipschitz conditions on $f'$, $g'$, and the following assumption on $h(t,x)$:

(B6) $h(t,0)=0$ for all $t \geqslant 0$, and:

$$\left\| \frac{\partial h(t,x)}{\partial x} \right\| \leqslant k \tag{4.41}$$

for all $t \geqslant 0$, $x \in B_h$.

This new assumption implies that $v(t,0)=0$, and:

$$\left\| \frac{\partial v(t,x)}{\partial x} \right\| \leqslant k' \tag{4.42}$$

for all $t \geqslant 0$, $x \in B_h$.

This condition is sufficient to guarantee Lipschitz conditions for the system (4.1), (4.2), given Lipschitz conditions for the system (4.32), (4.33). The theory developed earlier in this section can therefore be directly applied to systems of the form (4.32), (4.33). The key to the preceding transformation is the fact that the new state variable $y$ is truly a *fast* variable, so that the two time scales have been separated.

Although the theorems of this section allow for nonlinear terms in (4.1)-(4.2), (4.32)-(4.33), the linearity of the dominant term in (4.2) and (4.33) precludes the application of the theorems to adaptive control problems. In joint work with Anderson, Mareels, and Bitmead, we have recently shown that *Theorems* 4.1 and 4.2 hold for the larger class of systems

$$\dot{x} = \epsilon f(t,x,y) \tag{4.43}$$

$$\dot{y} = A(x)y + \epsilon g(t,x,y) \tag{4.44}$$

provided that $A(x)$ is a Hurwitz matrix (for $x \in B_h$), and provided that $\dfrac{\partial A(x)}{\partial x}$ is bounded. The extension to the mixed time scales system of the type (4.32)-(4.33) follows under the same conditions.

## 5. Two-Time Scale Averaging Applied to Model Reference Adaptive Controller

To apply the theory of Section 4 to model reference adaptive controllers we review the model reference adaptive system of Narendra, Valavani [18] for the relative degree 1 case (our notation is however consistent with Sastry [19]). Consider a plant with transfer function

$$\hat{p}(s) = k_p \frac{\hat{n}_p(s)}{\hat{d}_p(s)} = c_p (sI - A)^{-1} b_p \tag{5.1}$$

where $\hat{n}_p$, $\hat{d}_p$ are relatively prime monic polynomials of degree $n-1$, $n$ respectively and $k_p$ is a scalar (the representation in (5.1) is assumed minimal). The following are assumed to be known about the plant transfer function:

(C1) The degrees of the polynomials $\hat{d}_p$, $\hat{n}_p$ are known.

(C2) The sign of $k_p$ is known (say $k_p > 0$).

(C3) The plant transfer function is assumed to be minimum phase.

The objective is to build a compensator so that the plant output asymptotically matches that of a stable reference model $\hat{m}(s)$ with input $r(t)$, output $y_m(t)$ and transfer function

$$\hat{m}(s) = k_m \frac{\hat{n}_m(s)}{\hat{d}_m(s)}$$

where $k_m > 0$ and $\hat{n}_m$, $\hat{d}_m$ are monic polynomials of degree $n-1$, $n$ respectively (not necessarily relatively prime but both Hurwitz). If we denote the input and output of the plant $u(t)$ and $y_p(t)$ respectively, the objective may be stated as: find $u(t)$ so that $y_p(t) - y_m(t) \to 0$ as $t \to \infty$. By using suitable prefiltering of the reference signal if necessary, we may assume that the model $\hat{m}(s)$ is strictly positive real.

The scheme is shown in Figure 5.1. The dynamical compensator blocks $F_1$ and $F_2$ (reminiscent of those in Section 3) are identical one input, n-1 output systems, each with transfer function $(sI - \Lambda)^{-1} b$ ; $\Lambda \in R^{n-1 \times n-1}$, $b \in R^{n-1}$ where $\Lambda$ is chosen so that its eigenvalues are the zeros of $\hat{n}_m$. The pair $\Lambda, b$ is assumed controllable and, for ease of book-keeping (in the algorithm proof alone), we assume that they are in controllable form so that

$$(sI - \Lambda)^{-1} = \frac{1}{\hat{n}_m(s)} \begin{pmatrix} 1 \\ s \\ \cdot \\ \cdot \\ s^{n-2} \end{pmatrix}$$

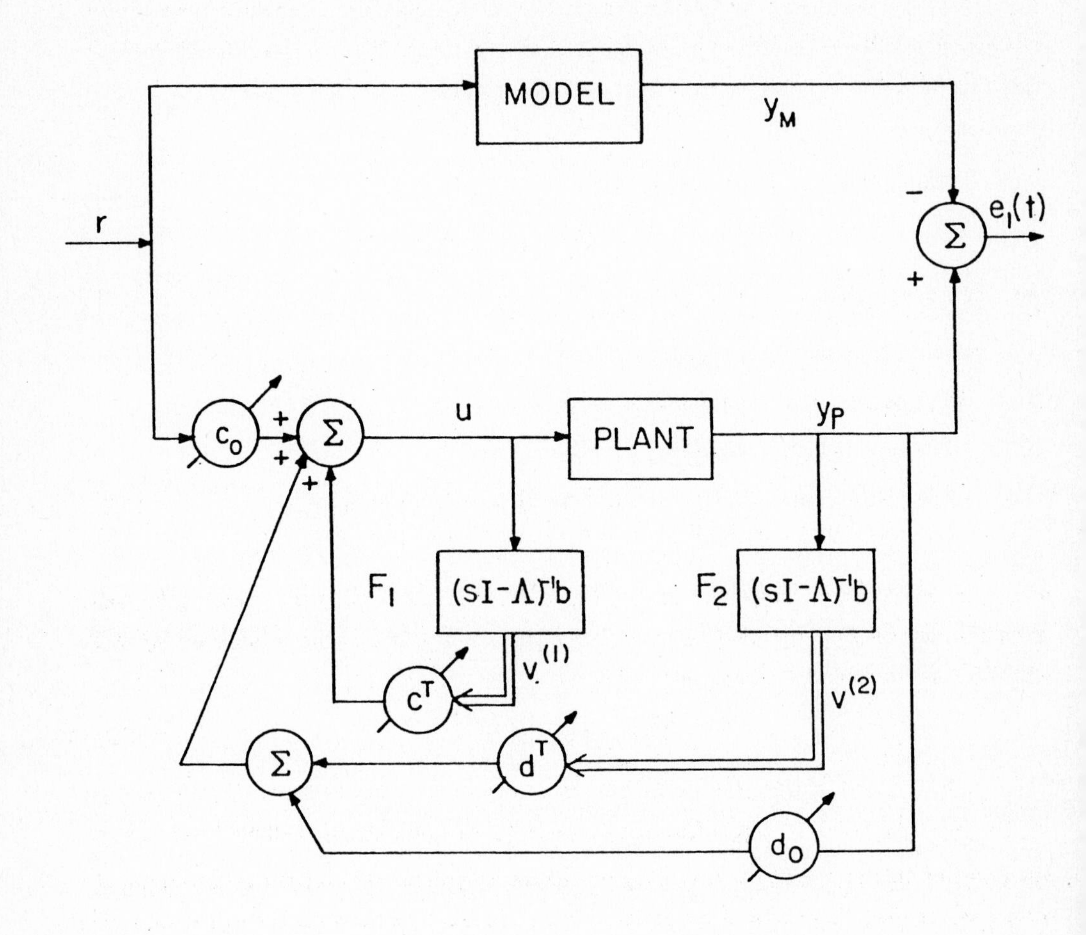


**Fig 5.1 Block diagram of model reference adaptive control system**

The parameters $c \in R^{n-1}$ in the precompensator block serve to tune the closed-loop plant zeros; $d \in R^{n-1}$, $d_0 \in R$ in the feedback compensator assign the closed loop plant poles. The parameter $c_0$ adjusts the overall gain of the closed loop plant. Thus, the vector of 2n adjustable parameters denoted $\theta$ is

$$\theta^T = [c_0, c^T, d_0, d^T]$$

with the signal vector $w \in R^{2n}$ defined by

$$w^T = [r, v^{(1)T}, y_p, v^{(2)T}]$$

The input to the plant is seen to be

$$u = \theta^T w$$

and the state equations of the plant loop are given by

$$\begin{bmatrix} \dot{x}_p \\ \dot{v}^{(1)} \\ \dot{v}^{(2)} \end{bmatrix} = \begin{bmatrix} A_p & 0 & 0 \\ 0 & \Lambda & 0 \\ bc^T{}_p & 0 & \Lambda \end{bmatrix} \begin{bmatrix} x_p \\ v^{(1)} \\ v^{(2)} \end{bmatrix} + \begin{bmatrix} b_p \\ b \\ 0 \end{bmatrix} \theta^T w \tag{5.2}$$

It may be verified that there is a unique constant $\theta^* \in R^{2n}$ such that, when $\theta = \theta^*$, the transfer function of the plant plus controller equals $\hat{m}(s)$. It can also be shown [18] that when $r$ is bounded and the parameter update law is given by

$$\dot{\theta} = -\Gamma e_1 w = -\Gamma (y_p - y_m) w \tag{5.3}$$

with $\Gamma \in R^{2n \times 2n}$, a positive definite matrix, all signals in the loop, i.e. $u, v, v^{(1)}, v^{(2)}, y_p, y_m$ are bounded. In addition, $\lim_{t \to \infty} e_1(t) = 0$ so that asymptotically $y_p(t)$ approaches $y_m(t)$. The proof of this fact used the following procedure: represent the model ( in non-minimal form ) as the plant loop with $\theta$ set equal to $\theta^*$. The state equations for the model loop are given by

$$\begin{bmatrix} \dot{x}_m \\ \dot{v}_m^{(1)} \\ \dot{v}_m^{(2)} \end{bmatrix} = \begin{bmatrix} A_p + b_p d_0^* c_p^T & b_p c^{*T} & b_p d^{*T} \\ bd_0^* c_p^T & \Lambda + bc^{*T} & bd^{*T} \\ bc_p^T & 0 & \Lambda \end{bmatrix} \begin{bmatrix} x_m \\ v_m^{(1)} \\ v_m^{(2)} \end{bmatrix} + \begin{bmatrix} b_p \\ b \\ 0 \end{bmatrix} c_0^* r \tag{5.4}$$

The 3n-2 $\times$ 3n-2 matrix in (5.4) is henceforth referred to as $\tilde{A}$, and the 3n-2 vector in (5.4) as $\tilde{b}$. Then, subtracting (5.4) from (5.2) with

$$e^T = [x_p^T, v^{(1)T}, v^{(2)T}] - [x_m^T, v_m^{(1)T}, v_m^{(2)T}]$$

we have that

$$\dot{e} = \tilde{A} e + \tilde{b} \phi^T w \tag{5.5}$$

and

$$e_1 := y_p - y_m = [c_p^T, 0, 0]\, e =: \tilde{c}^T e \tag{5.6}$$

where $\phi := \theta - \theta^*$ is the parameter error. Note from (5.4) that $\tilde{c}^T (sI - \tilde{A})^{-1} \tilde{b}\, c_0^*$ is equal to the model transfer function and that $c_0^* = k_p / k_m$ is the ratio of the high frequency gain (positive by assumption). Now the update law (5.3) is

$$\dot{\theta} = \dot{\phi} = -\Gamma w\, \tilde{c}^T e \tag{5.7}$$

To apply averaging, we consider $\Gamma = \epsilon I$ resulting in

$$\dot{\phi} = -\epsilon\, w\, \tilde{c}^T e \tag{5.8}$$

Equations (5.5) and (5.8) are superficially of the form (4.32), (4.33) with $h(t, \phi) := \tilde{b} w^T \phi$ and $f'(t, \phi, e) = -w\, \tilde{c}^T e$. Difficulty, however, arises from the fact that $w$ in (5.5), (5.8) is not independently and exogenously specified, but in fact depends on e. To show this dependence explicitly, we set

$$w_m^T := [\, r, v_m^{(1)T}, y_m, v_m^{(2)T}\,]$$

an exogenously defined 3n-2 dimensional vector that can be obtained either from $r(t)$ alone or as linear combinations of the state variables of (5.4), and rewrite

$$w = w_m + Qe \tag{5.9}$$

where

$$Q = \begin{bmatrix} 0 & 0 & 0 \\ 0 & I & 0 \\ c_p^T & 0 & 0 \\ 0 & 0 & I \end{bmatrix}$$

Using (5.9), the equations (5.5) and (5.8) are

$$\dot{e} = \tilde{A} e + \tilde{b}\, w_m^T \phi + \tilde{b}\, e^T Q^T \phi \tag{5.10}$$

$$\dot{\phi} = -\epsilon w_m\, \tilde{c}^T e - \epsilon Q\, e\, \tilde{c}^T e \tag{5.11}$$

With the exception of the last terms (quadratic in $e$ and $\phi$ ), equations (5.10), (5.11) are linear time varying equations describing the linearized adaptive control system, around the equilibrium $e = 0$, $\phi = 0$. In this section, we apply averaging to the linearized equations (5.10), (5.11) corresponding to small $e$ and $\phi$. Averaging of the full nonlinear equations (5.10), (5.11) is more subtle and is not considered here. We consider

$$\dot{e} = \tilde{A} e + \tilde{b}\, w_m^T \phi \tag{5.12}$$

$$\dot{\phi} = -\epsilon\, w_m\, \tilde{c}^T e \tag{5.13}$$

Since $r$ is bounded and $\tilde{A}$ is stable (its eigenvalues are the union of the zeros of $\hat{d}_m$, $\hat{n}_p$

and the eigenvalues of $\Lambda$), $w_m$ is bounded. Hence it is easy to see that the equations (5.12), (5.13) are of the form of (4.33), (4.32) with the functions $f'$ and $h$ satisfying the conditions of Section 4.

To establish the averaging results, we assume that $r$ is stationary. This implies, as has been shown in Boyd and Sastry [14], that $w_m$ is stationary. Its spectral measure is related to that of $r$ by

$$S_{w_m}(d\nu) = \hat{n}(j\nu)\hat{n}^*(j\nu)s_r(d\nu) \tag{5.14}$$

with

$$\hat{n}(s) := \begin{bmatrix} 1 \\ \hat{m}\hat{p}^{-1}(sI-\Lambda)^{-1}b \\ \hat{m} \\ \hat{m}(sI-\Lambda)^{-1}b \end{bmatrix}$$

an exponentially stable transfer function.

The function $v(t,\phi)$ of Section 4 for the system (5.12), (5.13) is

$$v(t,\phi) := [\int_0^t e^{\tilde{A}(t-\tau)}\tilde{b}w_m^T(\tau)\,d\tau\,]\phi$$

and the averaged $f$ is given by

$$f_{av}(\phi) = -\lim_{T\to\infty}\frac{1}{T}\int_s^{s+T} w_m(t)\tilde{c}^T[\int_0^t e^{\tilde{A}(t-\tau)}\tilde{b}w_m^T(\tau)\,d\tau\,]dt\;\phi \tag{5.15}$$

Since $w_m$ is stationary, the limit in (5.15) may be shown to exist as follows. Define a filtered version of $w_m$ to be

$$w_{mf}(t) = \int_0^t \tilde{c}^T e^{\tilde{A}(t-\tau)}\tilde{b}w_m(\tau)\,d\tau \tag{5.16}$$

Since $\tilde{c}^T(sI-\tilde{A})^{-1}\tilde{b} = \frac{1}{c_0}\hat{m}(s)$ is stable, it follows that $w_{mf}(t)$ is also stationary. The quantity inside the square brackets in (5.15) is

$$\lim_{T\to\infty}\frac{1}{T}\int_s^{s+T} w_m(t)w_{mf}^T(t)\,dt = R_{w_m w_{mf}}(0) \tag{5.17}$$

i.e. the cross correlation between $w_m$ and $w_{mf}$ evaluated at 0. Consequently, we may use (5.14) and (5.16) to obtain a formula for $R_{w_m w_{mf}}(0)$ as

$$R_{w_m w_{mf}}(0) = \frac{1}{c_0^*}\int_{-\infty}^{\infty}\hat{n}(j\nu)\hat{n}^*(j\nu)\hat{m}(j\nu)s_r(d\nu) \tag{5.18}$$

and the averaged system is a linear time-invariant system

$$\dot{\phi}_{av} = -\epsilon R_{w_m w_{mf}}(0)\,\phi_{av} \tag{5.19}$$

Since $\hat{m}(s)$ is strictly positive real, the matrix $R_{w_m w_{mf}}(0)$ is a positive definite matrix. Unlike the matrix $R_w(0)$ of Section 3, $R_{w_m w_{mf}}(0)$ need not be symmetric, and its eigenvalues need not be real. However, the real parts are guaranteed to be positive. When the reference input $r$ is almost periodic, i.e.

$$r(t) \sim \sum_k r_k e^{j\nu_k t}$$

a simple formula for $R_{w_m w_{mf}}(0)$ is given by

$$R_{w_m w_{mf}}(0) = \frac{1}{c_0^*} \sum_k \hat{n}(j\nu_k)\hat{n}^*(j\nu_k)\hat{m}(j\nu_k) r_k \tag{5.20}$$

As an illustration of the preceding results, we consider the following example of a first order plant with an unknown pole and an unknown gain :

$$\hat{p}(s) = \frac{k_p}{s + a_p}$$

The adaptive process is to adjust the feedforward gain $c_0$ and the feedback gain $d_0$ so as to make the closed-loop transfer function match the model transfer function

$$\hat{m}(s) = \frac{k_m}{s + a_m}$$

To guarantee persistency of excitation, we use a sinusoidal input signal of the form

$$r(t) = a \sin(\omega t)$$

Thus, equations (5.5), (5.7) become

$$\dot{e} = -a_m e + k_p(\phi_1 r + \phi_2 y_m)$$

$$\dot{\phi}_1 = -\epsilon\, e\, r$$

$$\dot{\phi}_2 = -\epsilon\, e\, y_m$$

where

$$\phi_1 = c_0 - c_0^* \quad , \quad \phi_2 = d_0 - d_0^*$$

Consequently, the averaged system defined in (5.19) now is

$$\begin{bmatrix} \dot{\phi}_{av\,1} \\ \dot{\phi}_{av\,2} \end{bmatrix} = -\frac{\epsilon a^2}{4}\frac{k_p}{k_m} \begin{bmatrix} \dfrac{18}{(9+\omega^2)} & \dfrac{18(9-\omega^2)}{(9+\omega^2)^2} \\ \dfrac{18}{(9+\omega^2)} & \dfrac{162}{(9+\omega^2)^2} \end{bmatrix} \begin{bmatrix} \phi_{av\,1} \\ \phi_{av\,2} \end{bmatrix}$$

using equation (5.18). With $a_m=3$, $k_m=3$, $a_p=1$, $k_p=2$, $a=3$, $w=2$, the two eigenvalues of the averaged system are computed to be $-3.10\epsilon$ and $-0.43\epsilon$, both real negative. Figs 5.2, 5.3 show the plots of the parameter errors of $c_0$ and $d_0$ for the original and averaged system, with three different adaptation gains. Fig. 4 corresponds to a higher frequency input signal $\omega = 4$ such that the eigenvalues of the matrix $R_{w_m w_{mf}}(0)$ are complex $(-0.49 \pm 0.30i)\epsilon$, and explains the oscillatory behavior of the original and averaged systems.

Using the results of Boyd and Sastry [14], it is easy to verify the following facts

(i) $R_{w_m w_{mf}}(0)$ is singular unless $R_w(0) > 0$, i.e. $w(t)$ is persistently exciting. Thus persistent excitation of $w$ is a necessary condition for exponential stability of (5.19).

(ii) If $\hat{m}(s)$ is strictly positive real and $w(t)$ is persistently exciting, then $R_{w_m w_{mf}}(0)$ is Hurwitz. Hence $\hat{m}(s)$ being strictly positive real is a sufficient condition for stability of (5.19), given that $w(t)$ is persistently exciting.

It is intuitive that if $w$ is persistently exciting and $\hat{m}(s)$ is close in some sense to being strictly positive real that $R_{w_m w_{mf}}(0)$ will be Hurwitz (in particular, this is the case if $\mathrm{Re}\,\hat{m}(j\nu)$ fails to be positive at frequencies where $\hat{n}(j\nu)$ is small enough). More specific results in this context are in [11,17].

In view of the results stated at the end of section 5, averaging can also be applied to the *nonlinear* system described by (5.10)-(5.11), with $A(x) = \tilde{A} + \tilde{b}x^T Q$. Consequently, the nonlinear time varying adaptive control scheme can be analyzed through the autonomous averaged system (a generalisation of the ideas of [24]). However, due to the nonlinearity of the system, the frequency domain analysis, and the derivation of guaranteed convergence rates are not straightforward.

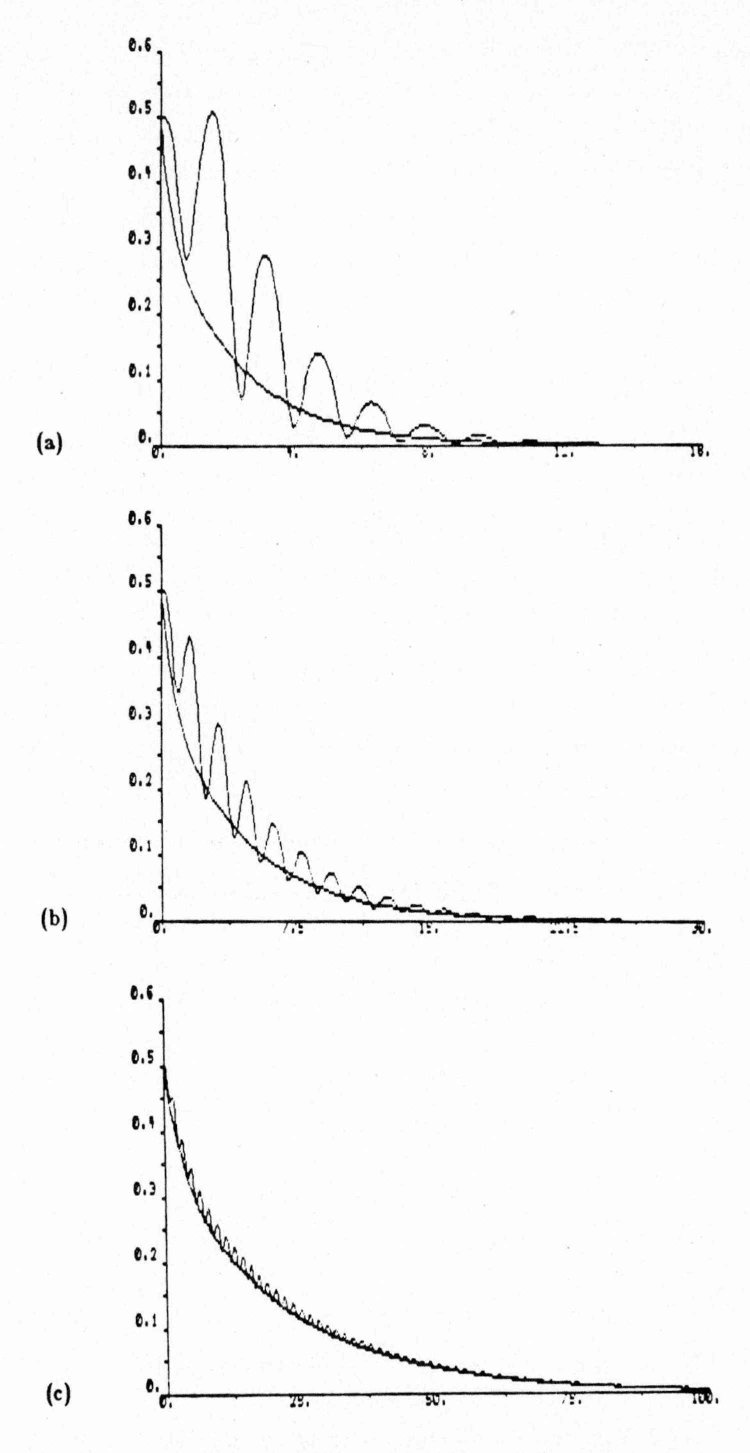


**Fig 5.2** Trajectories of parameter error $\phi_1 (= c_0 - c_0^*)$ and $\phi_{av1}$ with three adaptation gains (a) $\epsilon=1$ (b) $\epsilon=0.5$ (c) $\epsilon=0.1$

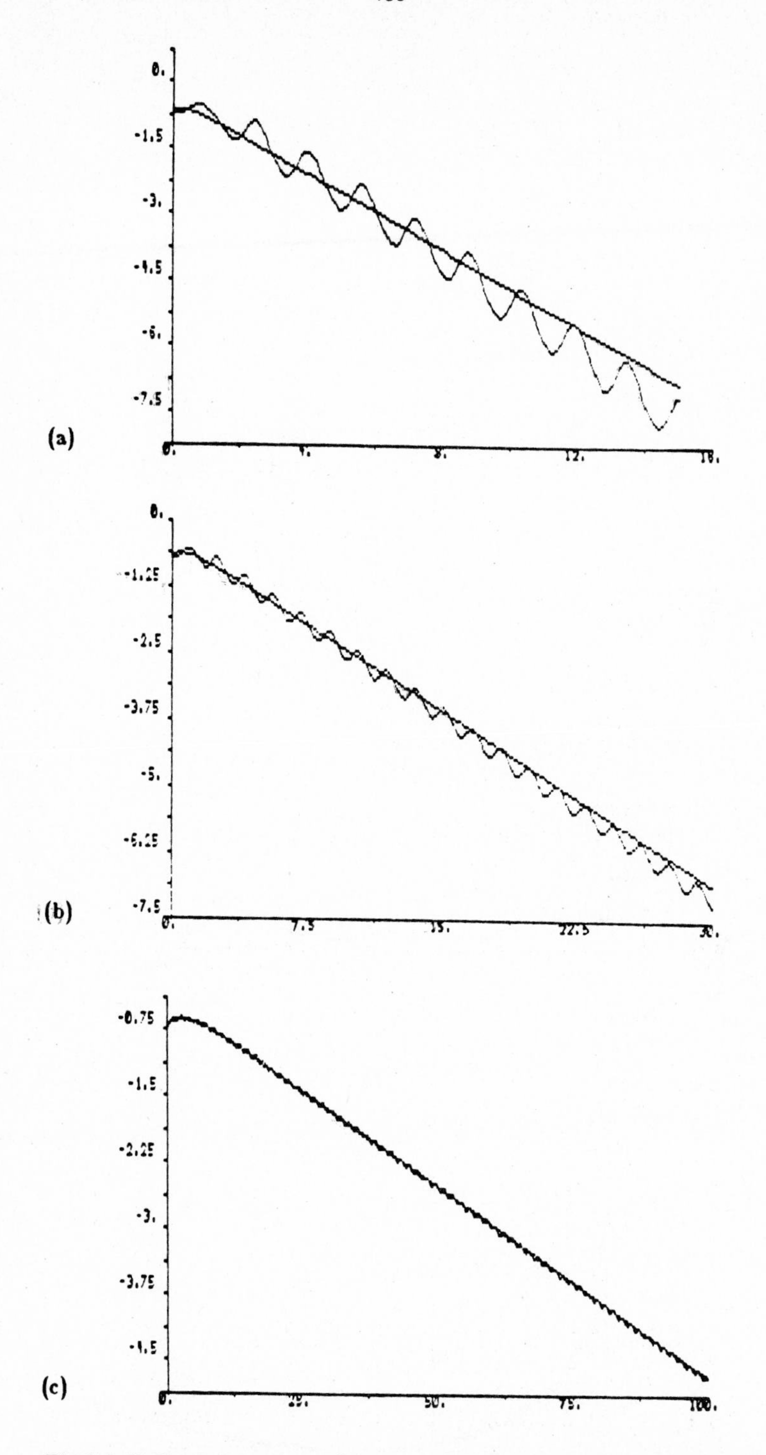


Fig 5.3 Trajectories of parameter error $\phi_2 (= d_0 - d_0^*)$ and $\phi_{av2}$ with three adaptation gains (a) $\epsilon=1$ (b) $\epsilon=0.5$ (c) $\epsilon=0.1$ using log scale.

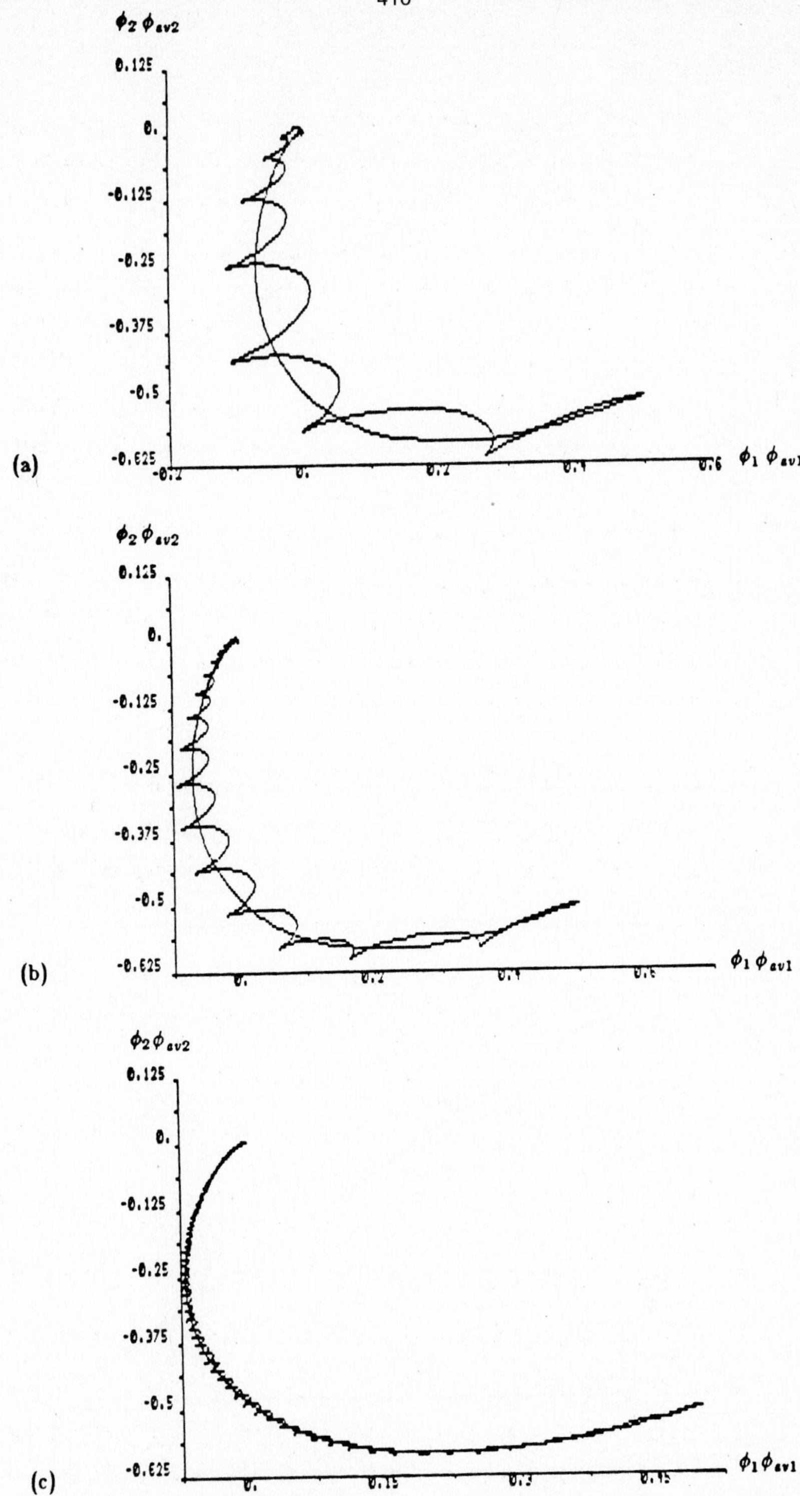


Fig 5.4 Phase plot for $\phi_2(\phi_1)$ and $\phi_{av2}(\phi_{av1})$ with three different adaptation gains (a) $\epsilon=1$ (b) $\epsilon=0.5$ (c) $\epsilon=0.1$ ; and $r=3sin(4t)$.

## 6. Concluding Remarks

We have presented in this paper new stability theorems for averaging analysis of one and two time scale systems. We have applied these techniques to obtain bounds on the rates of convergence of adaptive identifiers and controllers of relative degree 1.

We feel that the techniques presented here can be extended to obtain instability theorems for averaging. Such theorems could be used to study the mechanism of slow drift instability in adaptive schemes in the presence of unmodelled dynamics, in a framework resembling that of [10].

# APPENDIX

## Proof of Lemma 2.1

Define:

$$w_{\epsilon}(t,x)=\int_{0}^{t} d(\tau,x)\,e^{-\epsilon(t-\tau)}d\tau \tag{L1}$$

and:

$$w_{0}(t,x)=\int_{0}^{t} d(\tau,x)d\tau \tag{L2}$$

From the assumptions:

$$\| w_{0}(t+t_{0},x)-w_{0}(t_{0},x)\| \leqslant \gamma(t).t \tag{L3}$$

for all $t, t_0 \geqslant 0$, $x \in B_h$. Integrating (L1) by parts:

$$w_{\epsilon}(t,x)=w_{0}(t,x)-\epsilon\int_{0}^{t} e^{-\epsilon(t-\tau)}w_{0}(\tau,x)\,d\tau \tag{L4}$$

Using the fact that:

$$\epsilon\int_{0}^{t} e^{-\epsilon(t-\tau)}w_{0}(t,x)\,d\tau = w_{0}(t,x)-w_{0}(t,x)\,e^{-\epsilon t} \tag{L5}$$

(L4) can be rewritten as:

$$w_{\epsilon}(t,x)=w_{0}(t,x)\,e^{-\epsilon t}+\epsilon\int_{0}^{t} e^{-\epsilon(t-\tau)}(w_{0}(t,x)-w_{0}(\tau,x))d\tau \tag{L6}$$

and, using (L3):

$$\| w_{\epsilon}(t,x)\| \leqslant \gamma(t)\,t\,e^{-\epsilon t}+\epsilon\int_{0}^{t} e^{-\epsilon(t-\tau)}(t-\tau)\,\gamma(t-\tau)d\tau \tag{L7}$$

Consequently,

$$\| \epsilon w_{\epsilon}(t,x)\| \leqslant \sup_{t'\geqslant 0}\gamma(\frac{t'}{\epsilon})t'e^{-t'}+\int_{0}^{\infty}\gamma(\frac{\tau'}{\epsilon})\tau'e^{-\tau'}d\tau' \tag{L8}$$

Since, for some $\beta$, $\| d(t,x)\| \leqslant\beta$, we also have that $\gamma(t)\leqslant\beta$. Note that, for all $t'\geqslant 0$, $t'e^{-t'}\leqslant e^{-1}$, and $t'e^{-t'}\leqslant t'$, so that:

$$\| \epsilon w_{\epsilon}(t,x)\| \leqslant \sup_{t'\in[0,\sqrt{\epsilon}]}\left|\gamma(\frac{t'}{\epsilon})t'e^{-t'}\right|+\sup_{t'\geqslant\sqrt{\epsilon}}\left|\gamma(\frac{t'}{\epsilon})t'e^{-t'}\right|$$

$$+\int_{0}^{\sqrt{\epsilon}}\gamma(\frac{\tau'}{\epsilon})\tau'e^{-\tau'}d\tau'+\int_{\sqrt{\epsilon}}^{\infty}\gamma(\frac{\tau'}{\epsilon})\tau'e^{-\tau'}d\tau' \tag{L9}$$

This, in turn, implies that

$$\| \epsilon w_\epsilon(t,x) \| \leqslant \beta\sqrt{\epsilon} + \gamma(\frac{1}{\sqrt{\epsilon}})e^{-1} + \beta\frac{\epsilon}{2} + \gamma(\frac{1}{\sqrt{\epsilon}})(1+\sqrt{\epsilon})e^{-\sqrt{\epsilon}}$$

$$:= \xi(\epsilon) \tag{L10}$$

Clearly $\xi(\epsilon) \in K$. From (L1), it follows that:

$$\frac{\partial w_\epsilon(t,x)}{\partial t} - d(t,x) = -\epsilon\, w_\epsilon(t,x) \tag{L11}$$

so that both (2.6) and (2.7) are satisfied.

If $\gamma(T) = a / T^r$, then the right-hand side of (L8) can be computed explicitly:

$$\sup_{t' \geqslant 0} a\,\epsilon^r (t')^{1-r} e^{-t'} = a\,\epsilon^r (1-r)^{1-r} e^{r-1} \leqslant a\,\epsilon^r \tag{L12}$$

and, with $\Gamma$ denoting the standard gamma function:

$$\int_0^\infty a\,\epsilon^r (\tau')^{1-r} e^{-\tau'} d\tau' = a\,\epsilon^r \Gamma(2-r) \leqslant a\,\epsilon^r \tag{L13}$$

Defining $\xi(\epsilon) = 2a\,\epsilon^r$, the second part of the lemma is verified.

**Proof of Lemma 2.2**

Define $w_\epsilon(t,x)$ as in *Lemma 2.1*. Consequently,

$$\frac{\partial w_\epsilon(t,x)}{\partial x} = \frac{\partial}{\partial x}\left[\int_0^t d(\tau,x) e^{-\epsilon(t-\tau)} d\tau\right] = \int_0^t \left[\frac{\partial}{\partial x} d(\tau,x)\right] e^{-\epsilon(t-\tau)} d\tau \tag{L14}$$

Since $\frac{\partial d(t,x)}{\partial x}$ is zero mean, and is bounded, *Lemma 2.1* can be applied to $\frac{\partial d(t,x)}{\partial x}$, and inequality (2.6) of *Lemma 2.1* becomes inequality (2.10) of *Lemma 2.2*. Note that since $\frac{\partial d(t,x)}{\partial x}$ is bounded, and $d(t,0)=0$ for all $t \geqslant 0$, $d(t,x)$ is Lipschitz. Since $d(t,x)$ is zero mean, with convergence function $\gamma(T)\|x\|$, the proof of *Lemma 2.1* can be extended, with an additional factor $\|x\|$. This leads directly to (2.8) and (2.9) (although the function $\xi(\epsilon)$ may be different from that obtained with $\frac{\partial d(t,x)}{\partial x}$, these functions can be replaced by a single $\xi(\epsilon)$).

**Proof of Lemma 2.3**

The proof proceeds in two steps.

**Step 1:** for $\epsilon$ sufficiently small, and for $t$ fixed, the transformation (2.15) is a homeomorphism.

Apply *Lemma* 2.2, and let $\epsilon_1$ such that $\xi(\epsilon_1) < 1$. Given $z \in B_h$, the corresponding $x$ such that:

$$x = z - \epsilon\, w_\epsilon(t,z) \tag{L15}$$

may not belong to $B_h$. Similarly, given $x \in B_h$, the solution $z$ of (L15) may not exist in $B_h$. However, for any $x,z$ satisfying (L15), inequality (2.8) implies that:

$$(1-\xi(\epsilon))\| z \| \leqslant \| x \| \leqslant (1+\xi(\epsilon))\| z \| \tag{L16}$$

Define:

$$h'(\epsilon) = \min \left( h(1-\xi(\epsilon)), \frac{h}{1+\xi(\epsilon)} \right) = h(1-\xi(\epsilon)) \tag{L17}$$

and note that $h'(\epsilon) \to h$ as $\epsilon \to 0$.

We now show that:

- for all $z \in B_{h'}$, there exists a unique $x \in B_h$ such that (L15) is satisfied,
- for all $x \in B_{h'}$, there exists a unique $z \in B_h$ such that (L15) is satisfied.

In both cases, $\| x - z \| \leqslant \xi(\epsilon) h$.

The first part follows directly from (L16), (L17). The fact that $\| x - z \| \leqslant \xi(\epsilon) h$ also follows from (L16), and implies that, if a solution $z$ exists to (L15), it must lie in the closed ball $U$ of radius $\xi(\epsilon)h$ around $x$. It can be checked, using (2.10), that the mapping $F_x(z) = x - \epsilon w_\epsilon(t,z)$ is a contraction mapping in $U$, provided that $\xi(\epsilon) < 1$. Consequently, $F$ has a unique fixed point $z$ in $U$. This solution is also a solution of (L15), and since it is unique in $U$, it is also unique in $B_h$ (and actually in $R^n$). For $x \in B_h$, but outside $B_{h'}$, there is no guarantee that a solution $z$ exists in $B_h$, but if it exists, it is again unique in $B_h$. Consequently, the map defined by (L15) is well-defined. From the smoothness of $w_\epsilon(t,z)$ with respect to z, it follows that the map is a homeomorphism.

**Step 2:** the transformation of variable leads to the differential equation (2.16)

Applying (L15) to the system (2.1):

$$\left( I + \epsilon \frac{\partial w_\epsilon}{\partial z} \right) \dot{z} = \epsilon f_{av}(z) + \epsilon \left( f(t,z,0) - f_{av}(z) - \frac{\partial w_\epsilon}{\partial t} \right)$$

$$+ \epsilon \left( f(t,z+\epsilon w_\epsilon,\epsilon) - f(t,z,\epsilon) \right)$$

$$+\epsilon(f(t,z,\epsilon)-f(t,z,0))$$

$$:=\epsilon f_{av}(z)+\epsilon p'(t,x,z,\epsilon) \tag{L18}$$

where, using the assumptions, and the results of *Lemma 2.2:*

$$\| p'(t,z,\epsilon) \| \leq \xi(\epsilon) \| z \| + \xi(\epsilon) l_1 \| z \| + \epsilon l_2 \| z \| \tag{L19}$$

For $\epsilon \leq \epsilon_1$, (2.10) implies that $(I+\epsilon \frac{\partial w_\epsilon}{\partial z})$ has a bounded inverse for all $t \geq 0$, $z \in B_h$. Consequently, $z$ satisfies the differential equation:

$$\dot{z} = \left( I+\epsilon \frac{\partial w_\epsilon}{\partial z} \right)^{-1} (\epsilon f_{av}(z)+\epsilon p'(t,z,\epsilon))$$

$$= \epsilon f_{av}(z)+\epsilon p(t,z,\epsilon) \qquad z(0)=x_0 \tag{L20}$$

where:

$$p(t,z,\epsilon) = \left( I+\epsilon \frac{\partial w_\epsilon}{\partial z} \right)^{-1} \left( p'(t,z,\epsilon) - \epsilon \frac{\partial w_\epsilon}{\partial z} f_{av}(z) \right) \tag{L21}$$

and:

$$\| p(t,z,\epsilon) \| \leq \frac{1}{1-\xi(\epsilon_1)} (\xi(\epsilon)+\xi(\epsilon) l_1+\epsilon l_2+\xi(\epsilon) l_{av}) \| z \|$$

$$:= \psi(\epsilon) \| z \| \tag{L22}$$

or all $t \geq 0$, $\epsilon \leq \epsilon_1$, $z \in B_h$.

**eneralized Bellman-Gronwall Lemma (cf. [7], p 169)**

: $x(t)$, $a(t)$, $u(t)$ are positive functions satisfying:

$$x(t) \leq \int_0^t a(\tau) x(\tau) d\tau + u(t) \tag{L23}$$

for all $t \in [0,T]$, and $u(t)$ is differentiable.

**Then:**

$$x(t) \leq u(0) e^{\int_0^t a(\sigma) d\sigma} + \int_0^t \dot{u}(\tau) e^{\int_\tau^t a(\sigma) d\sigma} d\tau \tag{L24}$$

for all $t \in [0,T]$.

**References**

[1] Bogoliuboff, N. N. and Y. A. Mitropolskii, *Asymptotic Methods in the Theory of Non-Linear Oscillators*, Gordon & Breach, New York, 1961.

[2] Hale, J. K., *Oscillations in Non-linear Systems*, McGraw-Hill, New York, 1963.

[3] Hale, J. K., *Ordinary Differential Equations*, Krieger, Molaban (Florida), 1980.

[4] Sethna, P. R., "Method of Averaging for Systems Bounded for Positive Time," *Journal of Math Anal and Applications*, Vol. 41 (1973), pp 621-631.

[5] Balachandra, M. and P. R. Sethna, "A Generalization of the Method of Averaging for Systems with Two-Time Scales," *Archive for Rational Mechanics and Analysis*, Vol. 58, 1975, pp 261-283.

[6] Arnold, V. I., *Geometric Methods in the Theory of Ordinary Differential Equations*, Springer Verlag, 1982.

[7] Guckenheimer, J. and P. Holmes, *Nonlinear Oscillations, Dynamical Systems, and Bifurcations of Vector Fields*, Springer Verlag, New York, 1983.

[8] Hoppensteadt, F. C., "Singular Perturbations on the Infinite Time Interval," *Trans. American Math Society*, Vol. 123 (1966), pp 521-535.

[9] Krause, J., M. Athans, S. Sastry and L. Valavani, "Robustness Studies in Adaptiv Control," *Proc. 22nd IEEE Conf. on Decision and Control*, San Antonio, Texas (1983) pp 977-981.

[10] Astrom, K. J., "Interactions between Excitation and Unmodelled Dynamics in Ada tive Control," *Proc. 23rd IEEE Conf. on Decision and Control*, Las Vegas, Neva (1984), pp 1276-1281.

[11] Riedle, B. and P. V. Kokotovic, "A Stability-Instability Boundary for Disturba Free Slow Adaptation and Unmodelled Dynamics," *Proc. 23rd IEEE Conf. on De sion and Control*, Las Vegas, Nevada (1984), pp 998-1101.

[12] Ljung, L., "On Positive Real Transfer Functions and the Convergence of Some Rec sions," *IEEE Trans. on Automatic Control*, Vol. AC-22 (1977), pp 539-557.

[13] Rohrs, C. E., L. Valavani, M. Athans and G. Stein, "Robustness of Continuous-Tim Adaptive Algorithms in the Presence of Unmodelled Dynamics," *IEEE Trans. on Automatic Control*, Vol. AC-30 (1985), pp 881-889.

[14] Boyd, S. and S. Sastry, "Necessary and Sufficient Conditions for Parameter Convergence in Adaptive Control," *Proc. of the Ames- Berkeley Conf. on Non-Linear Dynamics and Control*, Math-Sci Press, Brookline, Massachusetts, 1984, pp 81-101.

[15] Bodson, M. and S. Sastry, "Small Signal I/O Stability of Nonlinear Control Systems: Application to the Robustness of a MRAC Scheme," Memorandum No. UCB/ERL M84/70, Electronics Research Laboratory, University of California, Berkeley, 1984.

[16] Riedle, B. and P. Kokotovic, "Stability Analysis of Adaptive System with Unmodelled Dynamics," to appear in *Int. J. of Control*, Vol. 41, 1985.

[17] Kosut, R., B.D.O. Anderson and I. Mareels, "Stability Theory for Adaptive Systems: Method of Averaging and Persistency of Excitation," Preprint, Feb. 1985.

[18] Narendra, K. and L. Valavani, "Stable Adaptive Controller Design-Direct Control," *IEEE Trans. on Automatic Control*, Vol. AC-23 (1978), pp 570-583.

[19] Sastry, S., "Model Reference Adaptive Control - Stability, Parameter Convergence, and Robustness," *IMA Journal of Mathematical Control & Information*, Vol. 1 (1984), pp 27-66.

[20] Hahn, W., *Stability of Motion*, Springer Verlag, Berlin, 1967.

[21] Luders, G. and K.S. Narendra, "An Adaptive Observer and Identifier for a Linear System," *IEEE Trans. on Automatic Control*, Vol. AC-18 (1973), pp. 496-499.

[22] Kreisselmeier, G. "Adaptive Observers with Exponential Rate of Convergence," *IEEE Trans. on Automatic Control*, Vol. AC-22 (1977), pp. 2-8.

[23] Goodwin, G. C. and R. L. Payne, *Dynamic System Identification*, Academic Press, New York, 1977.

[24] Riedle B. and P. Kokotovic, "Integral Manifold Approach to Slow Adaptation," Report DC-80, University of Illinois, March 1985.

# Lecture Notes in Control and Information Sciences

Edited by M. Thoma

Vol. 43: Stochastic Differential Systems
Proceedings of the 2nd Bad Honnef Conference
of the SFB 72 of the DFG at the University of Bonn
June 28 – July 2, 1982
Edited by M. Kohlmann and N. Christopeit
XII, 377 pages. 1982.

Vol. 44: Analysis and Optimization of Systems
Proceedings of the Fifth International
Conference on Analysis and Optimization of Systems
Versailles. December 14–17, 1982
Edited by A. Bensoussan and J. L. Lions
XV, 987 pages, 1982

Vol. 45: M. Arató
Linear Stochastic Systems
with Constant Coefficients
A Statistical Approach
IX, 309 pages. 1982

Vol. 46: Time-Scale Modeling of Dynamic Networks
with Applications to Power Systems
Edited by J. H. Chow
X, 218 pages. 1982

Vol. 47: P. A. Ioannou, P. V. Kokotovic
Adaptive Systems with Reduced Models
V, 162 pages. 1983

Vol. 48: Yaakov Yavin
Feedback Strategies for Partially
Observable Stochastic Systems
VI, 233 pages, 1983

Vol. 49: Theory and Application of Random Fields
Proceedings of the IFIP-WG 7/1
Working Conference
held under the joint auspices of the
Indian Statistical Institute
Bangalore, India, January 1982
Edited by G. Kallianpur
VI. 290 pages. 1983

Vol. 50: M. Papageorgiou
Applications of Automatic Control Concepts
to Traffic Flow Modeling and Control
IX, 186 pages. 1983

Vol. 51: Z. Nahorski, H.F. Ravn, R.V.V. Vidal
Optimization of Discrete Time Systems
The Upper Boundary Approach
V, 137 pages 1983

Vol. 52: A. L. Dontchev
Perturbations, Approximations and Sensitivity Analysis
of Optimal Control Systems
IV, 158 pages. 1983

Vol. 53: Liu Chen Hui
General Decoupling Theory of Multivariable
Process Control Systems
XI, 474 pages. 1983

Vol. 54: Control Theory for Distributed
Parameter Systems and Applications
Edited by F. Kappel, K. Kunisch,
W. Schappacher
VII, 245 pages. 1983.

Vol. 55: Ganti Prasada Rao
Piecewise Constant Orthogonal Functions
and Their Application to Systems and Control
VII, 254 pages. 1983.

Vol. 56: Dines Chandra Saha, Ganti Prasada Rao
Identification of Continuous
Dynamical Systems
The Poisson Moment Functional
(PMF) Approach
IX, 158 pages. 1983.

Vol. 57: T. Söderström, P. G. Stoica
Instrumental Variable Methods
for System Identification
VII, 243 pages. 1983.

Vol. 58: Mathematical Theory of
Networks and Systems
Proceedings of the MTNS-83 International
Symposium
Beer Sheva, Israel, June 20–24, 1983
Edited by P. A. Fuhrmann
X, 906 pages. 1984

Vol. 59: System Modelling and Optimization
Proceedings of the 11th IFIP Conference
Copenhagen, Denmark, July 25-29, 1983
Edited by P. Thoft-Christensen
IX, 892 pages. 1984

Vol. 60: Modelling and Performance
Evaluation Methodology
Proceedings of the International Seminar
Paris, France, January 24–26, 1983
Edited by F. Bacelli and G. Fayolle
VII, 655 pages. 1984

Vol. 61: Filtering and Control of Random
Processes
Proceedings of the E.N.S.T.-C.N.E.T. Colloquium
Paris, France, February 23–24, 1983
Edited by H. Korezlioglu, G. Mazziotto, and
J. Szpirglas
V, 325 pages. 1984

# Notes in Control and Information Sciences

M. Thoma and A. Wyner

. Analysis and Optimization
stems
ceedings of the Sixth International
onference on Analysis and Optimization
of Systems
Nice, June 19–22, 1984
Edited by A. Bensoussan, J. L. Lions
XIX, 591 pages. 1984.

Vol. 63: Analysis and Optimization
of Systems
Proceedings of the Sixth International
Conference on Analysis and Optimization
of Systems
Nice, June 19–22, 1984
Edited by A. Bensoussan, J. L. Lions
XIX, 700 pages. 1984.

Vol. 64: Arunabha Bagchi
Stackelberg Differential Games
in Economic Models
VIII, 203 pages, 1984

Vol. 65: Yaakov Yavin
Numerical Studies
in Nonlinear Filtering
VIII, 273 pages, 1985.

Vol. 66: Systems and Optimization
Proceedings of the Twente Workshop
Enschede, The Netherlands, April 16–18, 1984
Edited by A. Bagchi, H. Th. Jongen
X, 206 pages, 1985.

Vol. 67: Real Time Control of Large Scale Systems
Proceedings of the First European Workshop
University of Patras, Greece, Juli 9–12, 1984
Edited by G. Schmidt, M. Singh, A. Titli,
S. Tzafestas
XI, 650 pages, 1985.

Vol. 68: T. Kaczorek
Two-Dimensional Linear Systems
IX, 397 pages, 1985.

Vol. 69: Stochastic Differential Systems –
Filtering and Control
Proceedings of the IFIP-WG 7/1 Working Conference
Marseille-Luminy, France, March 12-17, 1984
Edited by M. Metivier, E. Pardoux
X, 310 pages, 1985.

Vol. 70: Uncertainty and Control
Proceedings of a DFVLR International Colloquium
Bonn, Germany, March, 1985
Edited by J. Ackermann
IV, 236 pages, 1985.

Vol. 71: N. Baba
New Topics in Learning Automata
Theory and Applications
VII, 231 pages, 1985.

Vol. 72: A. Isidori
Nonlinear Control Systems:
An Introduction
VI, 297 pages, 1985.

Vol. 73: J. Zarzycki
Nonlinear Prediction
Ladder-Filters for Higher-Order
Stochastic Sequences
V, 132 pages, 1985.

Vol. 74: K. Ichikawa
Control System Design based on
Exact Model Matching Techniques
VII, 129 pages, 1985.

Vol. 75: Distributed Parameter
Systems
Proceedings of the 2nd International
Conference, Vorau, Austria 1984
Edited by F. Kappel, K. Kunisch,
W. Schappacher
VIII, 460 pages, 1985.

Vol. 76: Stochastic Programming
Edited by F. Archetti, G. Di Pillo,
M. Lucertini
V, 285 pages, 1986.

Vol. 77: Detection of
Abrupt Changes in Signals
and Dynamical Systems
Edited by M. Basseville,
A. Benveniste
X, 373 pages, 1986.

Vol. 78: Stochastic
Differential Systems
Proceedings of the 3rd Bad Honnef
Conference, June 3–7, 1985
Edited by N. Christopeit, K. Helmes,
M. Kohlmann
V, 372 pages, 1986.

Vol. 79: Signal
Processing for Control
Edited by K. Godfrey, P. Jones
XVIII, 413 pages, 1986.

Vol. 80: Artificial Intelligence
and Man-Machine Systems
Edited by H. Winter
IV, 211 pages, 1986.